61st Annual Safety in Ammonia Plants and Related Facilities Symposium 2016

AIChE Technical Manual Volume 57

Denver, Colorado, USA
18 - 22 September 2016

ISBN: 978-1-5108-5321-8

Printed from e-media with permission by:

Curran Associates, Inc.
57 Morehouse Lane
Red Hook, NY 12571

Some format issues inherent in the e-media version may also appear in this print version.

Printed by Curran Associates, Inc. (2017)

AIChE
120 Wall Street, FL 23
New York, NY 10005-4020

Phone: (800) 242-4363
Fax: (203) 775-5177

www.aiche.org

Additional copies of this publication are available from:

Curran Associates, Inc.
57 Morehouse Lane
Red Hook, NY 12571 USA
Phone: 845-758-0400
Fax: 845-758-2633
Email: curran@proceedings.com
Web: www.proceedings.com

TABLE OF CONTENTS

Safe Operation in Chemical Plants with Stop Work Authority

Most companies say that their operators have (are fully empowered with) Stop Work Authority (SWA). It is generally easy to apply it in occupational health and safety (OH&S) area. When it comes to process safety, SWA is not well understood. In this paper we would review the current state of SWA as practiced in various major companies and would share typical guidelines for plant operation teams to act in safe shutdown of the plant to minimise safety and reliability issues such as delays in bringing down the plant as needed. Events, however, do emerge where prior foreseeability was not easily forthcoming or may have been inadvertently missed in the early hazard evaluations. During such times operator intervention is absolutely required and the expectation is that the operator will act appropriately, including the act of stopping production or shutting the plant down, in the interest of people and plant safety. Examples may include BFW excursions, leaking gas lines, loss of containment, vibrations of machines etc., where operators could have been given clear instructions (or policy/decision guidelines) to stop production or to shut down the plant. We would discuss some recent accidents where Stop Work Authority could have been applied. We will demonstrate the usefulness of such simple guidelines for risk judgement and corresponding prudent action by the operator in moments of perceived operational crisis.

Kawal Maraj,
Yara Trinidad Ltd

Raghava Nayak,
Process Safety Expert, Sydney

Venkat Pattabathula,
Incitec Pivot Ltd

Introduction

Chemical Plants are designed, constructed, operated and decommissioned within the Safe Operating Envelope as dictated by risk. The tolerable level of this risk is bounded and constrained strictly by the Principles of Deontology, viz a viz the science of the laws, codes, standards and the authorities. In addition to the employer's duty of care imposed by law, Stop Work Authority (SWA) policy requires that employers empower their personnel to intervene and "stop" unsafe work in a timely manner, to prevent injuries and harm. Similar intervention is expected to forestall an emerging threat or the escalation of a dangerous incident at the plant. Controlling a dangerous situation may require the costly, and sometimes unjustifiable, shut down of operations to maintain the desired level of safety. At such times, a SWA policy explicitly indemnifies personnel engaged in the stop work action.

SWA programs can be easily incorporated into existing management systems of assurance for continuing safe operations. These systems basically comprise the tenets of keeping plant and equipment in a state of "fitness for purpose"; systems, procedures and policies by which the plant and equipment are operated, maintained and managed safely; and by trained people who are competent with knowledge, skills and attitude, to operate, main manage the assets.

Management systems are then subjected to certification audits to verify the upkeep of the systems integrity and continuously improve as long as associated practices are conforming and compliant. Compliant behaviour is also expected to include the requirement to intervene and stop unsafe work and unsafe conditions as necessary, on a real time basis.

Whilst the Stop Work aspect regarding occupational health, safety and environmental matters may be well – understood, and to some extent, conscientiously practised [ref 3], the same may not hold true for process safety type events.

This paper seeks to explore and analyse the factors underlying the continuing trend of such unacceptable events despite the existence of well-established mandates of initiatives like Stop Work Authority and other Safe Systems of Assurance at the work place. It also proposes some practical ways by which the frequency and severity of impact of such events may be reduced, in both the short and long term.

Operational Risk

There exists numerous publications, reviews, analysis of trends regarding world episodic industrial events and their catastrophic consequences. We reviewed a few of these from various types of industry with special focus on the human intervention at the onset of an emerging crisis, during crisis and in the complex mitigating circumstances resulting from secondary failures and escalating events.

It is noted that there are twenty one (21) common elements of the anatomy of an accident [ref 4]. Out of these the following eight (8) are deemed relevant to the area of interest (operational risk judgement), of this paper:

a. Ignorance of Operating Equipment Characteristics.
b. Sense of Invulnerability (*loss of a sense of danger).*
c. Acceptance of Abnormalities.
d. Ignorance of Warning signs.
e. Capitulation to production (and time) pressure.
f. Inadequate Emergency (and Crisis) preparedness.
g. Insufficient Management oversight.
h. Degradation of operating limits.

Operational risk judgement and decision making requirements at the moment of imminent crisis also apply to other industry sectors such as aviation, shipping, pipelines, mining and nuclear.

Operational risk is defined as the risk of direct or indirect loss resulting from inadequate or failed internal processes, people, and systems or from external events [ref 8]. The definition also includes legal risks.

In practice, operational risk comprises all those risks that are not dealt with by any other specific control mechanism in a company. This makes it awkward to deal with as there is a tendency to push anything and everything into the category of operational risk whenever any of the other – more traditional risk areas (environmental, legal, design) do not appear to apply or fit the situation. In addition, there is another problem as operational risk does not have its own special champions [ref 9].

The default operational risk experts are the front line operating supervisors (with the support of local managers) who are also responsible for the plant's performance. Hence their actions depend on right balance between pushing production and minimising risk (maintaining the plant within the safe operating envelope) in a dynamic operating environment.

Abnormal Situations

An abnormal situation is an event that triggers the need for operator intervention. The situation can be either a perceived or actual event but in either case causes uncertainty and delays in identifying and responding to the root cause. Abnormal situations can be triggered by failures in equipment, people, processes, or some combination of the three.

A cascading approach for tackling the abnormal situations is suggested [ref 20]:

a. If operating parameters are well inside the safe operating envelope and the rate of change is steady, continue the normal operation;
b. If not, intervene to restore the process to normal operation;
c. Failing this, bring the process to a safe state;
d. If the timely achievement of safe state is difficult, consider actions to minimise severity of the potential incident.

Such a structured approach would most likely prevent the process operation rapidly growing out of control and escalating to major incident. Also this approach minimises the likelihood of inadvertent trips / last minute drastic actions.

Hazard studies such as HAZOP, Bow-Tie, safety integrity level studies and fire & safety study would have identified the abnormal situations, process safety time and required operator interventions. If these are incorporated in to the emergency procedures and the operator is well trained, the issue of the inadequacy of information, data suspicion and the perception of time pressure that seems to overwhelm the decision maker in an emergency is avoided.

Experience with SWA

The following four (4) cases cited from Yara Trinidad Ltd illustrates the application of Stop Work Authority in Abnormal Situation Management [ref 10].

a. 2010 September – The unit experienced a failure of the mixed feed coil in the convection section of the primary reformer when an inlet

motorized valve, without warning, went closed at the inlet of the high temperature shift converter causing an increase in plant front end pressure. The operating crew on being alerted to the abnormal situation engaged in intense troubleshooting during which time the mixed feed coil failed. The time elapsed during the troubleshooting effort was in excess of one hour.
This is a case of SWA not applied in time. The incident was investigated and lessons were shared.

b. 2015 January – One of the plants experienced the development of a hotspot on the primary reformer transfer line. The operator on observing the abnormal situation of the hotspot took immediate action to shut down the primary reformer and prevent consequential damages. The action by the operator was consistent with the expectation of Stop Work Intent. This is a case of a timely shutdown where zero time had elapsed.

c. 2015 – March – On one of the units, during a temperature excursion in the methanator, the supervisor on observing the temperature rising above the trip set point took immediate action and tripped the methanator. The trip action by the operator was without delay. The post trip investigation revealed that the trip setting had drifted away from the initial set point. Anyone familiar with a methanator runaway reaction in an ammonia plant knows that the exotherm is extremely rapid. As a result the supervisor knew that he would not have had time to attempt any alleviating action except to trip the unit. The incident was comprehensively investigated and the learnings were shared. The supervisor was also commended for his action.

d. 2015 May - The unit experienced a sudden rupture of the level bridle on the CO2 absorber. After fifteen (15) minutes of troubleshooting the inlet isolation valve to the absorber was eventually closed. The delay in this case was caused by the initial uncertainty regarding the real cause of the abnormal situation. The eventual closing of the isolation valve does of course mean that the plant was taken out of production.

Some recent incidents in other plants / industry are listed below:

e. On August 6, 2012, catastrophic pipe rupture and release of flammable vapours in the crude unit at Chevron refinery, Richmond USA engulfed 19 employees. Although none of the employees had major injuries, subsequent vapour cloud ignition and extensive smoke plume required large evacuation, resulted in major business interruption and caused health impacts.
When the leak was first discovered at 15:48 hrs, it was observed to be a minor leak and the risk of pipe rupture / auto ignition were not considered during leak response risk assessment. The perception that leak was small, stable leak reported to have influenced the risk assessment by the response team. Isolating the leak was not possible and two and half hours of attempts to stop the leak were not successful. Whist the employees were attending the leak,

the pipe ruptured and resulting vapour cloud and auto ignition caused major disruption in the surrounding areas. The CSB (Chemical Safety Board) report [ref 2] identified the need to improve Stop Work Authority process in this incident.

f. Tosco refinery: On February 23, 1999, a fire occurred in the crude unit at Tosco Corporation. Avon oil refinery in Martinez, California [ref 21]. Workers were attempting to replace piping attached to a 50 m (164 ft.) tall fractionator tower while the process unit was in operation. During removal of the piping, naphtha was released onto the hot fractionator and ignited. The flames engulfed five workers located at different heights on the tower. Four men were killed, and one sustained serious injuries. The CSB investigation noted that the SWA was not exercised.

g. Two severe explosions occurred at 22:48 hrs. June 3, 2014 at styrene-propylene oxide plant of Shell Moerdijik, The Netherlands [ref 17] during the warmup as the plant was being commissioned after the catalyst replacement. Contents of the reactor and its auxiliary vessel were released to a wide environment and reactor fragments were blown to hundreds of meters. Two people were hit by the pressure wave. Considerable smoke impacted a large area of Southern South Holland and crisis management authorities in the Southern South Holland region were put under severe stress.

   During the restart after catalyst replacement, the operators had resorted to manual heating of warm-up liquid (ethyl benzene) to hasten the rate of warm–up of the reactors. Side exothermic reactions (not known to the operations team at that time) increased the reactor pressure which triggered the automatic protection system (to prevent the liquid to flare). Continued warming up of the reactor snow balled the reaction rate, temperature and reactor pressures and resulted in explosions. The Dutch Safety Board investigations noted that the operators had erroneously decided to continue the process (warm-up) even though critical process boundaries were breached (Alarm and trip of automatic protection system) [ref 17].

h. An ammonia plant (confidential site) was shut down to carry out some maintenance work on steam piping and compressors. As a standard operating practice in NH_3 plants, all the catalytic reactors (HTS, LTS, methanator and NH_3 synthesis converter) are kept under nitrogen pressure. This plant also does keep all the reactors under N_2 and for some reason, the HTS reactor was not kept under N_2 this time. As part of work permit system, the steam system was depressured and drains and vents were opened on BFW/steam system which was downstream of HTS. But the HTS bottom bed temperatures went up slowly over 4-days well above the vessel design temperatures of 478DegC (892.4F) and beyond its maximum TI range of 516C (961F). As the vessel could have been subjected to much higher than its design temperatures, the plant team decided to unload the catalyst and carry out internal inspection of the vessel to confirm its mechanical integrity to operate it safely.

Preliminary investigation has revealed that because of a human error, drain and vent valves were opened on HTS reactor which resulted in to high temperatures in the bottom of the bed from exposure of reduced catalyst to air. Unfortunately, the high temperature alarms on HTS bed were also ignored during outage until plant resumed start-up activities. Timely operator intervention / SWA would have avoided the start-up delays.

i. Multiple gas explosions occurred in Kaohsiung, Taiwan about midnight on 31 July 2014 and early on 1 August 2014. The flame height reported to be over ~50 m (164 ft.) high. Miles of roads surface was broken and adjacent buildings were seriously damaged [ref 18, 19] as shown. More than 300 persons were killed, lost or injured. Underground propylene supply pipe to the petrochemical complex had leaked releasing a large quantity of flammable material (several tons) in a populated area which ignited causing the extensive impact.

The pipe line was connected to 3 firms namely the petrochemical company, the distribution company and the state run oil company. The petrochemical company bought propylene sometimes from the oil company and other times from the distribution company.

On the day of the blast the petrochemical complex was receiving the feed stock propylene from the harbor via the pipeline operated by the distribution company. When the petrochemical company noticed that it was not getting the correct amount of propylene the pipeline was shut initially around 8 PM but reopened about one and half hours later.

A pressure drop from normal pressure of 400 kPa (58 psi) to 130 kPa (19 psi) at the petrochemical plant was identified early, but the plant operator did not terminate the operation to allow checking and resumed the supply to the pipeline. The leakage continued and the pipeline was shut just 15 -20 minutes before the explosion at midnight.

Lessons learned

A review of above incidents reveals two basic elements. These are, the decision maker; and, the decision to operate through the uncertainty versus exercising the Stop Work. The decision to continue operations in the face of uncertainty is fundamental to events escalating out of control. In fact failure to recognize the abnormal situation going out of control and continuing the troubleshooting, unaware of the elapsed time and potential other developments result in major incidents.

The intent of Stop Work Authority (SWA) and associated Duty of Care legislation (e.g. OSHA act) is in fact to do exactly the opposite and to move the unit to a safe state.

Stop Work Authority programmes augment the concept of achieving a level of safety where risks are reduced to a standard described by the phrase "As Low As is Reasonably Practicable (ALARP)." This ALARP principle relies on the discretion of the decision making process to arrive at a satisfactory safe position. This ambiguity inadvertently augmented by both, the absence of prescriptive rules by SWA, and the sense of having legal latitude to judge the risks in the moment of crisis, lies at the heart of the dilemma faced by the individual.

Unsafe situations regarding occupational health and safety are more clear-cut and present less of an issue [ref 11].

In a nutshell, an operator / operations supervisor has to judge the precise timing of stop work action whilst wrestling with the urge to go forward keeping the plant online.

Clearly the plight of the decision maker (Shift Supervisor) is not necessarily helped by the vagaries of interpretation of any generalized statements regarding SWA policy, the Law, and internal work processes. The continuing occurrences of tragic and costly incidents calls for urgent management intervention with specific and prescriptive guidelines to forestall and mitigate incidents and lighten the burden of managing operational risks on the fly.

Managing operational risk in the context of safe operations with Stop Work Authority is a special case in point.

Before moving on to possible solution pathways, we undertake a quick review of current responses to abnormal situations in the chemical industry.

SWA – Summary of Practice

In his book, Alarm Management for Process Control, Rothenberg provides very good examples and guidance on how an abnormal situation can be prevented from getting out of control. Of special interest in this paper is the chapter on "Permission to Operate". Stop Work Authority policy, as intended and in spirit, can be easily seen to augment some of the solution paths suggested by Rothenberg.

There exists a plethora of practices to manage the abnormal situations including

a. application of ALARP coupled with "Grossly Disproportionate" rule [ref 23] for determining the limits of Reasonably Practicable.
b. application of precautionary principle [ref 22];
c. management assurance systems including Stop-Work Authority policies and procedures;
d. shared learnings from near-misses and accidents;
e. scenario specific emergency response procedures; and,
f. Regulatory requirements such as safety cases, emergency plans and procedures.

Despite the existence of the above noted practices including the Stop-Work Authority initiatives, the frequency of abnormal situations getting out of control remains a source of concern and regret. Whilst we continue to witness successful use of SWA in OHS incidents [ref 13], its' use in process safety incidents needs support and work place involvement.

As CCPS Guidelines for Risk Based Process Safety notes: "Workers are potentially the most knowledgeable people with respect to the day-to-day details of operating the process and maintaining the equipment and facilities and may be the sole source of some types of knowledge gained through their unique experience."

We would emphasize, however, that stop work authority is a less effective measure for incident prevention than good pre-planning, and that its success is contingent upon the existence of a "culture of safety" wherein workers are encouraged and empowered to advocate for their safety on the job. As the CSB [ref 21] noted in its investigation report of TOSCO Avon Refinery fire, stop work authority must often be exercised when pressures to get a job completed are significant, and delays may result in significant financial costs to the facility. In an environment where production pressures trump safety, this authority is often of limited value.

Recommendations

Thus, the SWA policy should be complemented by the following key guide posts which are linked to the lessons learned from process safety incidents described earlier:

a) Pre-accident investigations (*pre-mortem*) [ref 14] or analysing issues long before they develop into incidents. This guidepost links with the CSB finding of the requirement for more pre-planning and aptly relates to the recommendation for the development and use of checklist as an enabling instrument of the SWA Policy.

b) Training and documentation on diagnostics namely how to detect and appreciate whether the plant state is abnormal or out of control. If all operators had a sign that read, OUT OF CONTROL, every time their plant was abnormal and out of control, it's unlikely that they would ever try to keep it running in the vain hope of a miracle". All of the recommendations put forward by this paper, especially those requiring the establishment of suitable guidelines for detecting incipient crisis will require intense and repeated training to maintain the skill for handling operational risks and abnormal situations.

c) Employees to be made fully aware of operational risk potentially escalating to reputation and compliance risks during an abnormal situation; and recovery issues after the incident (maintain the licence to operate etc.)

d) Lessons learned on the application of SWA in the abnormal situations at site level and industry level (Including employees participating in safety seminars, annual user workshops etc.);

e) Develop effective discriminators for improving probability estimates. This aspect is covered comprehensively by Lindorfer [ref 15]; and [ref 16]. At Yara Trinidad this simple technique is used as a cross check for Job Hazard Analysis and Incident Investigations.

The above guide posts form the basis for four main recommendation categories listed below:

(a) Management intervention

Management should strengthen and augment its SWA Policy Statement with clear guidelines to enable and empower the decision maker in the moment of crisis, to act confidently, decisively and without fear of reprimand in his guided Stop Work Action. To achieve this level of specificity, the guidelines must comprise at least the following three (3) elements:

i. It must explicitly "Indemnify" the decision maker as long as his / her action is in keeping with the spirit and intent of the SWA Policy and the Policy guidelines.
ii. It must contain an unambiguous trigger mechanism that provides a virtual "License" to shift from a production mind set to a strategic Safety level. This "trigger" may take the form of a checklist that mandates

predetermined shutdown actions based on occurrence of certain specific conditions. In this paper we are dealing with situations that may develop unaware, unannounced and without any external clues including control room alarms.

iii. The pre-conditions of the checklist must be thoroughly contemplated and must carry the integrity equivalence of a typical "Process or Equipment Safety Trip" and be authoritative enough to lure and mandate the decision maker toward timely action away from the abnormal situation going out of control and toward a safer direction, even if it may result in a premature shutdown.

Chevron (ref 2), developed and implemented such a checklist following the August 2012 incident. This paper suggests a slightly modified version as follows:

1. Are we sure of the cause of the situation being observed? Yes / No
2. Do we have control over the incident? Yes/No
3. Do we know of any problem history relating to the situation? Yes/No
4. Can the initiating condition be detected on time? Yes/No
5. Do we have time to effectively intervene and correct in a simple way? Yes/No
6. Is the worst case impact on the situation tolerable? Yes/No
7. Are we sure that the present situation, if it worsens, will create no threat to overall business exposure (e.g. Environmental Liability)? Yes/No
8. Are we still in Compliance mode? Yes/No
9. Are we comfortable that we are still out of danger? Yes/No

If any one of the above answers is No, then it may be the sufficient basis to exercise Stop Work *and shutdown* in the interest of safety.

The above list is of course not all inclusive and merely serves as an example of the kind of thinking required for breakthrough solutions.

(b) Effective discriminator as a means of reducing uncertainty and communicating to management

An employer has a legal duty of care at the work place to maintain a safe system of work. One aspect in exercising its duty of care, is the implementation and enforcement of an SWA Policy. The standard of duty of care expected, is to be as safe as is reasonably practicable (ALARP). The concept of ALARP embodies the philosophy of dealing with risks that fall between an upper and lower limits. Above the upper limit where the risk is so great (or uncertain), that no start or instant stop is done. It is the ALARP region of uncertainty (or high risk) that SWA seeks to attend. The lower extreme, on the other hand, is where the risk has been made so small as to be insignificant (or certain and controlled).

ALARP requires that a suitable and sufficient Risk assessment be undertaken to justify preventive and mitigating action commensurate with the quantum of risk estimated. Abnormal situations with disastrous outcomes as reviewed in this paper, mostly originated in the risk region described by ALARP. Many of the scenarios appeared to have started off small, leading to assessed risk levels that were considered acceptable or tolerable at the time. This undesired and adverse development reveals an urgent need to improve the quality and reliability of risk estimates for such transient situations.

Risk (or the Risk Score) as we know, is a function of both the probability of occurrence, and the severity of the consequences. Estimations of probability on the practical scene is largely subjective, whilst severity is more easily quantifiable and more objective.

Against this back drop, this paper recommends the adoption and utilization of effective discriminators in (a) improving estimates of probability by reducing subjectivity [ref 15] in arriving at a reliable Risk Score and (b) the adoption of a suitable way to communicate to management, in management terms, the potential impact of the risk score. By this we are of the opinion that management will be interested in at least the following three (3) concern areas [ref 15, 16]:

1. The extent of impact of the abnormal situation (as it is assessed at the time), on the overall goal of production or start up or schedule.
2. The extent to which the cause of the situation is discoverable in time to avert the situation getting out of control. For instance; can it only be discovered after the situation is too far gone? And,
3. The impact on company reputation and compliance. For example, incidents widely quoted in the media, authorities revoking the licence to operate, litigations by regulators / public.

Couching it in the above terms for management attention will allow the decision maker to think in terms of the strategic risks (and not only the operational risks of avoidance of a shutdown in the moment). This will encourage a tendency to shift mind-set to longer term viability of the company instead of the short term gains in the moment of crisis [ref 24], – it is much better to avoid an incident altogether than to be skilled at dealing with it in real time".

(c) Always reduce human exposure and reduce energy levels when faced with an abnormal situation in incipient crisis –

This paper recommends that the following two key actions must always be performed at the onset of an abnormal situation going out of control:

1. Reduce exposure by removing personnel from the vicinity of any area of potential loss of containment, and
2. Reduce the energy level in the system by cutting back on plant rates etc.

The above actions are obvious responses to tentatively retreat to a safer situation until a more informed solution can be worked out.

This action immediately puts the operator in a frame of mind with the real possibility of a shutdown, and lessens any fear of him having to eventually shutdown, consistent with Stop Work Type action.

(d) Further developmental work and Research

(1) Plants should investigate their own abnormal events to learn from both, instances when SWA was invoked and from incidents when SWA was not invoked, and share the learnings. It will also assist in developing emergency scenarios to help in the early detection and timely diagnostics to be better prepared for handling potential similar abnormal situations.
(2) Plants should continue the research and development work in decision support system technology and in the area of narrowing the range (band) between residual risk and uncertainty

in way that is able to provide practical checklist type guidance for the decision maker in the moment of crisis. This latter aspect has the potential for premature shutdowns and must be properly balanced with the overall economics of the business.

Conclusions

This paper attempted to show that although Stop Work authority policy exists in the chemical industry, the exercise of stop work or plant shutdown action, prior to or during abnormal situations, requires major improvement. Stop Work Policy statements need to be augmented with clear and prescriptive guidelines for consistent and systematic actions.

We set out to show that providing a basis or an approach for consistent and systematic application of simple rules that we may be able to improve the decision making process and lighten the burden of the individual making the decision to shut down in abnormal situation.

The essential issue is the precise timing reflecting an acute balance of operational risks versus reputation and compliance risks. The decision maker in the moment of crisis, is unaware that he may have reached the point of impossible distinction between risk and uncertainty, when a decision has to be made for the better or the worse. On a day-to-day basis, this scenario plays out in the controlling of operational risks (in abnormal situation management), in the plant. In such times, every time, the reputation of the company hangs on how well the individual is equipped and supported to make the right decision and if not; to err on the side of safety. The best support he / she can have in such times is the unequivocal and express policy support of senior management indemnifying him / her from his / her shutdown action, which, only in the retrospect advantage of perfect hindsight may prove unjustified.

Both the strategic and the tactical approaches are crucial. Senior management involvement (by statement and visible demonstrated commitment of a sound SWA policy fortified by enabling checklist guidance), is most critical because the extent of operational risk exposure being taken may go unnoticed and unchecked until it reaches a level that represents an unacceptable threat to the company's operations and viability.

Ultimately we look for a system of assurance that consistently and systematically allows for the confident frontline execution of the SWA such that our accidents are prevented as well as inadvertent shutdowns are minimised. Furthermore, to seek buy-in; beyond compliance thinking; sound appreciation of operational risks and their control mechanisms is required. In other words, to forge a sound and sustained operational risk management culture.

References

1. Castle B, The Authority to Stop Work, Incident Prevention Magazine, June 2013
2. U.S. Chemical Safety and Hazard Investigation Board, Chevron Richmond Refinery- Pipe rupture and fire final report, August 2012
3. Zerarka S, Managing Risk through Combination of Hazard Observation and Stop-Work Authority Programs – Lessons Learned (Halliburton), December 2011

4. Howlett H C, The Industrial Operator's Handbook, Techstar 2001
5. Robert L. Cohn, Pilot Error, True stories behind general aviation accidents, 1994
6. Atherton J. and Gill F., Incidents that define Process Safety, 2013
7. Kletz T., Learning from Accidents, 2007.
8. Fragniere E and Sullivan G, Risk Management – Safeguarding Company Assets.
9. Chorafas D. N., Operational Risk Control with Basel II, 2004.
10. Unpublished, Internal Reports, Yara Trinidad Ltd.
11. Sutton I., Offshore Safety Management, 2013.
12. Rothenberg D. H., Alarm Management for Process Control, 2009
13. Kinney G.F., et al., Practical Risk Assessment for Safety Management, 1971.
14. Todd Conklin, Pre-Accident Investigations, 2012
15. Lindorfer J.H., Hazard Prioritization by the Numbers, 1987
16. Lindorfer J.H., Discrepancy Criticality by the Numbers, 1987
17. Dutch Safety Board, Summary – Explosions MSPO2 Shell Moerdijk, Report 2015
18. Smith E M, Out of the Ashes, Taiwan Business Topics, November 24, 2015.
19. Chow W K and Pang C L, Lessons learnt from the Great Kaohsiung Explosions and Disaster Management for Tsing Yi Island with Fuel Tanks, Fire Engineering, 2014
20. Bullemer P.T, Cochran et.al, "Managing Abnormal Situations II: Collaborative support for Operations Personnel", Honeywell Technology Center. https://www.asmconsortium.net
21. U.S. Chemical Safety and Hazard Investigation Board; Investigation Report, Refinery Fire Incident, Tosco Avon Refinery Martinez, March 2001.
22. Precautionary Principle, https://en.wikipedia.org/wiki/Precautionary_principle.
23. ALARP "at a glance", HSE UK webpage http://www.hse.gov.uk/risk/theory/alarpglance.htm
24. Andow P, Alarm Performance Improvement during Abnormal Situations ICHEME Symposium series No. 147.

Field Experience from Fiber Optic Ammonia and LNG Leak Detection Systems Installations

Distributed fiber optic temperature sensing system is a unique tool for the detection and accurate localization of small ammonia or LNG (Liquefied Natural Gas) leaks on pipelines that can reach ten kilometers or more. Such distributed temperature sensing (DTS) systems have the advantage of being easy to deploy over long pipeline sections and have been shown to detect leakage events with good accuracy and reliability. This paper will present the experience gathered from ten ammonia and seven LNG leak detection installations, including examples of real accidental leaks that were detected using the fiber optic system. Reliability and availability data based on the use of the ATTS (Automated Trip Testing System) will also be presented along with information on the rate of undesired alerts and results of a SIL 2 rating analysis. Based on those field experiences, we finally formulate recommendations on the installation of the sensing cable, the alarm settings and the deployment process, to guarantee reliable results while reducing interference with day-to-day plant operations.

Daniele Inaudi and Roberto Walder
SMARTEC SA, Manno, Switzerland

Todd Roberts
Roctest, Ltd, Denver, USA

Introduction

Recent developments in distributed optical fiber temperature sensing techniques provide a cost-effective tool that allows monitoring over long distances (up to 80km) with high spatial resolution (typically every meter). Using a limited number of very long sensors it is now possible to monitor the behavior of pipelines with a high measurement speed at a reasonable cost. Unlike electrical and point fiber optic sensors, distributed sensors offer the unique ability to measure temperature along their entire length. This capability allows the measurement of thousands of points using a single transducer. Using this technology, it is possible to detect leaks from liquefied ammonia and LNG pipelines by observing the characteristic temperature drop associated with such leaks. After a comprehensive testing and qualification phase [1, 2, 3] several systems are now permanently installed in real operating leak detection installations.

In this paper we will review the experience gained from those installations and provide recommendations for future similar installations.

Leak Detection with Distributed Fiber Optic Temperature Sensors

The most developed technology of distributed fiber optic sensors is based on Raman scattering as described in more detail in [3].

The typical components of a distributed temperature sensing system are the following:

- Sensing cable to be installed along the pipeline.
- Interrogator.
- Multiplexer to allow multiple fiber optic cables to be measured from one interrogator or to provide interrogation from both ends of the same cable for redundancy.
- Data analysis software for automatic detection of leaks.
- Relay module used to transfer alarm information to other plant control systems (e.g. to initiate automated emergency shutdown sequence).
- SCADA Interface.
- Automated Trip Testing System.
- User graphical interface that shows the exact location of a leak.

Figure 1: Fully redundant leak detection system

A typical rack cabinet containing two fully redundant is shown in Figure 1, while Figure 2 illustrates a typical cable installation. Figure 3 shows the graphical user interface, providing information about pipe route and leak location.

Figure 2: Red cable installation under pipe

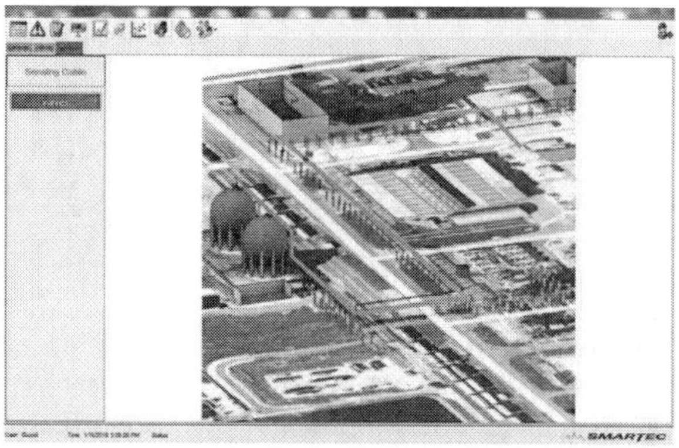

Figure 3: Example of user interface showing location of event, e.g. leak detection

The basic principle of pipeline leak detection through the use of distributed fiber optic sensing relies on a simple concept - when a leak occurs at a specific location along the pipeline, the temperature distribution around the pipeline changes. In the case of ammonia and LNG, we observe a rapid and localized drop of temperature

Automated Trip Testing System

One of the most effective tools to insure a high level of availability is the addition of an automated testing system. The ATTS (Automated Trip Testing System) is a device, fully independent from the data acquisition system, which can create an artificial leakage along the sensing cable and verify the correct response of the alert system (see Figure 4). Every hour, the ATTS automatically cools or heats 1m section of optical fiber at a rate similar to the one observed in the case of real leakage and observes the signal coming from the relay module to verify alarm triggering. A PLC (Programmable Logic Controller) is used to verify that for each testing cycle initiated by the ATTS, a corresponding alert is raised by the DTS. This configuration allows a higher level of self-diagnostic capability and insures that any issue in the system is rapidly detected.

This system configuration, including DTS, Multiplexer, sensing cable, relay module, leak detection software and ATTS, has received a SIL 2 certification with the following parameters:

Safety Integrity	SIL 2
"hazardous" failure rate (revealed)	7.73 10^{-6} per hour
"hazardous" failure rate (unrevealed)	0.12 10^{-6} per hour
"safe" failure rate (revealed)	0.47 10^{-6} per hour
"safe" failure rate (unrevealed)	0
Diagnostic Coverage	98%
Safe Failure Fraction	>98%

The failure rate was evaluated in respect of the following failure modes:

- In response to a valid temperature condition at the fiber cable, failure to deliver a loop disconnect output (open relay)
- Despite no valid temperature condition, deliver a loop disconnect output (open relay)

The rating assumes that the user can recognize the hourly "healthy" signal generated by the ATTS, by means of some logic function and to provide and act on a positive alarm in the event of no diagnostic signal. Thus, all failures within the ATTS, and those within the MUX, can only result in loss of hourly diagnostic function which, in itself, should be treated as a revealed failure. Failures not revealed by the use of ATTS are assumed to be revealed by an annual proof test which also covers the diagnostic features.

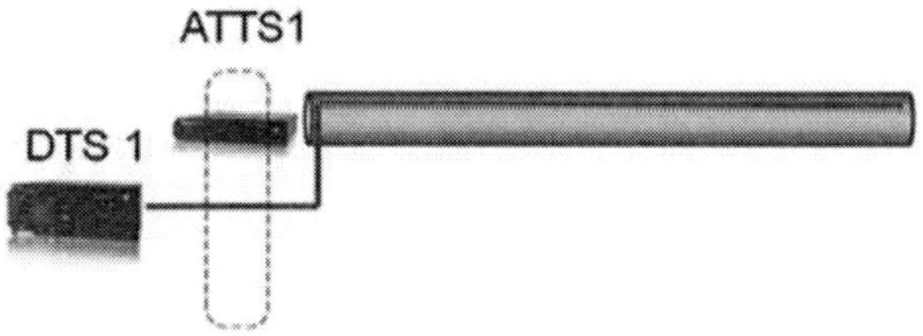

Figure 4: Example of simple leak detection system with additional self-testing capability

The overall integrity claim, which derives from the very high diagnostic coverage obtained, supports this safety-integrity assessment. As a consequence, both the unit and the simulated-based diagnostic unit must be subjected to the annual periodic proof test.

Field Installations

The leak detection system described in the previous chapters has been installed in a number of plants in Europe, as illustrated in the following tables.

Ammonia pipelines

Plant	Country	Year of install.	Approx. length of pipe	System configuration
Yara Ravenna	Italy	2006	5 km	2 x basic
Yara Le Havre	France	2010	10 km	2 x fully redundant
Yara Montoir	France	2013	2 km	1 x fully redundant
Yara Pardies	France	2013	1.5 km	1 x basic + ATTS
Yara Ambès	France	2014	2 km	1 x fully redundant
Borealis Grandpuits	France	2013	2.5 km	1 x fully redundant
Borealis Ottmarsheim	France	2010 pilot project	0.4 km	1 x basic (pilot test)
TOTAL			**23 km**	**14 systems**

4.2 LNG pipelines and tanks

Plant	Country	Year of install.	Approx. length of pipe	System configuration
LNG plant pipe	USA	2006	undisclosed	Ad-hoc
LNG Jetty pipe	USA	2007	2 km	Ad-hoc
LNG pipe	USA	2008	undisclosed	Ad-hoc
Grain Terminal LNG pipeline	UK	2008	4.5 km	1 x fully redundant
LNG transfer hose monitoring	France	2009	2 km	Ad-hoc
LNG tank	France	2015	1 km	1 x basic
LNG transfer line	France	2016	4 km	1 x basic + ATTS
TOTAL				**7 systems**

Learnings

For confidentiality reasons, the return of experience described in the following chapters will not make reference to a specific site, but rather summarize the lessons learned across all projects.

Detected leaks

Fortunately, none of the instrumented plants has suffered any major ammonia or LNG leak since installing the fiber optic detection system. There have been however two cases where a minor leak was detected.

The first instance concerned a very small leak from a bolted joint that was discovered during the commissioning of one of the systems. The leak was correctly identified by the system and was later confirmed by visual inspection. This leak was well below the minimum leak rate that requires immediate action.

A second occurrence concerned a coupling to a transportation vessel that leaked during an ammonia transfer operation. Also in this case the leak was correctly identified by the system and an alert was generated. The ammonia transfer was stopped and resumed after fixing the issue. Figure 5 shows the temperature evolution at the location of the leak. It is possible to observe how the temperature dropped from an initial value of about 20 °C (ambient temperature) to about -40 °C due to the leak. The location of the leak can be clearly identified at meter 887, but a larger section of cable is affected by the cooling effect of the ammonia evaporation cloud.

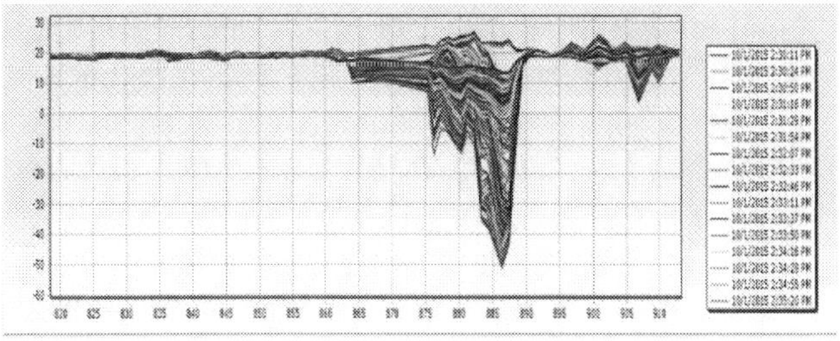

Figure 5: Temperature distribution along the sensing cable during a leak

The time evolution of the temperature at the leak location can be observed in Figure 6.

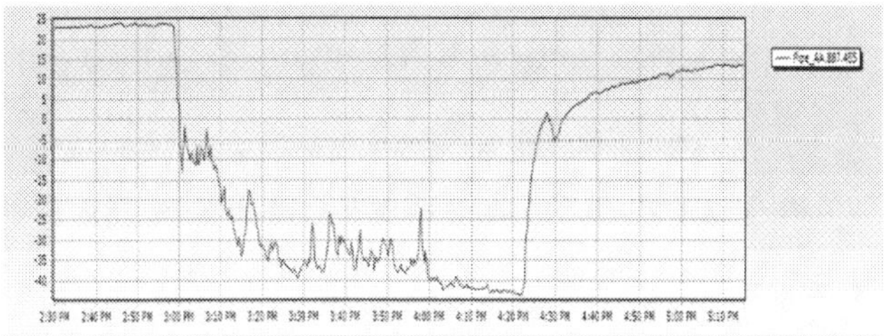

Figure 6: Temperature evolution at the location of the detected leak

Availability

Availability data from the sites were evaluated to quantify the reliability of the different components of the leak detection system. At the time of writing, the 14 systems that are in operation have added up 34 years of operation, with an average of 2.4 years per system. The recorded mean time between failures (MTBF) of the individual sensing components was evaluated to 8 years. The failure and/or maintenance of the individual sensing components (interrogator, ATTS, server) have prevented operation of the 34 systems for a total of 0.9 years, corresponding to an uptime of 97%. Considering only redundant system the uptime was 99% (with at least one of the two redundant systems available).

ATTS data

Figure 7 shows an illustration of temperature recording at the location of the cooling zone as a function of time. It can be observed that with an hourly frequency the apparition of the cooling peak is easily observable.

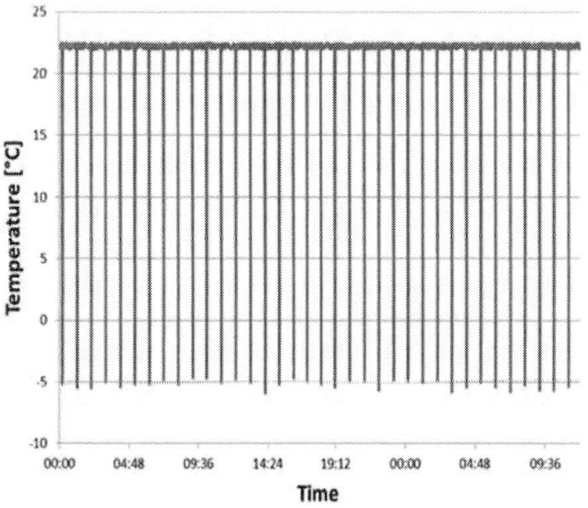

Figure 7: Example of ATTS temperature recording at the cooling location

Once hundreds or thousands simulated leakage events are generated, it becomes possible to calculate the probability of answer on demand (% of successful detections) and generate statistics on the reaction time (average and maximum delay). For example, a confidence level of 2 requires that 99% of the simulated leaks are correctly detected.

The data from the installed ATTS systems corroborates the availability data presented in the previous paragraph, respectively 97% for individual detection systems and 99% for redundant systems. When the systems were in service, the ATTS triggered an alert in 99.9% of the tests, demonstrating the high reliability of the ATTS device itself. Those results confirm experimentally the high availability expected from a SIL 2 certified system compliance.

False alerts

There have been no reported cases of false alerts, where an alarm was triggered without a corresponding temperature drop. There have, however, been numerous cases of false positives where unforeseen circumstances have triggered a "true" alert that did however not correspond to a real ammonia leak. The main causes of such false positives were the following:

- Trigger levels: temperature changes that would trigger an alert were initially set too aggressively and have generated false positive due to temperature variation induced by normal plant operation or weather conditions. After an observation period, the trigger levels were re-assessed and optimized to eliminate such false positives.
- Singular points: at some locations, extreme temperature changes were recorded due to external events. Examples include nearby fast cooling devices or vapor blows that would heat and cool the sensing cables. Those localized points were easily spotted by observing how false positives cumulated at the same location. It was then easy to correct those points by physically shielding the cable from the disturbance or adjusting the trigger levels of those isolated points.

Other encountered issues

The main obstacles that were encountered on the path to a successful deployment of ammonia leak monitoring system were organizational rather than technical. One common challenge in many sites concerned the transfer of system ownership from the project team to the operational team. In some cases this transfer did not occur with a sufficient training of the final operators. That caused misunderstandings on the way the system works and result in false expectations and the system sometimes became "orphan" since nobody was really taking ownership of it. In such cases it is possible that alerts generated by real leaks or warnings about failed components gets un-noticed. To avoid such misunderstandings, it is recommended to involve the production and maintenance teams from the beginning of the project and carry out in-depth training of all shifts to make sure the system functionality is properly understood.
In some other cases the opposite problem arose, when the system was transferred too soon to exploitation, before the initial false positives were completely eradicated, causing a general mistrust in the system that had to be reversed later on. In general, most safety systems that have no direct benefit on the day-to-day plant activities need to strike the right balance between staying out of the way of normal exploitation and not becoming forgotten and abandoned.

In two other sites, damage to the cable was caused by insufficient training of the maintenance personnel. They simply did not know how to handle the sensing cable if they need to service components supervised by the ammonia leak detection system. An early involvement of the maintenance staff is therefore highly recommended. At another site, the maintenance team was able to inform us about the most usual maintenance interventions that can affect the pipe and ancillary devices. It was then possible to install some extra length of cable at those locations to allow displacing the sensor and easing future maintenance.

Conclusions

The large number of field deployments of fiber optic based ammonia and LNG leak detection system has allowed learning valuable lessons on the technical and organizational sides. It was possible to improve the reliability and availability of the systems by implementing automated self-testing capabilities. The data gathered by the 14 systems currently in operation show an availability of 97%. All cases of non-availability were quickly identified so that they can be corrected rapidly or alternative safety measures could be implemented. The system is SIL 2 certified.

By monitoring false positives and understanding what operational or environmental condition would trigger unwanted alerts, it was possible to improve the leak detection algorithms and standard thresholds so that only real ammonia or LNG leaks would trigger alerts.

Finally, the process for deploying a system to a new site is improved by involving the right project and site team and by making sure that the system is finally owned by the people using it daily. Only one real leak of significant (although very small) size was recorded and correctly set off an alert.

Acknowledgements

We would like to sincerely thank Yara France, Yara Italy and Borealis France for the excellent cooperation demonstrated during the development and qualification phases of this innovative leak detection system and for letting us share the experience gained at their sites. We were particularly impressed by the excellent teamwork that could be established between competitors on topics related to process safety.

References

1. Inaudi D., Belli R., Walder R., 2008, "Detection and Localization of Micro-Leakages Using Distributed Fiber Optic Sensing", 7th International Pipeline Conference, IPC2008, Calgary, Canada.
2. Inaudi D., Glisic B., Figini A., Walder R., 2007, "Pipeline Leakage Detection and Localization Using Distributed Fiber Optic Sensing", Rio Pipeline Conference 2007, Rio de Janeiro, Brazil.
3. De Bont R., Inaudi D., Walder R., 2015, "Detection and Localization of Leakages in Toxic/Flammable Chemicals Pipelines using Distributed Fibre Optic Sensors", Nitrogen + Syngas, 28th International Conference proceedings.

Welding problems on ammonia plant primary reformer hot collectors

CF Industries operates an ammonia plant at their fertiliser manufacturing facility at Ince, Cheshire, UK. In May 2015 the primary reformer tubes were inspected as part of a general planned statutory inspection shutdown. The reformer is of Haldor Topsoe design – side fired with 160 vertical tubes. As a result of the inspection it was decided to replace 14 tubes due to excessive creep strain.

While attempting to weld the outlet pigtails into the sockets on the hot collectors, defects began to appear in the hot collector sockets. Multiple weld techniques and pre-heat temperatures were adopted without success. After discussion with the manufacturers and various consultants (metallurgists), it was decided to solution anneal the hot collectors. This action required removal of the hot collectors from the plant which resulted in a significant shutdown overrun.

The investigation concluded that the problems were caused by an unusually rapid cooling event (reformer trip) in 2013 very soon after plant start up with new hot collectors. This paper reviews the shutdown events and recovery/ start up. It then outlines the investigation that took place and reviews the findings.

John Hallett
CF Fertilisers UK Ltd, Ince, Cheshire, UK

Introduction

The primary reformer on CF Fertiliser's Ince site is of Haldor Topsoe design. It was originally installed in 1969 and was last fully re-tubed in 1992. It is side-fired with 160 catalyst tubes and 10 hot collectors. See figures 1 and 2.

The operating parameters for the tubes are as follows:-

Inlet temp/press = 520°C (968°F)/28 barg (420 psig)
Outlet temp/press = 765°C (1409°F)/26 barg (390 psig)

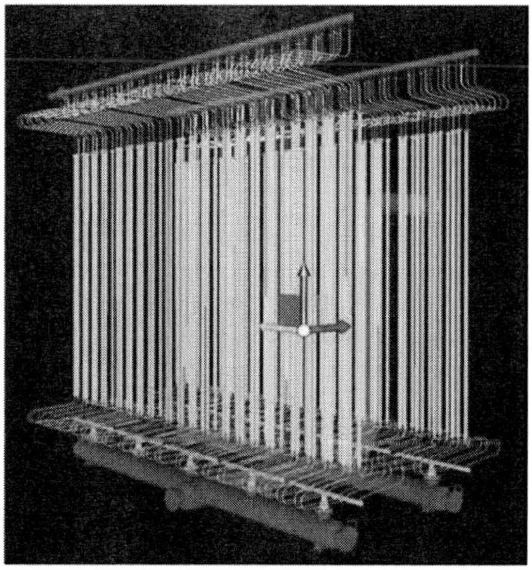

Figure 1 – Reformer view 1

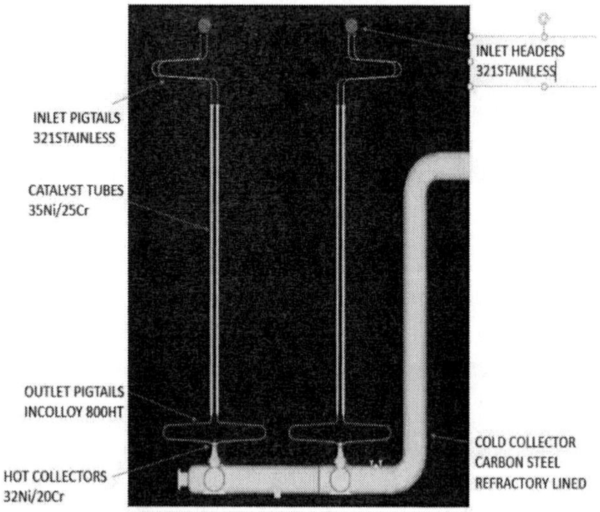

Figure 2 – Reformer view 2

In April 2015 a planned shutdown of the ammonia plant took place in order to carry out statutory inspections. These inspections will normally be done every 2 to 3 years. As part of the planned work the catalyst tubes in the reformer were examined for creep strain. This work was carried out by Quest integrity using their Mantis technique (this technique measures the OD of the tubes – the catalyst was not planned to be changed during this event).

Following the inspection it was concluded that 14 tubes would need to be replaced. This result was somewhat unexpected – it was thought that enough tubes had been replaced in 2013 to last until 2018 without further replacement.

In order to change the tube the bottom pigtail on each tube has to be removed. The top butt weld (tube to pigtail) and the bottom socket weld (pigtail to hot collector) both have to be removed.

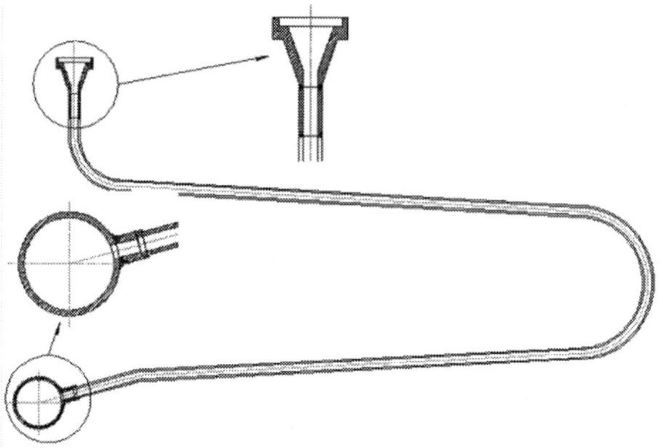

Figure 3 – Pigtail showing connections

This was done and the 14 tubes were replaced by lifting in/ out of the top of the furnace. The next task was then to weld the pig tails back in.

The hot collector arrangement is shown in figure 4.

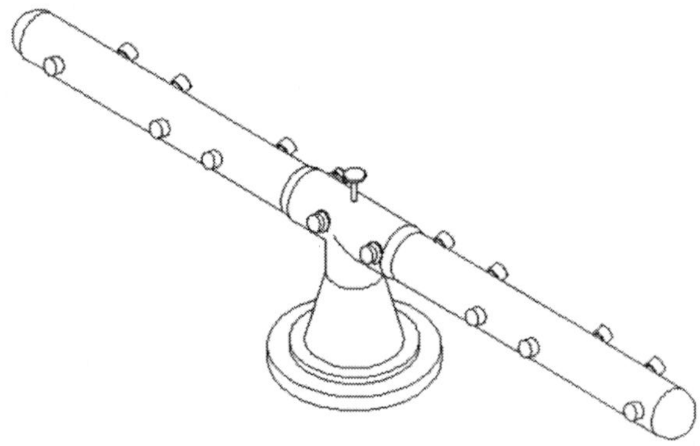

Figure 4 – Hot collector view

Welding of pigtails

The method of welding the pigtails back in can be summarized as follows:-

- Complete the butt weld onto the tube first. Argon purging can be established between the top of the tube and the bottom open end on the pig tail.

- Then weld the socket into the hot collector. Given this weld is essentially a fillet weld it does not need an argon purge.

The butt welds were completed successfully. However when welding to the hot collectors was attempted defects (cracks) were detected during the defined dye penetrant inspection (DPI). An example is shown in figure 5. The initial welding technique used was Orbital tungsten inert gas (TIG) with an Inconel 617 consumable. The weld geometry is shown in figure 6.

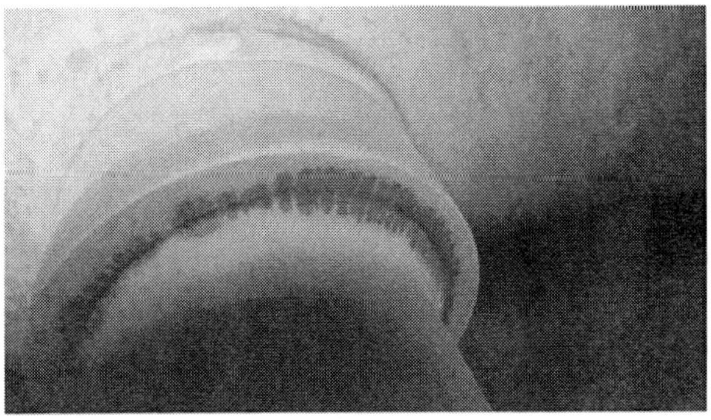

Figure 5 – View of weld defects

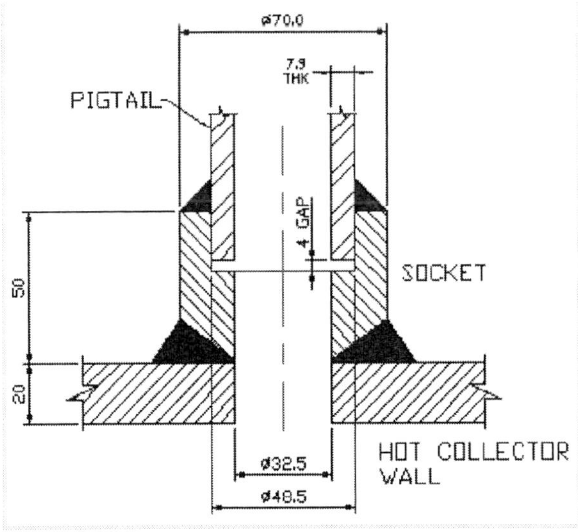

Figure 6 – Drawing of socket welds

Discussions then took place. Two courses of action were decided upon:-

(1) Attempts to weld using different welding techniques consumables and pre-heats.

(2) Metallurgical examination through replication.

Alternative welding techniques

A number of trials took place all without success. Examples of these were:-

An Inco 82 consumable was used rather than Inconel 617. This consumable would give a lower creep life for the completed weld but is more forgiving during the welding process.

Manual TIG and MMA (Manual metal arc) were also tried as welding processes.

At various stages during the trials different pre-heats were adopted (482°F, 932°F and eventually 1472°F). The 1472°F (800°C) pre-heat was eventually adopted for two reasons –

- To increase the ductility of the material ahead of welding
- To move the metal temperature above the critical 1100°F (600°C) temperature at which Incoloy 800H changes phase and becomes more ductile. See figure 7 which shows the properties of Incoloy 800H at 1100°F (600°C).

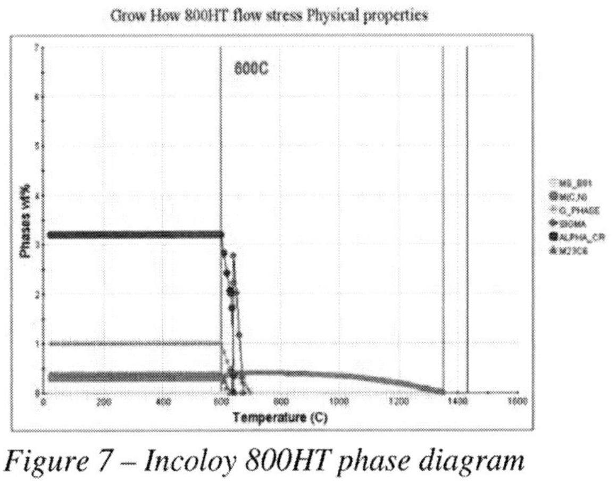

Figure 7 – Incoloy 800HT phase diagram

As well as welding to the socket attempts were made to put down a weld run on the surface of the hot collectors with similar defects being generated.

It was also noted during the trials that some of the defects took a very long development time to detect. We discovered this when a particular weld was passed clear and the developer left on accidentally overnight!

In summary though all these attempts to weld the material failed.

Results of metallurgical examination

Many replicas were taken during the welding trials. Figures 8 and 9 below are typical. The material had significant chromium carbide precipitates at the grain boundaries and twinning within the grains. Consultants commented the material looked significantly aged. ***It should be noted at this point that the hot collectors concerned had been installed in April/ May 2013 and were therefore only approximately 2 years old.***

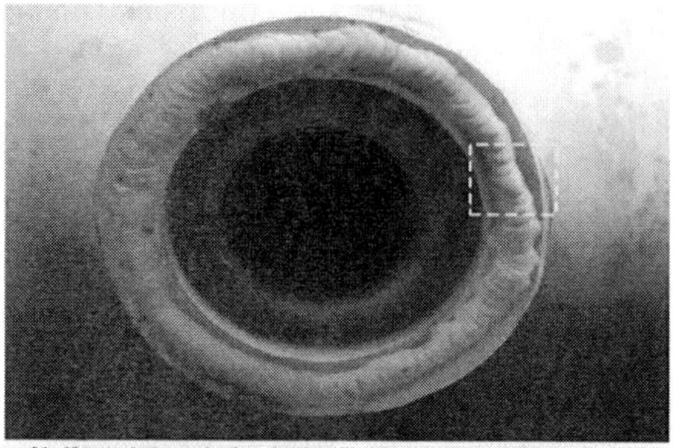

Figure 31: Macro photograph of socket 9 welded face showing dye pen indications, a replicated is shown.

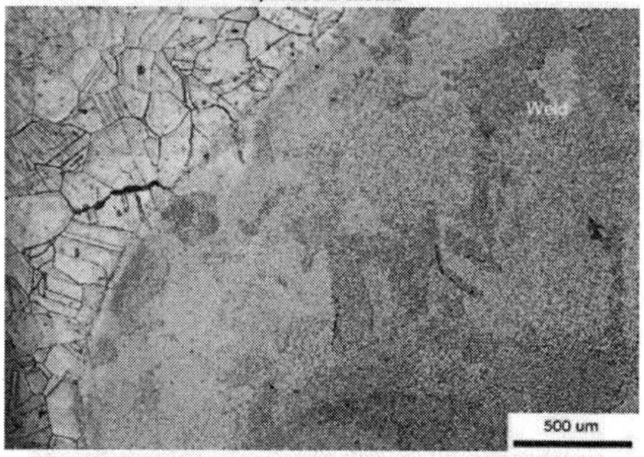

Figure 32: Showing the weld and the heat affected zone with a small tear.

Figures 8 and 9 – Microstructure of damaged material

Figure 10 shows a more normal structure:-

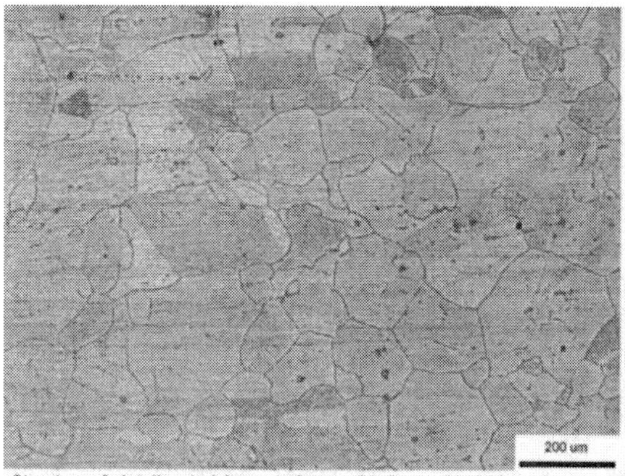

Structure of pigtail material remote from welded area. Variable grain size with few coarse precipitates visible.

Figure 10 – Microstructure of "normal" material

Discussion/ investigation

The nature of the replicas and the short operating life of the hot collectors suggest that the hot collectors had suffered from high levels of stress early in the life cycle. Investigation using the plant DCS history reminded us of a plant trip on 3 June 2013 approximately 4 weeks after the plant was put back on line after a planned shutdown. Figure 11 below shows the rate of cooling.

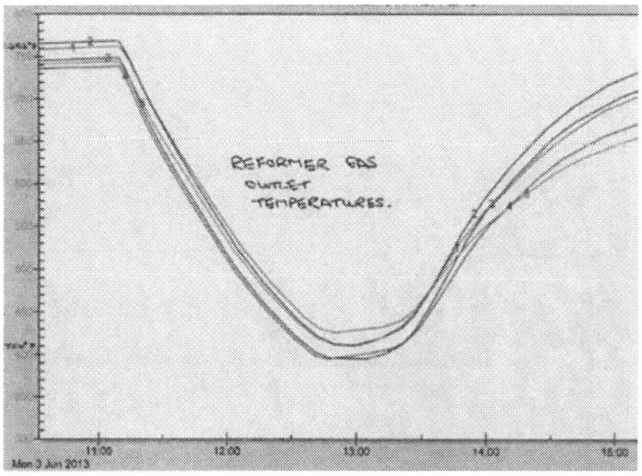

Figure 11 - June 2013 trip incident rate of cooling

The rate of cooling can be seen to be approximately 250°C in one hour. This rapid cooling would have generated considerable thermal stress. This stress would have introduced the formation of the "twins" (parallel grain boundaries) in the grains. Subsequently - as carbide precipitation had not fully taken place in the new material - the high stresses allowed chromium and titanium carbide formation at grain boundaries.

This combination serves to embrittle the material such that any further stresses applied – eg from welding – leads to cracking along grain boundaries.

It should be noted that while encountering the problems welding the pigtails to the sockets in 2015 there were no problems welding on to the pigtails. These pigtails were significantly older than the hot collectors and it was concluded the carbides had had time to disperse within the material.

Recovery actions – solution annealing

After discussion it was concluded that the only option was to remove the hot collectors and solution anneal them at 2000°F (1100°C) for 48 hours with 210°F (100°C) per hour heating and cooling rates. This course of action had several disadvantages associated with it:-

(1) The need to cut pigtails out of sockets and subsequently re-weld 144 back in rather than 14 (one hot collector was of a different vintage and was unaffected)
(2) Time taken to carry out the cycle
(3) The process is known to decrease the creep life of the resultant material. The extent is difficult to determine without destructive testing (which was later carried out – see below)
(4) The process will also reduce the strength of the material. Again the extent of this cannot be determined without destructive testing.

Following the decision to remove the hot collectors for solution annealing it was also decided to replace all the sockets on the headers for the following reasons:-

(1) Our inspections had shown the sockets to be in a worse condition than the spun cast tube.
(2) The socket to tube welds were also seen to be in a particularly poor condition.
(3) The socket to tube weld can be carried out in workshop conditions
(4) This approach ensured that virgin socket material would be welded in the more difficult site conditions.

(5) Time saved water cutting sockets rather than carefully machining out to allow re-welding.

A summary of the repair method eventually used is therefore:-

- Remove hot collector from plant. Sockets cut by HP water to reduce heat input/ further damage
- Solution anneal headers at 2000°F (1100°C) 48 hours
- Grind off remaining part of socket and remove socket to tube weld and HAZ (approx. 3mm depth). See figure 12.
- Weld on socket. Figures 13,14 and 15.
- Heat treat at 1710°F (950°C)
- Return to site and weld in pigtails
- Final NDT and inspection

Figure 12 – Hot collector with sockets and weld removed and "buttered" ready for new sockets

Figure 13 – Socket root and cap

Figure 14 – Repairs general view

Figure 15 – Completed socket replacement

Fitness for purpose assessment for resumed plant operation

At this point it was necessary to satisfy ourselves and our inspection authority that the plant was fit for safe and reliable operation. We had retained one of the solution annealed headers to carry out strength tests and creep rupture tests. However we would not have these results for some weeks (for the creep tests in particular). Resumption of production was justified on the following basis:-

- Replication of the annealed material showed a grain structure suitable for service (according to our consultants) but with significantly reduced creep life. A maximum of 2 years was agreed subject to further testing (see below).
- Further destructive tests on the "spare" solution annealed header were agreed to and progressed as quickly as possible.
- A rigorous inspection regime had been adopted in the repairs and re-fitting of the headers
- The situation would be reviewed following any plant trips or upsets. Detailed instructions were issued regarding actions to be taken in the event of a trip

The plant was brought into service on 1 July some 44 days later than planned.

Metallurgical (destructive) tests on solution annealed hot collector

Although production was resumed it had been decided to carry out tensile tests and stress rupture (creep) tests as follows:-

- Tensile tests - 3 samples from each of five areas – tensile test at room temp/ 400°C (752°F)/600°C(1112°F)/750°C (1382°F). ie 45 tests in total.

- Stress rupture (creep) tests - 2 samples from each of three areas. These were selected following the results of the tensile tests. These were done at a temperature of 780c and an applied stress of 54.7MPa.

The positions of the various test pieces are shown on figure 15 (more samples were cut then eventually tested)

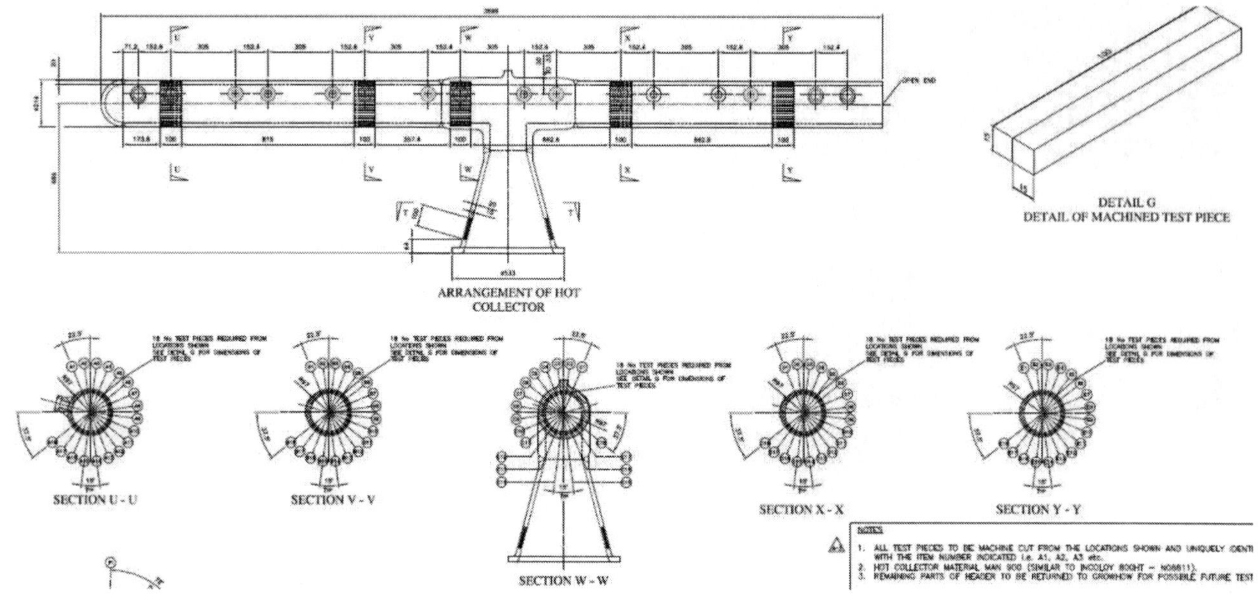

Figure 16 - Hot collector – locations of destructive test pieces taken

The results from the tensile tests gave some cause for concern. For example on the room temperature tests we had Ultimate tensile strength (UTS) as low as 332MPa and Rp0.2% as low as 41 MPa. We did however carry out specialist finite element analysis to verify the material was strong enough for the design conditions.

The creep tests were also disappointing. At the selected stress and temperature new material would be expected to last approximately 2000 hours. Our six samples lasted between 314 hours and 517 hours (average was 383 hours). Attempts were made to estimate a remaining life based on available Larson Miller curves but we concluded too many assumptions were needed and the result would be unreliable.

As a result of these tests – completed in November 2015 - it was concluded that the plant could not be operated for long enough to reach a planned shutdown in 2018 as had been hoped. It was therefore taken off line in Feb 2016 coincident with the planned shutdown of a downstream acid plant and the hot collectors were replaced.

Subsequent events

When the hot collectors were replaced in Feb 2016 problems were experienced when welding to the pigtails! No problems had occurred welding onto the pigtails in 2015! A typical defect pattern is shown in figure 16 below.

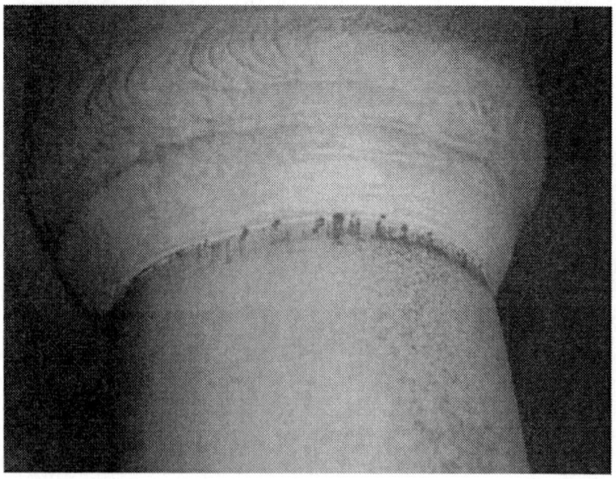

Figure 17

The approach adopted to solve this problem was similar to events outlined above. The ends of the pigtails were solution annealed and then welding proceeded. Replications showed the material was in better condition than the hot collectors the previous year (albeit unweldable) so a shorter annealing time was adopted (2 hours)

The reason for the 2016 pigtail welding problems is not fully explained at time of writing. However a similar trip to that of 3 June 2013 had occurred just before the planned shutdown in Feb 2016.

It is noted that the trips in June 2013 and Feb 2016 both had trivial causes and yet major consequences! In 2013 a scaffolder hit a fan stop button. In 2015 a cable tie failed resulting in a cable melting against hot pipework and tripping the same fan.

Conclusions and learning points

(1) **Minor events or failures can have major consequences**. Attention to detail essential.

(2) **Rate of reformer cooling can have a significant effect on metallurgy.** We were not aware of the potential and had successfully welded these items several times over the years.

(3) **To solution anneal or not?** We were forced to do so on this occasion but on talking to colleagues at our American sites they will sometimes pre-empt possible problems of this type by solution annealing up-front. Solution annealing up-front has the advantage of giving maximum confidence in successful welding with the disadvantages of up front time used and a certain degrading of the material.

(4) **The age of the material matters.** The structure of new material put into service will change in the first few weeks/ months of service. In particular carbide will be distributed within the material structure rather than concentrated at grain boundaries. Trip events early in material life therefore have the potential to cause increased damage

(5) **Material condition is likely to remain "hidden" until welded upon.** At high operating temperatures material will be ductile with no issues. Welding induces stresses leading to cracking.

(6) **Some defects only appear after a long period of DPI development in these circumstances.**

Topsoe Furnace Manager Technology Benefits for Ammonia Producers

Topsoe Furnace Manager (TFM) provides firebox images and data 24/7 monitoring burner flames and tube temperatures, to your organization locally and remotely without human interaction with the firebox.

Scott W. Sexton
Haldor Topsoe Inc., Houston, Texas

Introduction

Global food demand is expected to increase by 15% by 2030. Ammonia and fertilizer demand will increase accordingly in response to increased food demand. Higher expectations will be placed on existing and new plants in terms of efficiency, reliability, and compliance. However, safety of the plants remains the highest priority to be able to achieve all other expectations.

Personnel safety is very dependent upon personal behaviors. Increased focus on personal awareness and behavior-based safety programs has occurred throughout industry over the past 25 years. Also, process safety is dependent upon equipment, process design, hazard identification, and subsequent safeguards installed or implemented to mitigate the hazards' impact. These safeguards largely fall into two categories—administrative or engineered. Administrative safeguards typically require human interaction with the process, and engineered safeguards do not require human interaction. Engineered safeguards are typically considered more reliable, but costly.

The goal is to install safeguards that will economically mitigate process and equipment hazards--in an operating environment where additional demands are placed on personnel to administratively prove compliance with operating permits and a myriad of regulations.

A good safeguard provides information to drive action, or directly takes action, to mitigate the most severe hazards, and is reliable. A good safeguard is easy to use and understand. Pressure relief valves are good safeguards with audible feedback used extensively in industry to mitigate overpressure hazards, for example.

In ammonia plants, the primary reformer operation encompasses several hazards. Fire, loss of primary containment of process fluids, uncontrolled combustion, mechanical equipment failures, and process upsets are a few important hazards, or causes of hazards. Traditional hazard mitigation for primary reformer operation has required plant personnel to directly interface with the firebox to verify equipment and process status. This interaction has inherent risks, especially during potentially hazardous events. When personnel interact with a firebox to determine the status inside, they become human data collectors. It is recognized that this interaction is important, and tools have been developed to facilitate the best use of this interaction, including specialized personal protective equipment (PPE) and handheld data collectors. Even under the best of circumstances, the firebox interaction is typically a relative short duration, with only about 1000 hours or less of data collection annually. The amount of data collected in terms of temperatures and imagery is restricted by the duration of the personnel interacting with the firebox. To increase the duration of the interaction to improve data collection an increase in exposure to hazards occurs. Therefore, ammonia producers are faced with a decision to sacrifice firebox data acquisition to reduce hazard exposure. Extrapolation and leveraging of a limited amount of firebox data is the result of the decision. This extrapolation of limited data is typically historical based, as is the operation and maintenance of the firebox. Notable firebox failures occur from gaps in historical knowledge.

Topsoe Furnace Manager (TFM) provides an alternative. (Figure 1) TFM is an array of permanently installed image collectors acquiring images every second, 24/7. TFM collects millions of data points annually, with deviation alarms, stored in an easy-to-use historian, and is remotely available to the organization outside of the plant. TFM monitors your firebox with minimal safety risks. The virtual elimination of human-firebox interaction allows personnel to perform higher value work, such as image and data analysis, collaboration, and/or minor maintenance and turnaround planning.

Topsoe Furnace Manager (TFM) System:

TFM is a system-of-systems including electronics, optics, data handling, data analysis, placement design, visual display, self-protection, compliance, and human-machine-interface. Image acquisition is at the heart of the system. The image collectors send the images to the data collectors, which organize, transmit, store and analyze the data. Alarm safeguards (Figure 2) are immediately available to the plant operators, and remotely to the organization outside of the plant supporting operations.

Figure 1: Topsoe Furnace Manager Image and Data Acquisition Units

TFM Capabilities:

Image collection: Over 500,000 images per image collector per year

Data collection: Over 50,000,000 data points per collector per year

Alarms: Metal temperatures and burner flame quality, such as impingement, jetting, and poor mixing.

Benchmarked images of good examples and bad examples of burners and tubes

Historian: Time and date stamped images with data

Remote access to images and data

Wireless computer interface with firebox

24/7 persistence of the entire system

Self-protection for reliability

TFM Process Safety Benefits:

Process safety management mandates certain important areas of focus for industrial plant operations. Process safety represents best practice, and is a compliance requirement. TFM provides critical information for several process safety areas of focus.

Mechanical Integrity and Asset Protection: Asset protection is a very high priority in ammonia plant operations. Inside the firebox, catalyst tubes experience "creep." Creep is a metallurgical deformation with dependence upon stress, time-in-service and temperature-in-service of metals exposed to high operating temperatures. (Figure 3) Historically, reformer tube creep progression has been measured by reformer tube diameter growth over time, with measurements recorded during major firebox outages, such as during a catalyst change. These spot creep measurements are good indicators of remaining life, excluding consideration of significant events (power outages, startups, shutdowns, process upsets) which add a higher level of stress in a short period of time. Essentially, current creep prediction is calendar-based for an entire set of reformer tubes. Autopsy of reformer tubes typically indicates that not all tubes experience the same creep progression, and tube sections can also progress differently. This variation in creep progression for tubes with identical time-in-service is the result of variations in temperature-in-service (Figure 4), excluding local stress contributors. Temperature-in-service variations result from burner and combustion variations, combined with process variations. TFM provides images and temperature data to correlate combustion and process variations within the firebox. The resulting analysis and correlation provides specific mechanical integrity information that can

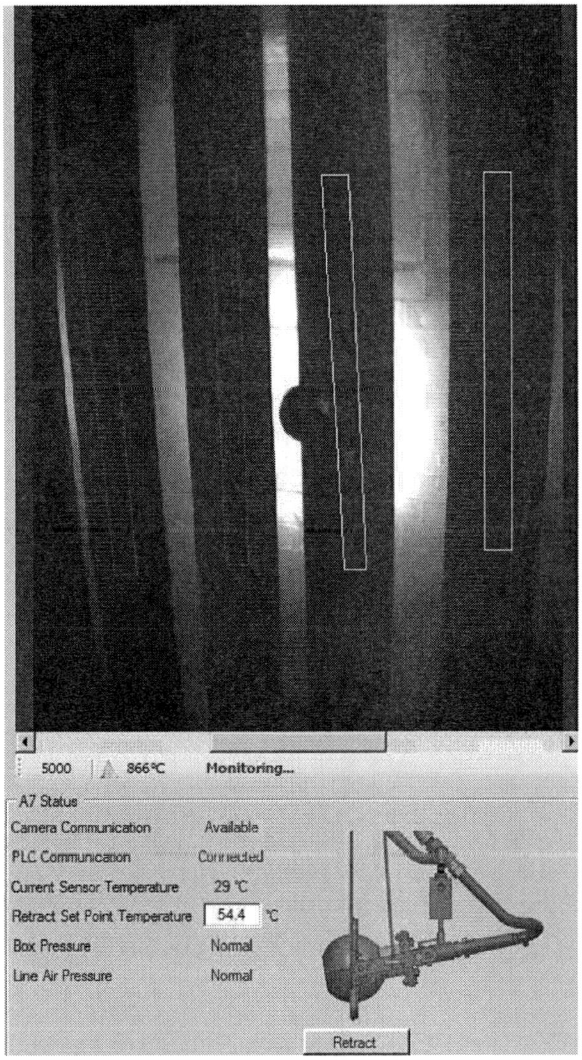

Figure 2: Hot banded tube with alarm

be acted upon by mechanical integrity and process subject matter experts (SME's) to reduce variation and improve overall tube mechanical integrity. TFM imagery and data also drive action on other aspects of firebox components, such as burner failure, refractory failure and structural compromise.

Asset Protection Economics: TFM pays for itself in an outage of less than a week if the outage is avoided utilizing TFM's temperature alarm function, and burner quality information;

based on TFM cost of $1,000,000, and $200/ton ammonia profit margin for a 1000 ton/day plant. For example, outage avoidance with TFM is the result of literally seeing flame impingement causing hot catalyst tubes, and making corrective actions before irreversible damage and an incident occurs. Additionally, if an outage can be avoided due to new operators' misunderstanding of firebox process conditions, such as upsets caused by draft swings, fuel swings, rate changes, or even weather disturbances, TFM pays for itself quickly. TFM enables new operators to see cause-and-effect as it happens in the firebox.

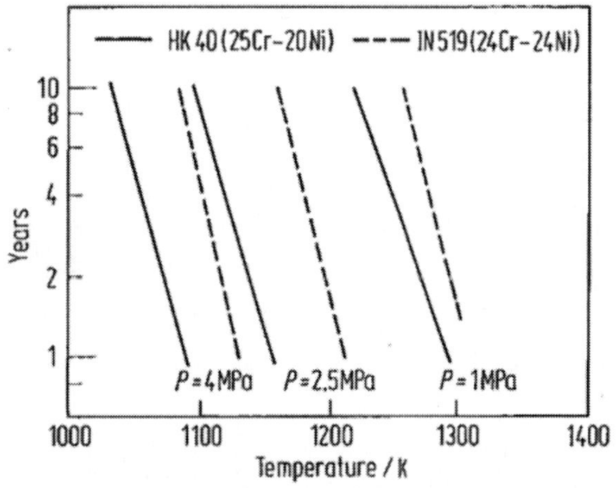

Figure 3: Example tube life and temperature correlations. A 10K temperature increase can decrease tube life by 30%

TFM's asset protection capabilities are especially useful on startups and shutdowns of the primary reformer, and TFM is working during normal operation (98+% of the time) when flame scanners are not active by design once auto ignition temperature is reached. Although unusual, loss of flame does occur at normal operating temperatures, and routinely observed on furnaces where TFM has been installed. Without alarms on these flame-outs, they go undetected.

Administrative Safeguard and Alarm History: TFM's alarm and historian capabilities also support the overall intent of process safety management by providing immediate feedback for operational mitigation, and historical capture of important equipment status. This duality of immediate information and historical data capture enables safe day-to-day operations, and continuous improvement through engineering over time.

Furnace Burner Balancing: Topsoe Furnace manager will provide the data and the tools to effectively balance firing to minimize the temperature variation across each row of tubes, as well as, throughout the furnace. (Figure 5) Based on previous reformer balancing studies, fuel gas consumption improvements of 1-2% are possible at the same production rate with savings estimated up to $100,000 per year. Burner balancing may provide a production increase excluding other bottlenecks. Balancing tube skin temperatures will result in longer tube and catalyst lives while significantly reducing the risk of tube failures and costly downtime. Expected efficiency improvements pay for TFM during a reformer catalyst life cycle.

Reliability Economics: TFM's burner balancing also improves reliability that results in savings over time due to longer catalyst tube life. TFM is paid for in a tube harvesting protocol based on creep measurements when about 10% of the reformer tubes avoid replacement at the first outage. 100% of the tubes in a world scale reformer with TFM installed documented creep of less than 1% at mid-life (about 45,000 hours) at normal operating temperatures.

Incident Investigation: Sudden incidents (such as power failures) on firebox component integrity have been difficult to correlate. Overheating events during startup and shutdown have been difficult to capture without increased exposure of personnel to firebox conditions. Subjectivity permeates most firebox incident investigations resulting in potentially inaccurate conclusions. TFM images and stored data show exactly when an event began, and the consequences of that initial event. TFM information is readily available to the entire incident investigation team, even remotely, in a matter of minutes so that the root-cause-analysis can begin quickly, and the next incident can be avoided.

Training: Training of new and existing personnel on firebox operations has traditionally involved a steady-state model with a minimal amount of information available regarding firebox behavior during transitions. With reformers, there are several dynamic scenarios that impact burner operation and catalyst tube temperatures, especially during transitions such as startup and shutdown. TFM captures images with data to provide actual transient information that can be implemented in training, including dynamic training simulators. Additionally, training benchmarks are captured, and can be inserted into training documentation to establish best practices.

Process Hazard Analysis: Mitigation of firebox overheating scenarios and consequences typically rely on safeguards such as fuel pressure, firebox draft, and spot process temperatures. These process measurements represent historical safeguards based on the capability of prior furnace process measurement technology. TFM provides additional firebox process measurements. Actual flame intensity and flame shape; and actual catalyst tube temperatures over a large surface area are now available to utilize as administrative safeguards without reliance on personnel directly interacting with the firebox.

Standard Operating procedures (SOP's): SOP's traditionally have relied upon historical operational information collected and documented based on past best practices. For firebox operations, SOP's typically require a person to physically inspect the firebox condition on a recommended frequency. This inspection frequency can be unrealistic, especially during startup and shutdown when other operational priorities take precedence. TFM provides continuous firebox inspection, and image targets with data targets can be captured and documented to provide new best practice examples. These examples can be imbedded within the SOP's, whether or not electronic, and provide direct comparison with real-time images and data captured by TFM during operation.

Compliance audits: Self-auditing of firebox compliance has traditionally relied upon written tube temperature and firebox inspection reports generated by operations personnel. The fundamental basis of the measurements is difficult to correlate. For example, tube temperature measurements are typically reliant upon a map generated by engineering or operations management specifying locations on the tubes and locations of the person performing the measurement, with specification of the measurement device (camera), and direction to capture results manually on a log sheet. This series of information handoffs is vulnerable to inaccuracy, even with the best efforts of trained personnel. TFM information is captured and stored with locations, device, images and data specifications held constant. TFM captures firebox burner imagery and data, which has historically been absent. Better audits result from better information and data.

Employee Participation: Employee participation in safety regarding fireboxes has typically involved PPE specification, SOP's for firebox interaction, permitting, lock-out-tag-out protocol, and other best practices heavily weighted in personnel safety. Process safety considerations typically have involved enforcement of SOP's and safety system calibrations. Process limits and trip points may not be well understood. Actual process safety limits regarding overheating of reformer tubes and poor burner performance resulting in flame impingement and localized tube overheating, afterburning, or flameouts are not fully considered due to lack of information. TFM provides that missing process information necessary for enhanced employee participation and understanding of firebox process safety.

Pre-Startup Safety Reviews (PSSR's): Firebox PSSR's typically focus on a review of process setpoints, process and instrumentation diagrams (P&ID's), safety system functionality, hazard review, and process transitions over an expected timeline. Human interaction with the firebox to collect data and interpret firebox condition during a period of constant change over at least 12 hours during a reformer startup is difficult to evaluate in a PSSR, especially for key personnel. TFM reduces or eliminates the level of physical interaction required during startup, and distributes the human interpretation element to a group utilizing TFM remote access capabilities. The natural intent of the PSSR to involve a group review with collaboration is enhanced and supported by TFM.

Additional Mechanical Integrity Information: Reformer firebox mechanical integrity information has historically consisted of design specifications, mechanical drawings, P&ID's, operational limits, and general operational guidelines provided by the technology supplier. TFM provides actual operating data in the form of images with catalyst tube metal temperatures and burner intensity and flame patterns. Combustion-side operational information can be correlated with process information to bridge the knowledge gap. Process and combustion upsets can be correlated with TFM information. SOP's can be tailored to specific furnaces based on their operational fingerprint. Training, including simulators, can be tailored to specific furnaces based on their actual behavior. A deeper understanding of the interaction between combustion and process side heat exchange is possible with analysis of TFM images and data.

TFM's Role in Personnel Safety

Evaluations, recommendations, and programs on industrial plant behavior improvement have become a big business for behavior professionals. Recognition of the importance of the human side of overall industrial performance is long standing. TFM mitigates many of the human issues associated with furnace operation. TFM improves individual performance through behavioral safety improvements.

Firebox Interaction: TFM reduces, or potentially eliminates, direct personnel interaction with firebox. Even when direct human interaction is required, it can be well planned, permitted with job safety analysis (JSA) evaluation, directed with specific information requirements identified, and minimized in terms of radiation exposure. In actual field use of TFM, furnace operations have continued with small process problems supported with strict enforcement of limited personnel interaction with the firebox. This enforcement included access restriction to the firebox upper deck during the time period the process issue persisted. By restricting personnel access to the firebox, and enabling a safe operation until proper materials were received, business risk was also reduced. The capability to view the problem and continuously compare its progression remotely provided transparency and confidence to all personnel.

Firebox Data Collection: TFM reduces frustration resulting from repeated firebox, visual data collection requests by non-plant colleagues. Collaboration is a great operational lever, and is highly valued in today's business environment. When data is remotely available, engineers can engineer, planners can plan, and managers can manage. When data is not remotely available, and with today's communication channels, such as cell phones, firebox information requests can overload those residing in closest proximity to the firebox. Often the task of firebox data collection falls on the furnace operators, or junior plant furnace image engineers. This can be a good training activity, but it also can lead to very high levels of frustration from those collecting the data. This frustration can lead to unpredictable behaviors in a plant environment, which is inherently dangerous, complex and frustrating by nature. There is high value by not adding any degree of frustration to a plant operating environment. TFM provides images and data remotely regarding firebox status. Requests for firebox information can then be directed toward TFM functionality rather than human firebox data collection.

Figure 4: Firebox color-coded temperatures

Furnace Stair Climbing: Although arguably good for health, physical exertion and stair climbing, especially on modern super-sized furnaces, is difficult. The design of many modern-day furnaces exasperates the exertion requirement. The furnaces are larger, taller, and more complex than in the past. The exertion required to climb up and down a large furnace is better spent ensuring that TFM remains on-station collecting important firebox images and data rather than looking through many peep doors and manually recording data.

Furnace Problems: TFM reduces confusion and anxiety during periods of time with furnace issues. Problems can be shared within the organization to make plans and provide guidance quickly. Furnace issues are nearly inevitable due to the nature of the operation. The ability to handle any problem is made easier with information. Without information, confusion, anxiety, and frustration build up to an unacceptable level. The remote viewing capability of TFM, supported by the capture of images and data stored in the historian, enable communications, planning and organizational coordination during periods of furnace problems.

Figure 5: Image of side fired burner rows

Furnace Guidelines and Documentation: TFM reduces confusion regarding linkage between hazard reviews, firebox SOP's, firebox critical safety systems, and actual operation. It is difficult and time consuming to 'connect-the-dots' between all sources of required process safety information. It is especially difficult to link process information if fundamental process parameters are not available. Reformer furnace tube surface area is one of the largest heat transfer surfaces in the ammonia plant. It is also one of the least monitored for data due to the harsh operating environment. TFM's non-contact temperature data and image acquisition provides the information needed to link process information so that it makes sense and isn't confusing. (Figures 6 & 7) This is especially important in fundamentally understanding critical safety system intent.

Key Performance Indicators (KPI) Attainment: TFM provides process documentation resulting in attainment of production KPI's while maintaining firebox components within their safe operating window. Ideally, all organizational KPI's are perfectly aligned and in support of an overall strategy that moves the business forward. In reality, there are typically some natural KPI conflicts, such as cost control and cost of repairs. These conflicts require human collaboration. Especially in the case of furnaces representing high cost operation, conflicting KPI's can lead directly to conflict. The conflict is best addressed with data and not emotion. TFM provides the data so that a collaborative decision-making process can take place.

Personnel Risk Exposure: TFM supports remote location of personnel with reduced risk exposure frequency. For example, a major, instantaneous reformer piping incident occurred on a reformer with TFM installed that resulted in no injuries because personnel were remotely located away from the furnace. Some older, existing plants were built with a lower emphasis on risk and exposure to personnel. Modern risk evaluations, including QRA's, actually quantify personnel exposure and risk.

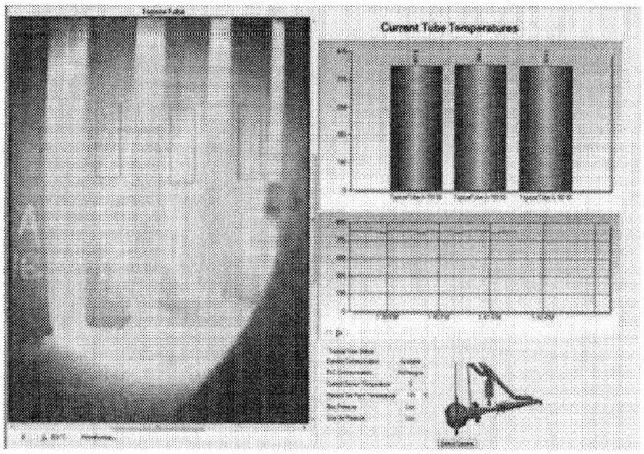

Figure 6: Reformer tube image with data

The quantified risk matrix considers work patterns, which are a function of equipment inspection frequencies, or "rounds." Furnace rounds have typically been historically set by previous habits and patterns established many years ago. With TFM, the furnace rounds are not required on the same frequency as historically understood to be required. Just as the industry trend has been to move personnel away from operations unless necessary, personnel within the plant itself are at less risk when exposed less frequently to the furnace firebox.

Figure 7: Raw reformer firebox image

Burner Evaluations: TFM eliminates frustration associated with human burner evaluations. Modern day, low NOx, industrial furnace burners are difficult to see with the human eye. When fired on a mixed fuel consisting of a significant amount of hydrogen, the low NOx burners develop a flame that cannot truly be evaluated with the human eye. TFM's capability to monitor flames which the human eye cannot see, and to provide alarms when the flame is under-performing, eliminates the need for personnel to visually decipher flame quality. TFM provides a flame intensity measurement which is charted and graphed continuously, so that the impact of rate changes and fuel/air changes is captured. When the flame is not behaving correctly, TFM will alarm to notify the organization that there is a burner issue. Sun burnt eyes from looking inside a firebox at flames too long are not necessary with TFM. Eliminating a source of physical pain is a very tangible improvement for personnel. TFM encourages burner tuning. Proper burner operation contributes directly to proper furnace operation. With TFM's ability to adjust view angles and regions-of-view, burner adjustments and burner performance improve with increased levels of understanding. Additionally, TFM captures the images to benchmark exactly how burners should appear during various process scenarios.

Weather: Operations personnel monitoring and managing large reformers face interesting and challenging scenarios involving weather events. Rain, snow, heat, cold, and wind all impact the ability to physically monitor a reformer. Weather compounds any difficulties associated with specialized PPE that is typically required for furnace interaction. Face shields and fire retardant clothing can be cumbersome during normal activities such as stair climbing. Weather events make the situation even more difficult, which can lead to personal safety events, such as tripping or falling. Importantly, serious furnace monitoring needs to occur during a weather event. Critical management decisions to continue operation during weather events depend on operations personnel climbing stairs and looking into the reformer firebox. TFM can eliminate the need, or greatly reduce the frequency, of physical human reformer inspections during weather events. This capability provides time for higher value plant activities while reducing personal safety risks.

Permitting: Lockout-tagout (LOTO), confined space, and vessel entry are permitted activities during furnace turnarounds (TAR's) and outages. These activities represent some of the most complex job permitting activities in an industrial plant. Furnace LOTO is a difficult process due to the process energy isolation and combustion energy isolation required. Air quality inside a firebox can also become contaminated if the firebox is part of a larger industrial complex. Continuous air monitoring could be required during firebox work on a TAR. And, confined space/vessel entry procedures are typically required during firebox work. Elevated work from scaffolding, and overhead hazards from crane equipment are also typically present during firebox work. Considering that firebox work during TAR's is also an infrequent activity, the more information and time to plan this activity inherently makes it safer to execute. Visual images of the firebox (Figure 7) from TFM are helpful to create understanding of where and how work will be performed in the firebox.

TFM's Role in Process Control Schemes: TFM provides data to establish actual equipment-operating-envelopes for tubes and burners. TFM data can be integrated into DCS. Tube temperatures and burner quality can be mapped and characterized for use in advanced firebox analysis.

TFM's Role in Operational Excellence: Support groups collaborate using the same TFM images and data as reference for discussions. Historical information is available "on demand" without relying on the operations group for retrieval. Compliance data is available at all times. Subject matter experts have remote access to furnace operations with time-and-date stamped images. (Figure 12)

TFM's Role in Furnace Lifecycle: Furnace equipment performance and repair history are captured with time and date stamped images. Temperature data is correlated with equipment performance, metallurgical analysis and repairs. Visual images are available for operations and reliability team training. Historical tube and burner performance is captured with images and data. (Figures 10 and 11)

Conclusion: Topsoe Furnace Manager is a proven personnel and process safety system for ammonia plant primary reformers, and has immediate economic benefits by avoiding outages, improving long term reliability, and enhancing operational excellence.

References:

1. FAO United Nations, "World Agriculture Towards 2030/2050", June 2012
2. J.R. Rostrup-Nielsen, "Catalytic Steam Reforming", 1984

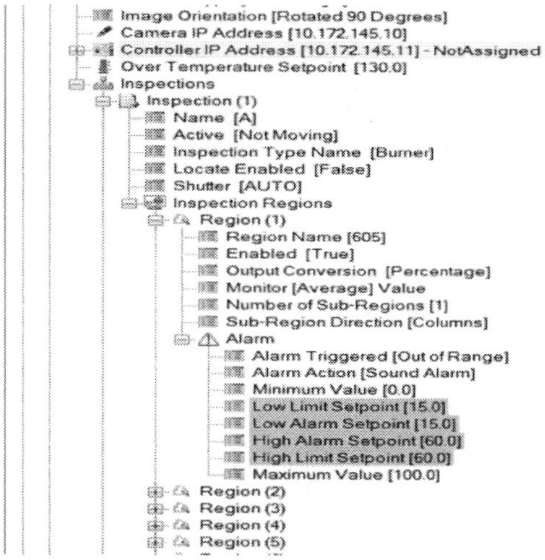

Figure 8: Data file tree organization

	A	B	C	D	E	F
1	Time Stamp	Furnace Load	Burner 24	Burner 23	Burner 22	Burner 21
2	Jul 31 2009 12:01:11 AM	75	96.33	88.35	70.23	74.05
3	Jul 31 2009 12:16:11 AM	75	97.14	88.01	68.97	76.53
4	Jul 31 2009 12:31:11 AM	75	96.32	86.08	69.09	79.05
5	Jul 31 2009 12:46:11 AM	75	95.19	86.03	72.75	77.25
6	Jul 31 2009 1:01:11 AM	75	96.77	87.26	70.56	74.86
7	Jul 31 2009 1:16:11 AM	75	96.38	88.4	69.67	75.13
8	Jul 31 2009 1:31:11 AM	75	95.98	89.18	72.81	77.61
9	Jul 31 2009 1:46:11 AM	75	96.72	87.28	70.69	77.73
10	Jul 31 2009 2:01:11 AM	75	97.21	90.35	73.3	77.55
11	Jul 31 2009 2:16:12 AM	75	95.78	87.17	70.62	76.6
12	Jul 31 2009 2:31:12 AM	75	96.48	85.72	69.47	76.05
13	Jul 31 2009 2:46:12 AM	75	96.93	87.94	69.94	75.95
14	Jul 31 2009 3:01:12 AM	75	97.81	89.08	71.21	77.3

Figure 11: Burner intensity historical data

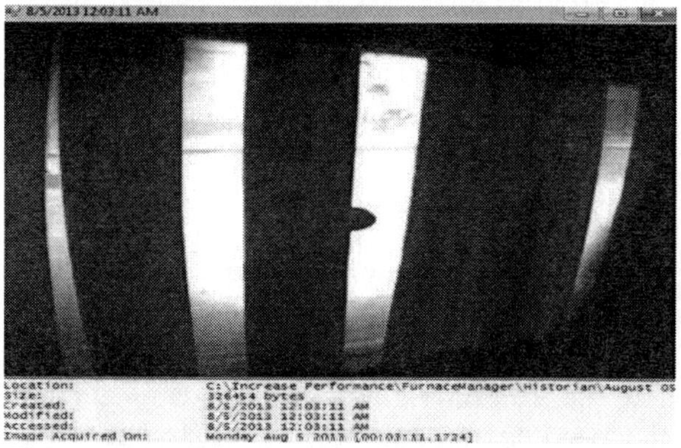

Figure 12: Time and date stamped image

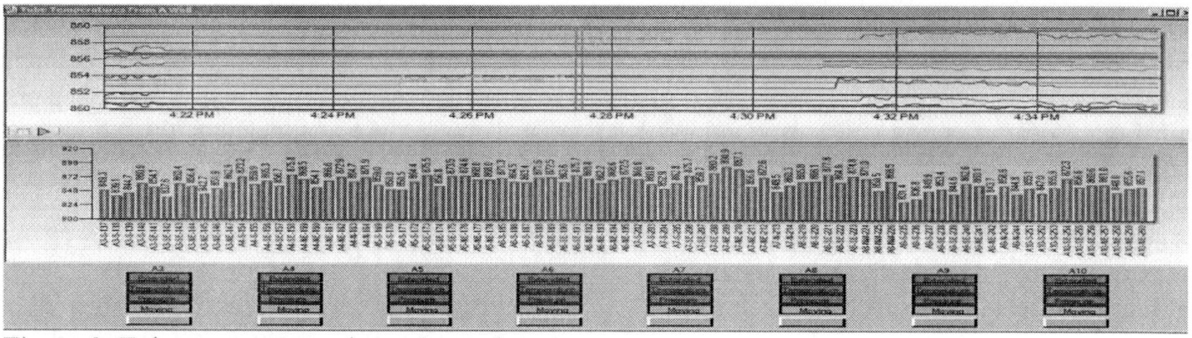

Figure 9: Tube temperature charts & graphs

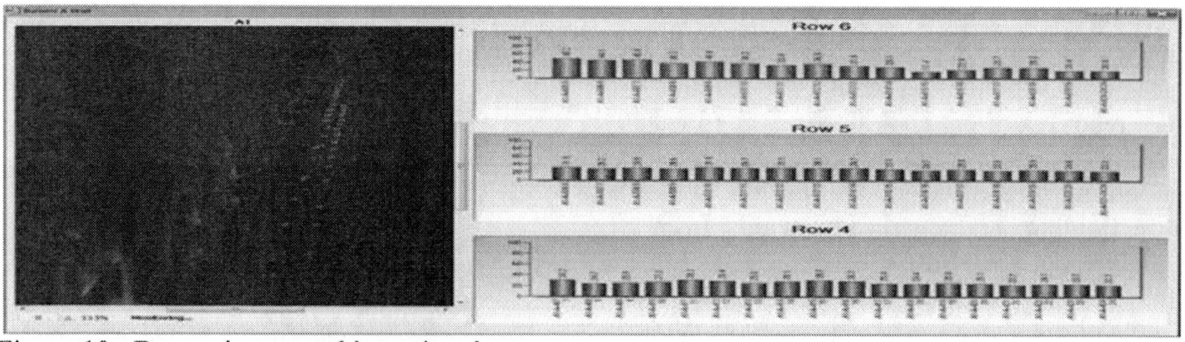

Figure 10: Burner image and intensity chart

FAILURE OF SECONDARY REFORMER DOME AND ITS REPAIR

A high pressure drop across the inlet and outlet of the secondary reformer in the Moranbah ammonia plant suddenly occurred on Friday 10th May 2013. The plant was shut down to determine the cause. After the catalyst and support alumina balls were removed it was found that parts of the refractory support dome had collapsed and that refractory balls and catalyst had filled the bottom of the vessel blocking the outlet transfer line.

Examination of the failed dome revealed that the bricks had primarily sheared at an angle close to 90 degrees to the major axis of the dome. It was concluded that the refractory dome failed as a result of high thermal stresses introduced as a result of the different thermal expansion between the refractory brick dome and the water cooled vessel shell. The design had no allowance for thermal expansion.

The dome was replaced with a novel design using readily available alumina blocks. The plant was returned to service 30 days after the incident.

Nia Aliabadi, John McGrath, David Keen and Venkat Pattabathula
Incitec Pivot Ltd, Australia

David Firth
Quest Integrity, Wellington, New Zealand

Introduction

A high pressure drop across the inlet and outlet of the secondary reformer in the Moranbah ammonia plant suddenly occurred on Friday 10th May 2013 when the plant was being restarted. Excessive metal temperature (700°C (1300°F)) was also noted in the refractory lined transfer line. The plant tripped and was taken completely off line to determine the cause. It was found that parts of the refractory support dome had collapsed and that refractory balls had filled the bottom of the vessel blocking the outlet transfer line.

This paper outlines the findings from the root cause analysis and the innovative modifications made to rapidly return the plant to operation.

Plant Background

The Moranbah ammonia plant designed by ICI using their LCA process, produces 450 MTPD of ammonia that is then converted to explosive grade ammonium nitrate. The ammonia plant was originally first commissioned in Yazoo, USA in 1998 and operated there until 2003. It was then purchased by Incitec Pivot moved to Moranbah and recommissioned in September 2012.

Moranbah in Central Queensland is roughly 1000km (600 mile) from Brisbane in a remote location. It is strategically located in the heart of one of the world's largest coal mining provinces to supply explosives.

Secondary reformer design, history and normal operational conditions

The plants reformer/secondary reformer system differs from a standard design as the primary reformer is a heat exchanger type system where the exhaust gas from the secondary reformer is used to heat the outside of the catalyst tubes in the primary reformer. However, the secondary reformer operates in an identical manner to one in a typical ammonia plant. The secondary reformer details are given in Figure 1 and Table 1. The refractory design details are shown in Figure 2.

Figure 1. The Moranbah secondary reformer.

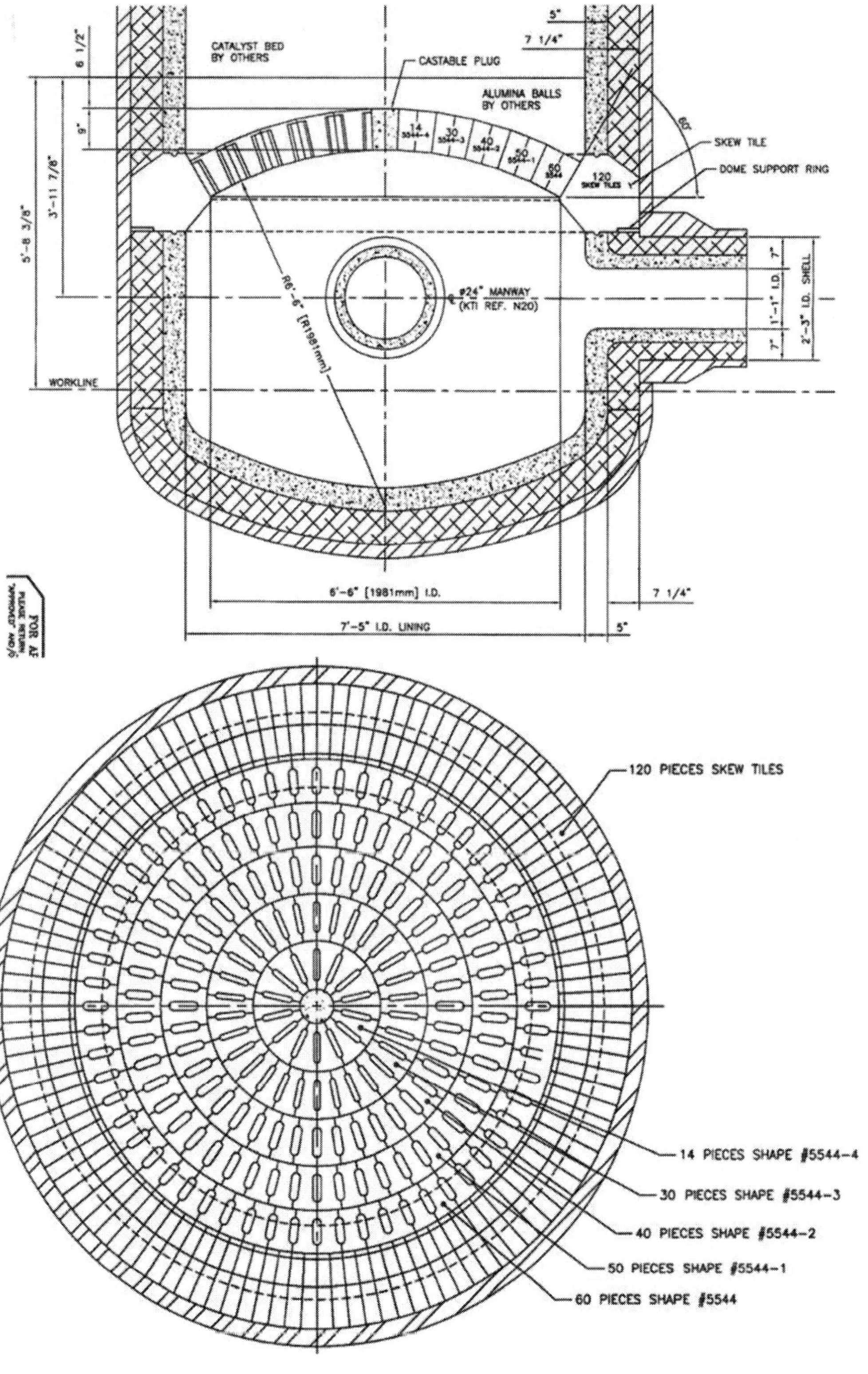

Figure 2. Layout of refractory in the lower portion of the vessel

Table 1. Secondary reformer details (dimension for lower catalyst section)

Internal Diameter	2883mm (9ft 5½ inch)
Wall thickness	76mm (3 inch) bottom
Shell material	SA516 Gr 70 carbon steel
Refractory in shell	184mm (7¼ inch) of insulation + 127mm (5 inch) of hard face high alumina refractory.
Water jacket	From bottom dome up the shell fed by boiler feed water nominally at 100°C (212°F)
Outlet temperature	Normally 950 to 985°C (1742 to 1805°F). At failure the temperature rapidly rose to 1080°C (1976°F). Note, in a normal ammonia plant the outlet is about 950°C.
Operating pressure	3400kPag (493psig)
Pressure drop	Normally 115kPa (17psi) At failure 900kPa (130psi)
Reformer outlet gas composition	11% CO, 6% CO_2, 0.6% CH_4, 30% N_2 and 50%H_2
Dew point	Approximately 177°C (350°F)
Support dome material	High alumina refractory bricks - Unicast 98, skew tiles - Greencast 95
Support dome layout	120 skew tiles supported on a steel shelf welded to the shell. Single layer of tapered bricks in a dome with a 1981mm (6ft 6 inch) bottom radius and 229 mm (9 inch) thick consisting of the followings rows of bricks . • Row 0: 60 positioned next the skew bricks. • Row 1: 50 • Row 2: 40. • Row 3: 30 • Row 4: 14 row • Central plug
Catalyst bed	About 10 tonne of catalyst + alumina balls

The bottom of the row 0 bricks sit on the skew blocks (also called skew tiles) and are mounted at 35 degree to the horizontal. The skew blocks are designed to fit against the inside of the vessel wall and sit on a steel support shelf.

The operational/history details available for the secondary reformer prior to being installed at Moranbah were very limited. However, in 2002 at a turn around and catalyst change, the dome was found to be damaged in some manner and was replaced.

After the shipping to Moranbah, the second dome was found to have collapsed. A new dome was installed in accordance with the original vessel design prior to recommissioning in August 2012.

After recommissioning in 2012, the Moranbah plant had numerous issues that resulted in the plant having 42 restarts before the 10th May 2013 incident. These typically were unrelated to the reforming section of the plant and ranged from Electrical and Instrumentation faults to rotating equipment failure. Every stoppage of the reformer area involves a rapid automated depressurization and 2 hour purge with nitrogen. Typically this would reduce the dome temperature to 650°C (1200°F)

Typically the secondary reformer operates at steady conditions. During start up and shutdown the secondary reformer is heated at a rate of approximately 100°C/hour (180°F/hour) from 450 to 670°C (840 to 1240°F) then 230°C/hour (414°F/hour) from 670 to 925°C (1240 to 1700°F) and cooled at a rate of 315°C/hour (567°F/hour) from 980 to 750°C (1800 to 1380°F) then 25°C/hour (45°F/hour) from 750 to 550°C (1380 to 1000°F), see Figure 3. The operational data was reviewed and there were no trends present such as an increased pressure drop that were observed.

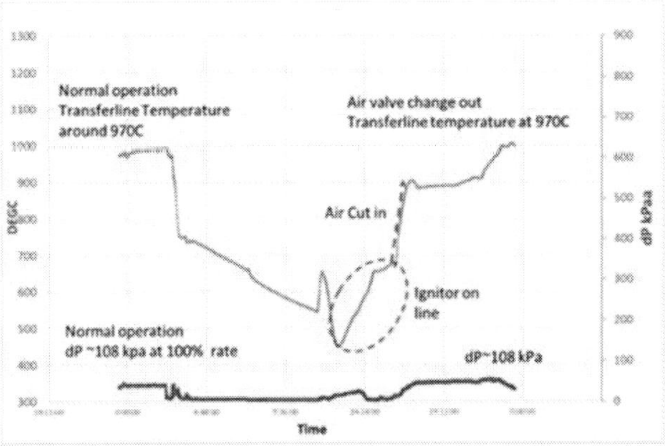

Figure 3. Normal temperature and pressure drop profile during start-up of the secondary reformer.

The incident on 10th May and observations

The plant had been running in a stable condition from 5th of May when it tripped on 10th of May at 2:11 am due to natural gas compressor discharge temperature high high. At 9:30am plant start-up commenced. At 17:09 pm after air cutting to the secondary reformer, the pressure drop suddenly increased from 104 to 825kPa (15 to 120psi) in 8 mins and to 909kPag (130psi) in 5 mins, see Figure 4. The plant tripped during the start-up due to secondary reformer outlet temperature high high (1080°C (1950°F)).

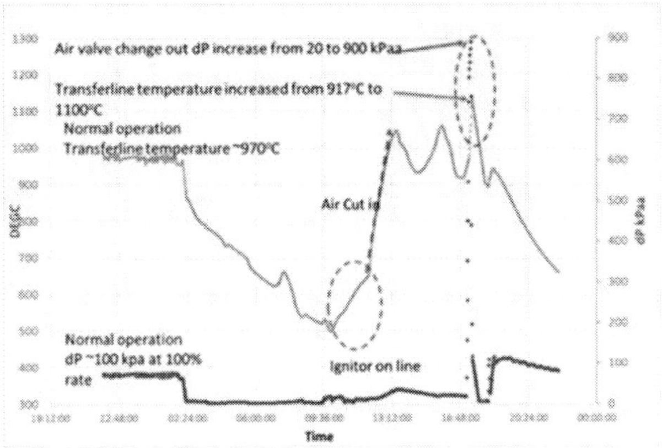

Figure 4. Pressure drop increase from 20 to 900kPa and temperature during the start-up on 10th May 2013.

When the bottom manway, below the refractory support dome, was opened it was found that the dome had failed in a number of locations and that alumina balls and catalyst had run into the area below the dome and blocked the outlet, see Figure 5. This also showed that the lower portion of the majority of bricks in row 0 and row 1 had broken off.

Figure 5. Below the dome.

The catalyst and alumina balls were removed from the vessel. Examination from above revealed that bricks had completely failed and were missing in three locations. These were in the West, East (see Figure 6) and South East Locations. In total 14 bricks in row 0, 7 in row 1, 2 in row 2 and 2 in row 3 had fallen out of the dome after failing. The bricks from the dome were removed. Examination of the skew blocks revealed that a large number had cracked and section had fallen off. After the removal of the skew blocks no damage was found to the inside of the shell or the steel support row which was present below the skew blocks. The shell diameter was measured and found to be within ±5mm (0.2in). Examination of the broken pieces of the dome showed that the failure of all the bricks was consistent with the dome trying to expand and being restrained from expanding; see Figure 7, i.e. the resultant force was radial.

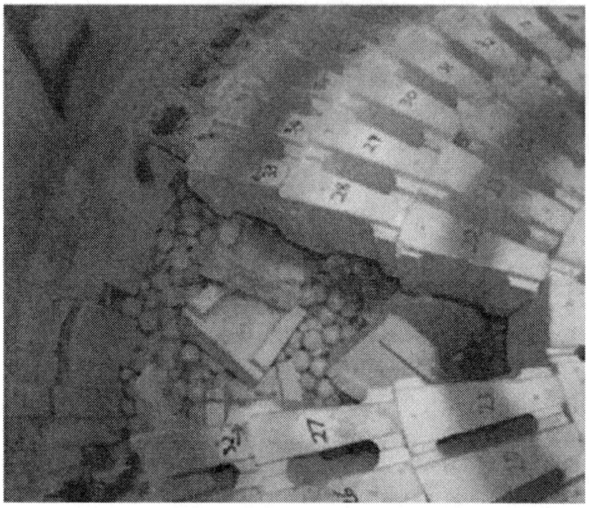

Figure 6. The refractory dome from above showing the East failure – cracked skew blocks and missing bricks from row 0, 1, 2 and 3.

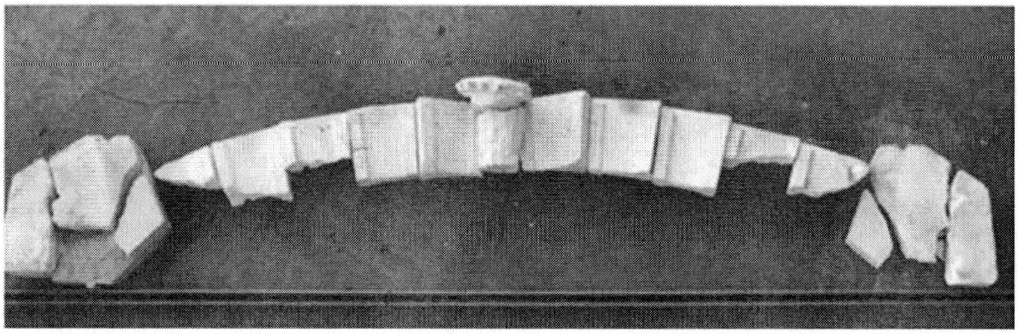

Figure 7. Typical cross section of the dome after the failure aligned in comparison with marker to show the horizontal plane.

The damage in the rows 0, 1 and 2 bricks is all consistent shear loading across the bricks with the failures running near to parallel with the horizontal. The damage to the skew bricks is consistent with high point loading where contact had occurred with top edge of the row 0 bricks, i.e. the loading was no longer evenly spread across the face of the skew bricks. The centre plug failed as a result of the compression point loading at the contact points with the tops of the row 4 bricks.

High alumina bricks are hard and not readily crushed. Whenever they are designed to be used in the high temperature condition the design should suitably allow for thermal expansion. For the design of the arch no thermal expansion had been accommodated for in the design. Typically this could include a softer mortar that is crushable, expansion gaps or a flexible outer portion such as alumina balls.

The refractory dome and the outer surface of the steel will expand or contract by different amounts as a result of start-up/shutdown. For a length of 2883mm (9ft 5in), the design inside diameter of the secondary reformer steel shell, it is calculated that this will increase in diameter by 3.5mm (0.14in) when it is heated from ambient temperature of 20°C (70°F) to water jacket temperature of 120°C (248°F) assuming an expansion rate of 12.5x10^{-6}mm/mm/°C (7 x10-6in/in/°F). In comparison a high alumina refractory brick of the same length of 2883mm (9ft 5in) will increase in length by 22.5mm(0.9in) when it is heated from ambient temperature to 950°C (1740°F) (operating temperature) assuming an expansion rate of (8x10-6mm/mm/°C (13 x10^{-6}in/in/°F). The difference in movement is 19mm. For a dome to be suitable for this application this movement of 19mm then needs to be compensated by a rise in the height of the dome. However, the movement is restricted by the weight and friction of the alumina balls/catalyst and the inability of the 9in (229mm) thick rectangular bricks to rotate. The design of the dome in a parallel (cylindrical) walled shell is not ideal as the shell acts as solid immovable anchor. If the shell is tapered at the location of the dome this may allow movement up and down the taper. Alternatively if the skew bricks had been at the more typical angle of 45 degree the forces may have been lower.

There are a number of different designs used for supporting the catalyst in secondary reformers. These include:

- A pyramid cone of bricks with each row of bricks sitting on top of the row below and each row being in compression from the weight of the catalyst supported on alumina balls. This design allows for thermal expansion by movement in the alumina balls.
- Multi-layered brick arch. This structure is far less stiff than a single layer of bricks and allows for shear forces between the bricks.
- A single solid plate of alumina mounted on a row a bricks. This design is similar to the pyramid cone as the expansions are taken up by movement of the alumina balls.

There are very few complete failures of secondary reformer domes that have been reported to have occurred. However, a failure occurred in a single layer brick design dome in 1982 at the Columbia Nitrogen fertilizer plant Augusta Georgia USA in an identical manner [1] to the Moranbah event. However, the cause of the failure was not identified. In addition, it was reported that damage occurred to the dome of the Moranbah vessel when it was operated in Yazoo.

Incitec Pivot operates four secondary reformers worldwide and this case was the first incident of a dome failure or any significant damage being seen to the domes.

Autothermal reformer after 4 years of operation

Partial damage to refractory domes is not unknown as can be seen by this case study. The relevant data is shown in Table 2 and Figure 8.

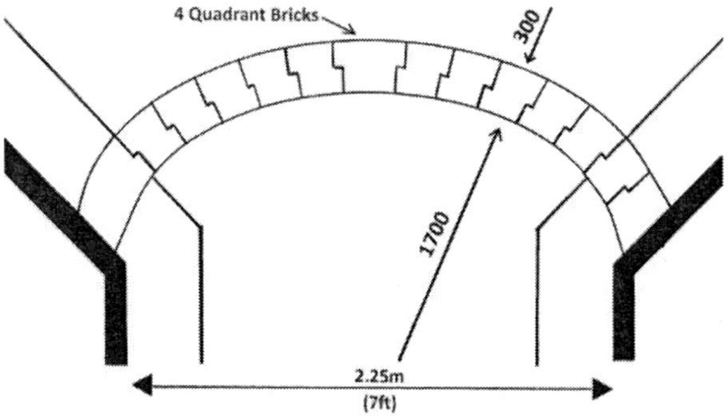

Figure 8. ATR dome design

Table 2. ATR

Age	4 years
Shell diameter	2.25m (7ft) at base of dome
Outlet temperature	950°C to1050°C
Shell temperature	Water cooled at about 100°C (212°F)
Support dome	Single layer of 300mm (12in) high alumina Z bricks with dome radius of 1700mm (5ft 6in). The centre plug was made from 4 quadrant bricks
Load supported by dome	Approximately 20 tonnes static load and 2 barg differential pressure dynamic load

A shutdown inspection revealed that all four of the quadrant centre bricks had fractured horizontally across the Z step line. The lower halves were lying on the floor below. The catalyst was removed so that the dome could be inspected. Note, health and safety requirements forbid the entry under the dome with the catalyst in place. No significant damage was found to any of the other bricks. However, it was found that the base of the dome was 25mm (1in) off being horizontal due to a variation in the fabrication of the steel shell and the original installation. The fractured centre plug bricks were removed and a new insert was cast in place.

Returning the Moranbah secondary reformer to service

When the incident occurred at Moranbah the plant had suffered numerous issues and a spare refractory dome was not available. For a dome the bricks designs vary between each row and as result it is not possible to get delivery of high alumina bricks urgently off the shelf. In addition, it was considered highly risky replacing the dome with the same design considering the history of the vessel. After the failure in Augusta they used a "goal" post design in front of the outlet. This was made from an Inconel 800 frame.

The Incitec secondary reformer at Gibson Island (GI) Brisbane Australia has a catalyst support structure made from alumina blocks spaced in lattice to allow gas flow. This had been in service for 10 years and no major issues had been seen.

A number of alternative designs were considered at Moranbah. It was decided to create a structure out of alumina blocks similar to that at GI using readily available material. The design used is shown in Figure 9.

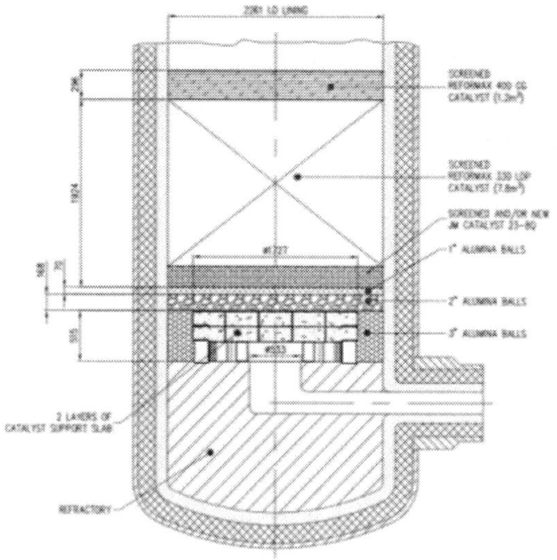

Figure 9. The new design for catalyst support for the Moranbah secondary reformer.

The new installed dome and repaired refractory required a very delicate dry out process this included four major steps as shown in Figure 10.

The catalyst support structure was installed, new catalyst was used and the plant was returned to service 30 days after the incident. Since installation the pressure drop has been within normal design window ranging from 101 to 108kPa (14.6 to 15.6psi).

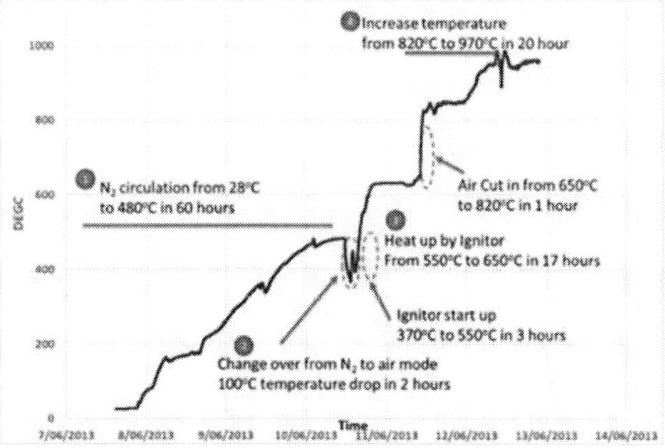

Figure 10. Refractory dry out and initial start-up of the secondary reformer after the installation of the new catalyst support structure over a five day period.

The support structure was inspected in March 2014. Upon inspection by a refractory expert it was seen that the refractory was in serviceable condition with no areas of spalling or significant cracking. Minor repairs were required to the refractory on the associated transfer piping to remove minor hotspots.

Conclusions

The Moranbah catalyst support dome failed as a result of high thermal stresses introduced as a result of the different thermal expansion between the refractory brick dome and the water cooled vessel shell. The design had no allowance for thermal expansion and high number of thermal cycles.

The dome was replaced with a novel design using readily available alumina blocks and the plant was returned to service after being shutdown for 30 days.

References

1. George E. Connaughton and Robert W. Clark Columbia Nitrogen Corp. Secondary Reformer Catalyst Support Dome: Failure and Repair, AICHE 1983.

Failures in Aging Ammonia Plants

In this paper some failures are described that occurred in the aging ammonia plants of OCI Nitrogen in Geleen. Three failures cases affected the reforming/waste heat recovery section of the plant. These failures could be related to the aging of the plants in combination with an underestimation of the impact of certain failure mechanism.

Jack Stoffels
Sitech, Geleen, The Netherlands

Introduction

OCI Nitrogen in Geleen The Netherlands (former DSM Agro) operates two ammonia plants, each with a capacity of 1550 mtpd of ammonia. One plant (AFA2) was designed and constructed by Bechtel; the plant was commissioned in 1971. The other plant (AFA3) was designed and constructed by M.W. Kellogg based on Kellogg's reduced energy ammonia technology; the plant was commissioned in July 1984.

In both ammonia plants a number of aging failures appeared in a relatively short period (the last 3 years).Aging is not only determined by the age of the plant, but is also about what is known about its condition. In this paper three failures are described, the actions taken to repair and the lessons to be learned to prevent these failures in the future.

Cases described in this paper are the following:

1. Failure of reformer inlet pigtails in the AFA2 plant
2. Failure of one of the steam risers between the waste heat boiler and the steam drum of the AFA3 plant
3. Crack indications at the transition between the reformer/riser tubes and the weldolet of the outlet header of the AFA3 plant.

Case 1. Failure of reformer inlet pigtails in the AFA2 plant

Introduction

After the Turnaround (April 2014), two leakages occurred just after the start–up of the ammonia plant (AFA2) at the inlet pigtails of the primary reformer.

Details

The inlet pigtails (in total 396 pieces) connect the inlet manifold(s) with the catalyst tubes (see figure 1).

The pigtails should absorb the movements between:

- Inlet manifold: displacement in horizontal or length of the manifold from the T-piece in the middle of the manifold (expansion at the ends is about 10 cm (4 inches)); and the
- Catalyst tubes: displacement in vertical direction as a result of thermal expansion of the tubes (movement about 20 cm (8 inches)).

There are 'long' and 'short' inlet pigtails because of the staggered row set up in the reformer. The material of the inlet manifolds and inlet-pigtails is 10CrMo 9.10 (2¼ Cr-1Mo steel), grade 22.

Process parameters

	P [barg/psi]	T [°C/°F]
Design	41/595	613/1135
Process	~ 33/479	~ 591/1096

Table 1. Process parameters at the inlet pigtails of the reformer

The age of the pigtails is 43 years. The pigtails were still from the original installation in 1971. During these 43 years it is estimated that these pigtails have had 43 x 5 = 215 start/stops.

The process gas flowing through the inlet pigtails is coming from the pre-reformer and contains about 19 (vol.)% CH_4, 75 % H_2O, 1,6 % CO_2, 4 % H_2 and small amounts of CO, N_2, Ar and He. The partial hydrogen pressure (pH_2) is about 1.43 bar-abs (20 psi), which is too low to cause degradation due to Nelson hydrogen attack. (See table 1)

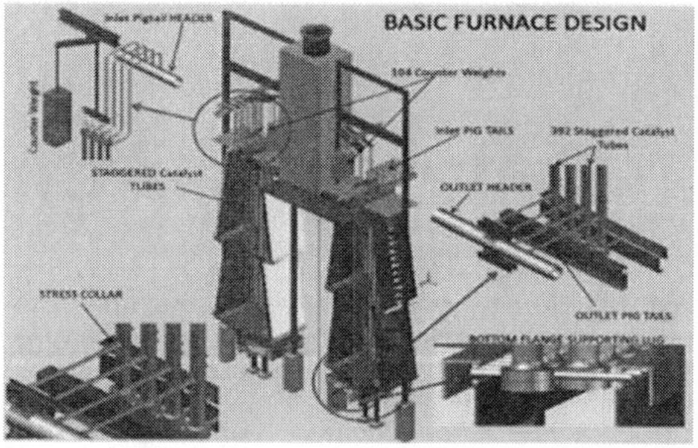

Figure 1. Basic Furnace Design

Visual inspection

After the detection of the leakages a visual inspection was performed and it was found that at two locations a crack was visible at the end of the weldolet close to the welded joints between the weldolet and pigtail (see figure 2).

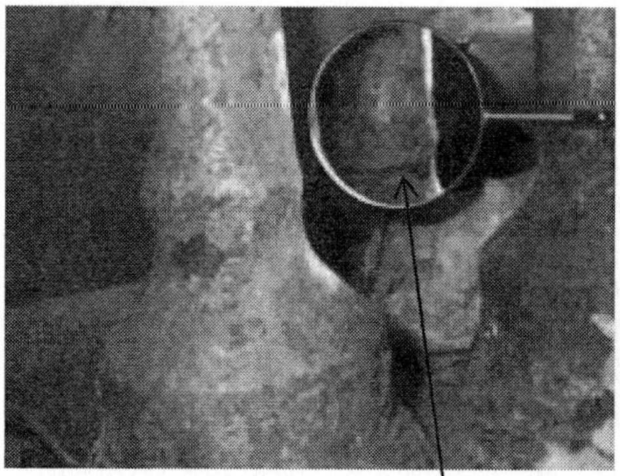

Figure 2. Indication of a Leakage

Failure investigation

A failure investigation has been carried out together with a third party Element Materials Technology on the removed tube part of one pigtail (see figure 3).

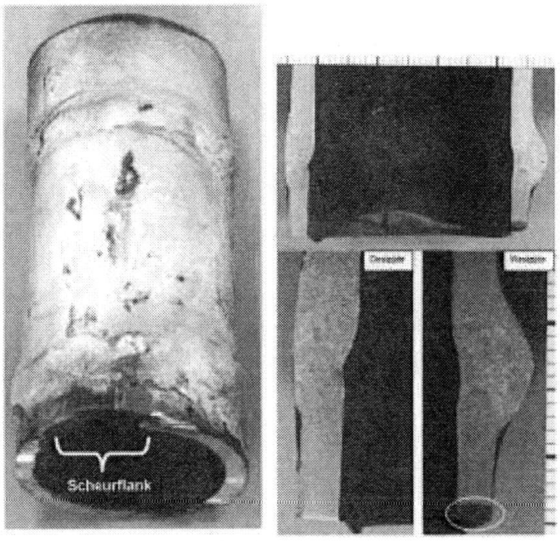

Figure 3.Removed tube part

Figure 4. Oxidation product

The oxide layer was present on both the in- and the outside of the pipe surface (figures 4 and 5)

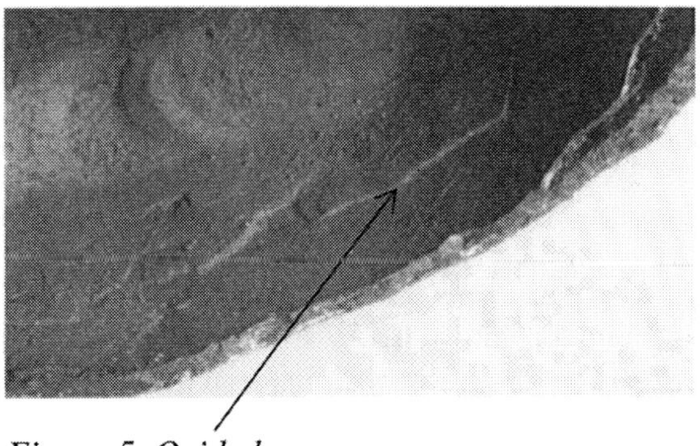

Figure 5. Oxide layers

The original wall thickness of the welding end of the weldolet is 4.0 mm (0.15 inch). while the original wall thickness of the pigtail is 4.85 mm 0.19 inch). (see figure 6)

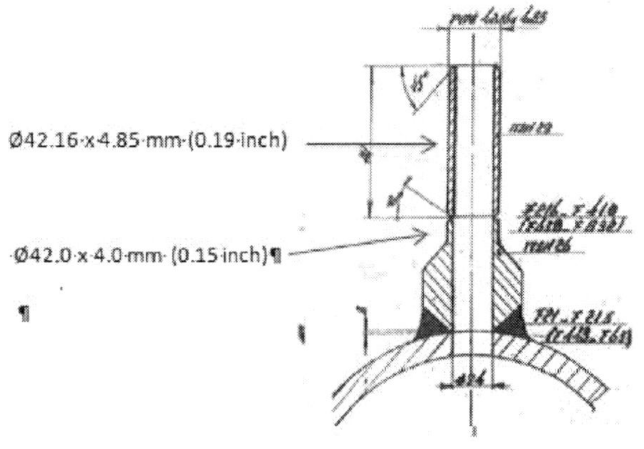

Figure 6. Welding end and pigtail

The wall thickness of the pigtail was decreased at the location of the cracked area till only about 1.2-2.5 mm (0.05- 0.1 inch) remained due to oxidation (both inside and outside).

From the failure investigation it was found that the crack was caused by a combination of low cycle fatigue, creep damage and oxidation

The crack flanks of the examined pigtails show also heavy oxidation product. (see figure 4) This indicates that cracks were already present for a longer time.

Discussion

The pigtails must compensate for the movement between the inlet manifold and catalyst tubes. This is accompanied by a changing load (every start/stop or hot/cold). Load cycles due to minimum firing scenarios of the reformer (hot-steam stand-by) will lead to bending of the pigtails tubes but these changes are much smaller than the hot/cold-swings caused by start/stop.

Because of the fact that this bending takes place under an internal pressure (35 bar, 500 psi) and at high temperature (600 °C, 1112 °F), there has been an interaction between creep and Low Cycle Fatigue (creep/LCF). This is also confirmed by the failure investigation done by Element Materials Technology.

Things that can negatively affect creep and LCF include:

High-temperature oxidation

This failure mechanism reduces the wall thickness. The high-temperature oxidation occurs both on the inside and the outside. On the inside by natural gas and steam; on the outside by air (oxygen). In the API RP-571[1] the oxidation rates are found in relation to the temperature is (in theory):

- 0,075-0,10 mm/0.003-0.004 inch/ year (till 1997, T= 570°C, 1058°F)
- 0.15-0.20 mm/ 0.006-0.008 inch/year (after 1997, T= 600°C, 1112°F)

In our NH3 plant we observed about 0,5-0,75 mm/0.02-0.03 inch in total since start-up (in/outside wall).

Temperature increase

Due to a revamp of the ammonia plant in 1997, the temperature at the inlet manifold of the reformer has been increased from about 570°C / 1058°F to about 600°C/1112°F. Due to this temperature increase, it is estimated that the creep/fatigue processes and the high-temperature oxidation are about three times faster after the revamp of 1997.

Smaller wall thickness

A smaller wall thickness of the weld end of the weldolet compared to the inlet pigtail caused a significant lower resistance to bending. The smaller wall thickness of the weld end was part of the design. Due to the thinner wall thickness the cracks/leaks occurred at the end of the weldolet close to the welded joints between the weldolet and pigtails (see figure 7).

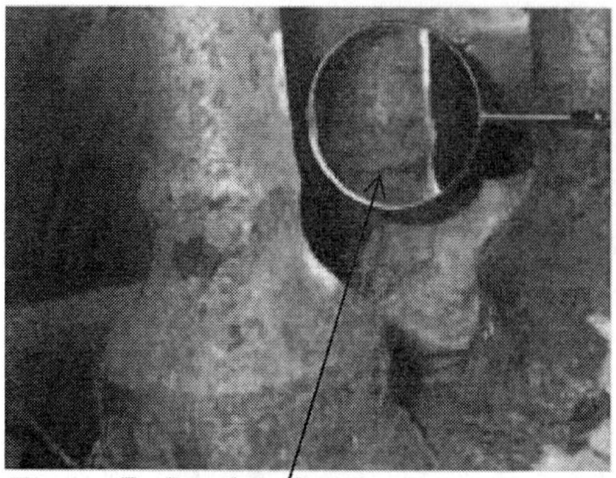

Figure 7. Crack indication

Negatively influence of the NDT inspections

The grinding of the surface area as part of the preparation for the NDT examination by magnetic particle examination, lead to further decrease of the wall thickness (see figure 8).

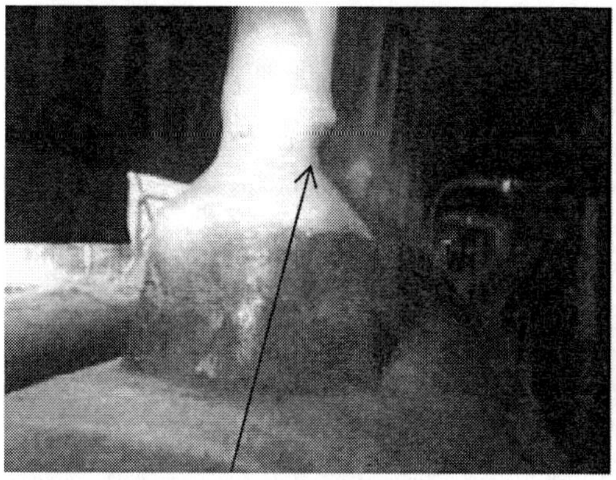

Figure 8. Excessive ground area

NDT Inspection of all inlet pigtails after this Failure

On-stream radiography

Due to the influence of the oxide layer on the surface (in/outside) it was found that discrimination based on on-stream radiography was not possible.

NDT examination, at the critical areas.

Knowing that the crack's started from the outside, all pigtails have been inspected by magnetic particle examination (black/white). No further crack indications have been observed.

Conclusions and Advice

<u>Remaining life time calculation based on oxidation and creep</u>

A remaining lifetime calculation was made by a third party company M.C. Know-How. Due to the increased temperature (30 °C, 54 °F) the high-temperature oxidation is about three times faster after the revamp of 1997. The outcome of the life time calculation showed that at the thinner part of the weldolet (4,0 mm/ 0.15 inch) the life time consumption was almost doubled compared to the pigtail (4,85 mm/0.19 inch). The results are shown in figures 9 and 10.

Life time calculation of the pigtail (d = 4,85 mm/0.19inch)

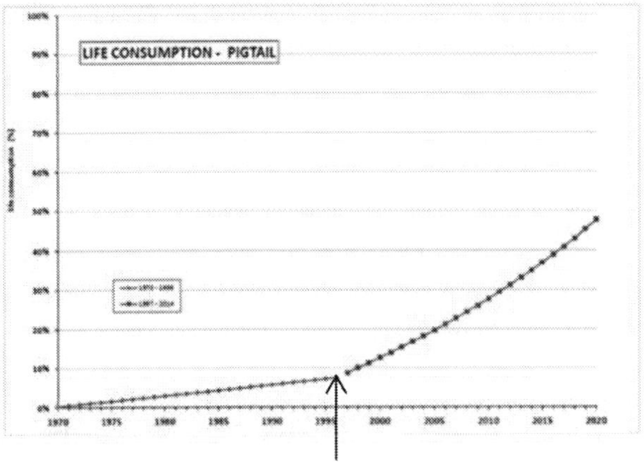

Figure 9 1997

Life time calculation on the end of the weldolet (d = 4,00/0.15 inch)

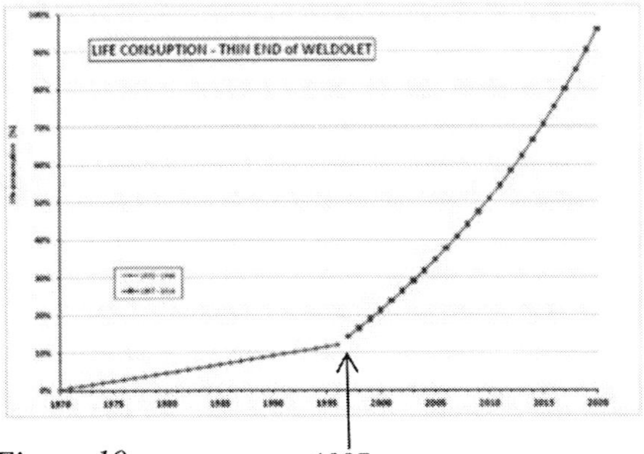

Figure 10 1997

Note: X-axis, lifetime consumption/ Y-axis , operating time

- The inlet pigtail leaks are caused by an interaction of creep and low cycle fatigue and high temperature oxidation after 43 years operating time. The above mentioned failure mechanisms have been accelerated due to the increased operating temperature with 30°C since 1997.
- The leakage occurred in the weld end of the weldolet with the smallest nominal wall thickness (4,0 mm/ 0.15 inch). The crack started at the outside.
- In a lifetime calculation it was shown that the combined effect of creep and wall thickness-decrease by oxidation resulted in almost <u>double life consumption</u> at the end of the weldolet.
- It was advised to replace the inlet pigtails in the next turnaround given, because the pigtails are at the end of life.
- For the repair/replacement it is recommended to use a stainless steel type (for instance 347H/ 304H).

Lessons to be learned

- High temperature oxidation is a slow process but due to the aging of the plants this failure mechanism must be included in the inspection program.
- When carrying out NDT inspections, attention must be given to the surface preparation. It should be carried out in such a way that wall thickness does not decrease.
- Temperature increase and oxidation reduce the life of the pigtails and has to be implemented in the life time calculations.

Case 2.Failure of a riser between the waste heat boiler and the steam drum, AFA3 plant

Introduction

In 2015 one of the four steam risers cracked between the steam drum and the waste heat boiler. During the inspection the crack was observed in Northern Riser (DN150 riser tube).The cracked riser was removed for further failure investigation.

Details

Between the steam drum and the waste heat boiler there are four risers, one with a Ø169,3/6.67 inch x10 mm/0.4 inch and three with Ø 360/14.17 inch x 36mm/1.42 inch. The most northern riser (red pipe with green circle in figure 11, location with the smallest diameter and wall thickness) ruptured between the waste heat steam boiler and the steam drum (see figures 12 and 13).

Pipe material:15NiCuMoNb5-6-4,
material identification no. 1.6368 (WB36)
Medium: water/steam
Process pressure: 125 barg/1813 psi
Process temperature.: 327°C/620°F

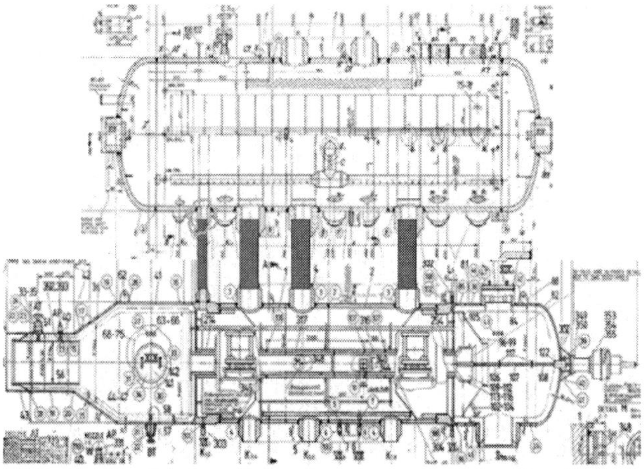

Figure 11. Steam drum (top), waste heat boiler (bottom), red, with green circle the cracked riser. Blue the other 3 risers

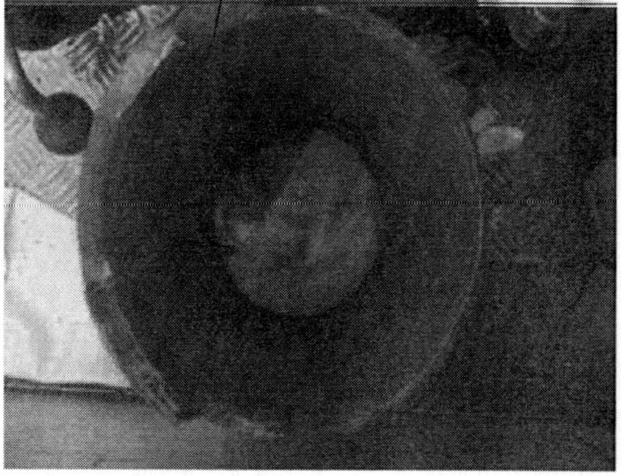

Figure 12. Ruptured riser

Figure 13. Failure piece, (black oxide layer is visible). Also in the cracked part.

Failure investigation

The failure investigation of the ruptured riser pipe has been carried out together with a third party, Element Materials Technolgy.

The conclusions of this investigation were:

- The welded seam of the riser was already cracked through about half the wall thickness before it failed due to mechanical overload (see figure 14). The crack started from the steam side.

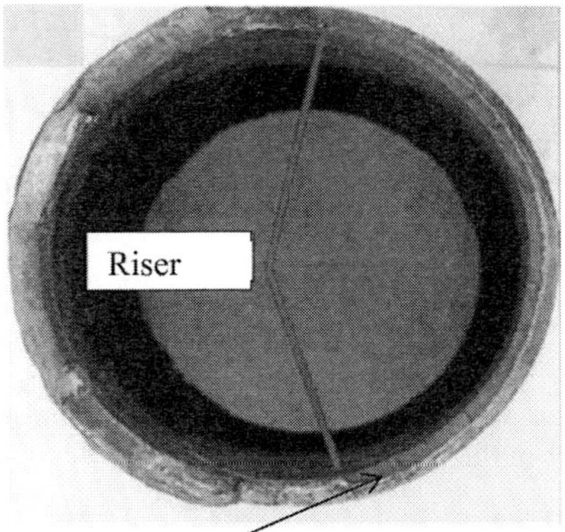

Figure 14. Oxide layer

- The presence of an oxide layer on the existing crack indicates possible step-by-step crack propagation in the circumferential direction. (see figure 15).

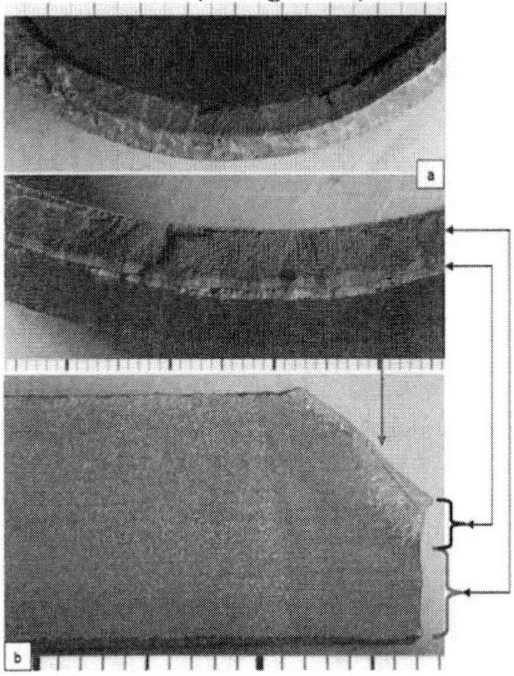

Figure 15. Step by step propagation

- Because the surface was oxidized it was not possible to detect the crack initiation location.
- The existing older crack was relatively small compared to the total cracked surface. The rupture of the pipe is not only caused by the existing crack but can only be explained by a very high stress level at the time of the failure (overload).
- The orientation of the older (primary) oxidized crack (in the length axis of the cracked gas steam boiler and steam drum) is possible an indication for up-set conditions and/or startup/shutdown.
- There are no indications found in the current micro structure of the material which could have had a negative influence on the observed cracking. The chemical composition and measured hardness of base and weld material are adequate for the applied materials.

Discussion

Based on the results of this investigation it can be stated that the failure mechanism that caused the initial cracking has been initiated by corrosion due to strain related cracking (or distortion induced corrosion) of the protective magnetite layer, also known as corrosion fatigue.

This failure occurs as follows:

The normally protective oxide film (magnetite layer) on C-steel in steam systems can crack by elastic deformation of the underlying steel. This oxide layer cracks due to high mechanical stresses; for example by start-stops, pressure changes, temperature changes, quality of the formed oxide layer (density, adhesion). Cracking occurs especially where the magnetite layer is under high tension.

Where the magnetite layer is cracked during a start/shutdown of the plant, a corrosion pit will form. In this pit a new magnetite layer will be formed (see figure 16).

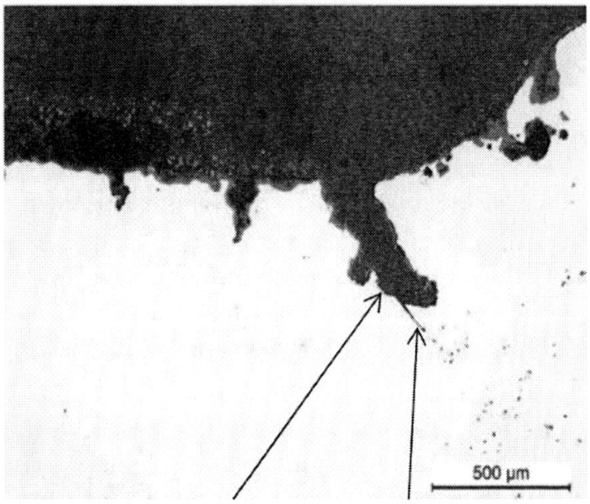

Figure 16. Corrosion pit and crack indication

This magnetite layer will crack at those places where the stress concentrations are the highest. In this case this will be in the crevice (weld neck) between the base material and the weld seam of the smallest riser.

Thus, every mechanical cycle can result in a further fatigue crack growth. The material of the riser 15NiCuMoNb5-6-4 (WB36) is a type of steel with enhanced yield strength which is especially vulnerable to this failure mechanism in high pressure steam systems.

This failure was not predicted because it was assumed that the steam system was protected by an excellent water treatment (good quality magnetite layer).

The aging of the plant for about 30 years plays a big role. The amount of start/stops has a negative effect on this failure mechanism.

A combination of corrosion fatigue (low cycle fatigue) and finally an overload due to process up-set conditions caused the failure of this riser. The cracked riser has a smaller diameter and wall thickness (10 mm, 0.4 inch) compared to the other three risers (36 mm, 1.4 inch). This results in more tensile stresses from the start/stop cycles. The waste heat boiler has a fixed position compared to the steam drum. The steam drum has moved sideways in the direction of the boiler water feed inlet. Due to process upset conditions. This corresponds with the observed crack orientation.

<u>Solutions to prevent this failure mechanism:</u>

1. Provide a good thin magnetite layer.
In this case, a new water treatment based on film forming amines in the steam system has been in place for the past two years. This conditioning is known that it forms a better protective, thinner and denser magnetite layer. This should prevent the cracking of the magnetite layer in the future.

2. The operation conditions are changed to avoid overload/up-set conditions in the future. Change in the design of the system is not an option. The riser that has cracked now remains the location where this phenomenon could occur again. Due to this failure the inspection plan was modified by including NDT examination from the outside by Phased-Array ultrasonic inspection of the welds of this riser and the three others risers.

Conclusion and advice

- The final cause of the cracking of the riser is a sudden overload due to process up-set conditions.
- Cracking has initiated by corrosion fatigue of the protective magnetite layer.
- Advice: Make sure a thin denser magnetite layer will be formed.
- Advice: Avoid mechanical and temperature shocks in the steam drum and the waste heat boiler.

Lessons to be learned

- Apply an optimum water treatment in these high pressure steam systems, e.g. by using film forming amine treatment.
- Take the correct process precautions to avoid sudden overload.

- Update the inspection program and perform the correct NDT inspections.

Case 3. Crack indications at the transition between the reformer/riser tubes and the weldolet of the outlet header

Introduction

During the Turnaround 2012 a NDT dye penetrant check on the riser and reformer tubes indicated a crack at the transition of the weld between the weldolet and one of the riser tubes in the primary reformer of the AFA3 Ammonia plant. (see figure 17)

Details Reformer/riser tubes

Tube material: Spun cast, 25Cr35NiNb
Weldolet mat.: Alloy 800H, 35Ni20CrAlTi
See also fig.17.
Reformer tube: Ø115,9/4.56 inch x13,5 mm/0.53 inch
Riser tube; Ø163,1/6.42 inch x19,3 mm/0.76 inch
Life time: 32 Years
Number of reformer tubes: 364
Number of risers: 7

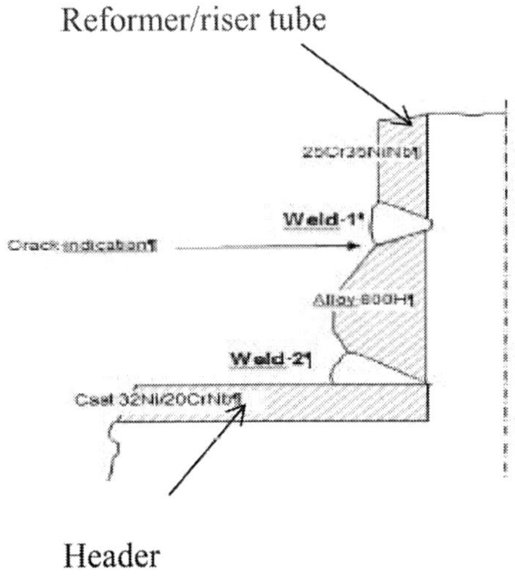

Figure 17. Detail tube-weldolet

By milling the crack depth has been determined. For the riser 7 milling was stopped at a depth of 7- 8 mm/0.27- 0.31 inch and a boat sample was taken to perform a failure investigation (see figure 18).

The depth of this boat sample was about 12 mm/0.47 inch and the crack was still present.

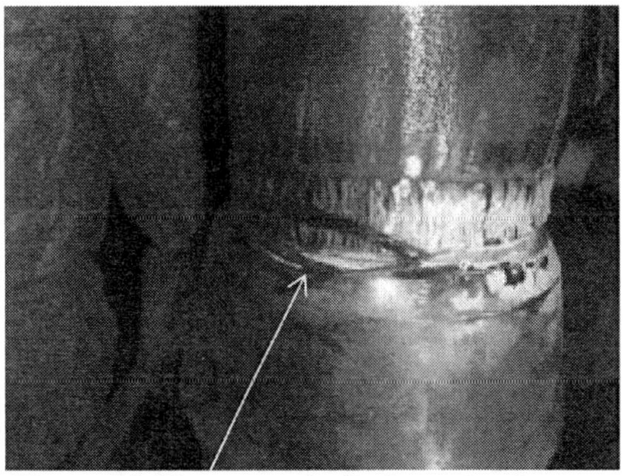

Figure 18. Boat sample for investigation

Failure investigation

The failure investigation has been performed together with a third party, Element Materials Technology. This showed that the crack was present in the weldolet base material (Alloy 800H) close to the heat affected zone along the weld material Alloy 625 (see figure 19).

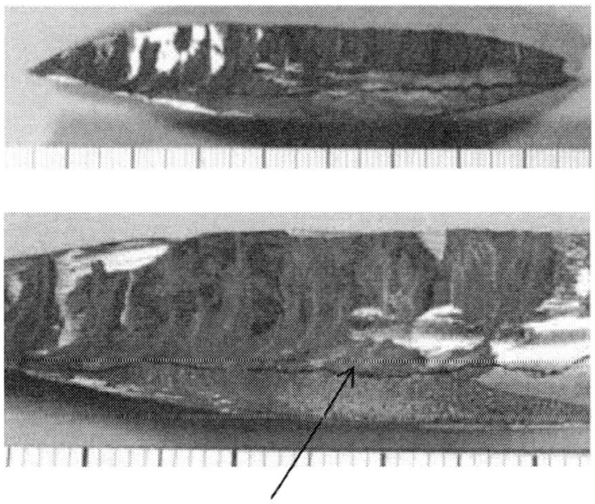

Figure 19. Crack indication

The crack initiated from the outside. Along the crack almost no creep cavities were present. In the cracks, oxides and nitrides are presence, which indicates that the crack already has been present for a long period. (See figure 20).

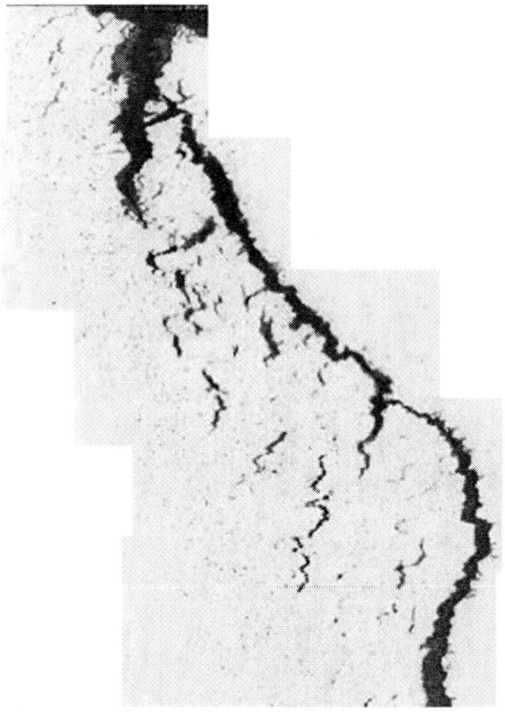

Figure 20. Overview of the crack in the weldolet weld connection

<u>Material identification</u>

The PMI examination performed identified that the weld metal was Alloy 625 and the weldolet material was Alloy 800H.

<u>Conclusion of the failure investigation</u>

The cause of this failure is low cycle fatigue (LCF) as a result of the difference in thermal expansion coefficient between the Ni-base weld metal (Alloy 625), and the austenitic Alloy 800H base metal. It is further concluded that the crack propagation is very slow and grows preferably during thermal cycles (starting/stopping) of the reformer.

Discussion

All risers (7 pieces) and adjacent reformer tubes (14 pieces) have been checked by dye penetrant testing. (for an overview see figure 21).

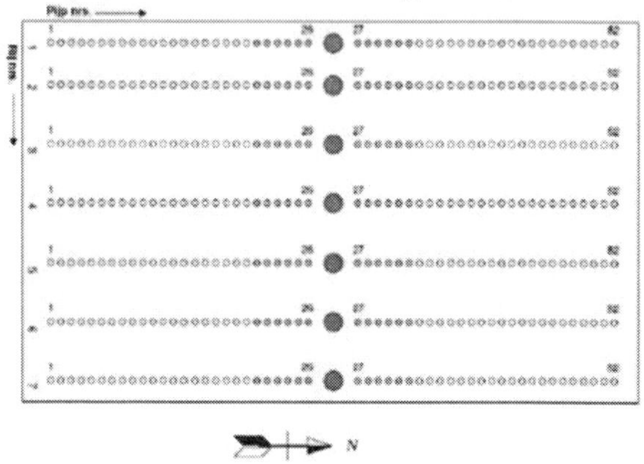

Figure 21. overview reformer and risers tubes

The investigated welds showed the same defects (cracks) (See also figures 22 and 23).

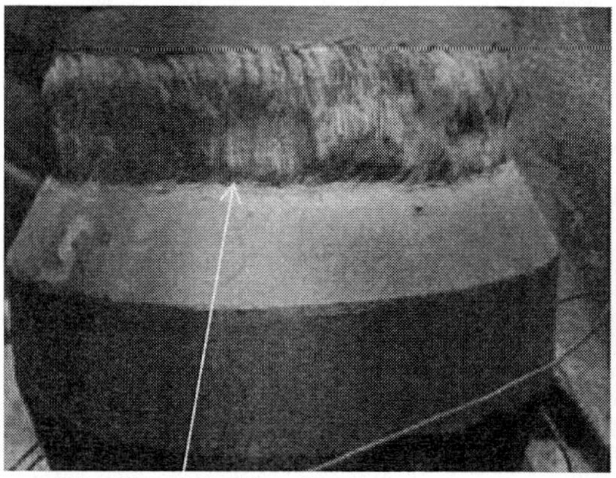

Figure 22. Crack indication

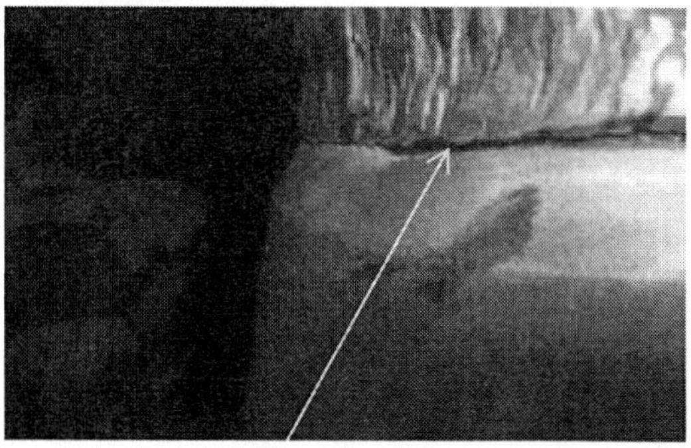

Figure 23. Crack indication Riser

JSC inspections performed third party NDT inspections by eddy current measurements in order to determine the crack depth of the risers and adjacent reformer tubes.

The results were as follows:
- all reformer tubes:max.1-2/0.039-0.078 mm/inch crack depth
- risers 1,2,3,4,5,6 max. 3-5/0.12-.197mm/inch crack depth
- riser 7 : 5 /0.2 – 10/0.4 mm/inch depth, locally about 19/0.75 mm/inch depth.

Original diameter/wall thickness
Reformer tube: Ø115,9/4.56 x13,5/0.53 mm/inch
Riser tube; Ø163,1/6.42 x19,3/0.76 mm/inch

Fitness For Service Analysis (FFS)

A Fitness for Service analysis has been performed by Sintra Engineers (Third party) to determine the allowable defect size. Because of the nature of the defect (most probably caused the difference in thermal expansion) the cyclic loading of the construction has an influence on the crack growth.

Two types of assessments have been performed:

- Limit load analysis
- Creep crack growth/limiting crack size

The defects in some of the risers have been ground/milled out.

To determine the effect of this local wall loss a limit load analysis has been performed to calculate the Remaining Strength Factor (RSF). Based on this calculation, it was concluded that a local wall loss of maximum 13/0.52 mm/inch is acceptable in the weld between the weldolet and the riser.

Since not all defects have been removed, a fracture mechanics analysis also has been performed. Based on this analysis, the maximum allowable defect depth in the riser is 17/0.67 mm/inch. This crack growth analysis was only based on creep, based on the expectation that crack growth due to cyclic behavior within a period of 4 years will be limited. From the creep crack growth analysis, it was concluded that a crack size of maximum 5/0.197 mm/inch will not grow due to creep within a period of 4 years. After 4 years, new inspections are required.

Further investigation for the influence of the cyclic nature of the loading has been recommended.

Repairs performed

The measured crack depth in the risers and reformer tubes (except for the riser in row 7) is much less than the allowable reduction in wall thickness as determined by the FFS-analysis. Because it is known that the crack propagation is very slow, it was decided not to carry out any repairs. Only the weldolet of the riser in row 7 has been repaired. (See figure 24).

The weldolet/manifold and riser have been solution annealed locally and the crack has been removed completely by grinding.

During this repair it appeared that the crack was present over approximately 2/3 of the circumference and was cracked almost through the wall.

The weld has been repaired using Ni-based filler metal (UTP068HH). Matching welding consumables are preferred, but cannot be used in combination with the existing Ni base consumables.

Figure 24. Repaired weld

As follow up to the observed cracks in these welds of the risers and reformer tubes, an inventory has been performed of all the Ni base weld metals to find out whether there are more potential damaged locations in this ammonia plant. No additional dissimilar weld connections which are not already in an inspection plan were found.

Conclusions

- The cracks initiate at the outside, in the Alloy 800H base material near to the fusion line with the alloy 625 weld metal.
- The alloy 800H weldolet material shows hardly any creep.
- The cracks are oxidized and the adjacent 800H base metal shows nitrides. This indicates that the cracks has been present for a long period of time.
- The saw tooth pattern of the crack indicates appearance of Low Cycle Fatigue (LCF).
- The failure mechanism of this LCF is the difference in thermal expansion coefficient between the Alloy 625 Ni-base weld metal and the austenitic Alloy 800H weldolet material.
- Crack propagation is (very) slow and grows preferentially due to thermal cycles (starting / stopping) of the reformer.

Lessons to be learned

- In the design stage, avoid a high difference of thermal expansion coefficients of dissimilar materials (Ni base welds vs austenitic base material).
- It is preferred to use matching welding consumables for these welds, such as the 35Ni25Cr and 32Ni20Cr types that are available.
- Perform NDT inspections on these dissimilar welds on a regular basis and monitor the crack propagation.

References

[1] API RP-571 Damage Mechanisms Affecting Fixed Equipment in the refining Industry

Successful Project Execution and Commissioning of Largest Capacity KRES™ Ammonia Unit

For the 2015 ammonia plant capacity expansion project at Lima, Ohio, PCS Nitrogen, elected to use KBR's Reforming Exchanger System (KRES) technology for increasing capacity of the reforming section. The KRES unit was licensed and engineered by KBR to increase the ammonia plant capacity and contains Clariant's novel ReforMax® catalyst. This paper will discuss how PCS, KBR, and Clariant collaborated on the project execution, installation and commissioning of the KRES unit at the Lima site. Completing a licensed technology based project requires specific technology focused proven project execution methodology. Engineers from PCS, KBR, Clariant, and contractors were on-site to make sure this new technology was safely and successfully implemented and commissioned. The Lima site's current KRES operations as well as lessons learned will also be discussed. The KRES unit in Lima is the largest capacity application of reforming heat exchanger technology. While this is the fourth KRES unit in operation, this is the first domestic unit within the USA.

Timothy Kearns
PCS Nitrogen Ohio, L.P., Lima, OH, USA

Kim Taniguchi and Akhil Nahar
KBR, Houston, TX, USA

Tim Rembold
Clariant, Louisville, KY, USA

Introduction

PCS Nitrogen operates an ammonia/urea complex in Lima, Ohio. The Bechtel engineered ammonia plant was constructed in 1971 with a Foster Wheeler Primary Reformer. The original nameplate capacity of the plant was 1360 MTPD (1500 STPD); just prior to this revamp the operating capacity was 1678 MTPD (1850 STPD) ammonia in summer conditions. The targeted capacity for the current revamp was 1955 MTPD (2150 STPD) ammonia; however, equipment was sized to accommodate a future revamp capacity.

The KBR Reforming Exchanger System (KRES) is KBR's proprietary reforming exchanger technology. It has been successfully implemented in commercial operations in ammonia plants for Methanex (Canada) since 1994, for Liaohe[1] (China) since 2003, and Chambal Fertilizers and Chemicals, Ltd.[2] (CFCL) (India) since 2009.

The KRES system can be implemented in three different process schemes which are described in papers given previously[3][4]. The process scheme utilized for the PCS Lima plant is similar to that used for the CFCL plant in India.

In this process scheme, KRES utilizes heat from the secondary reformer to produce additional synthesis gas. Mixed feed, consisting of desulfurized natural gas and steam, is preheated and fed to the KRES tube-side channel (top head) in parallel with the primary reformer. The feed passes through the tubes containing Clariant's specially formulated ReforMax® reforming catalyst and exits the tubes at the bottom of the shell. Secondary reformer effluent is introduced at the bottom of the shell where it mixes thoroughly with the tube-side effluent. The gas mixture then passes over the outside of the tubes containing catalyst and supplies heat for the reforming reaction (see Fig. 1 above, 2, and 3 below). Implementing a KRES retrofit with this process scheme increases capacity with minimal impact on the existing primary reformer while reducing the service severity for the Reformed Gas Waste Heat Boilers (RG WHBs).

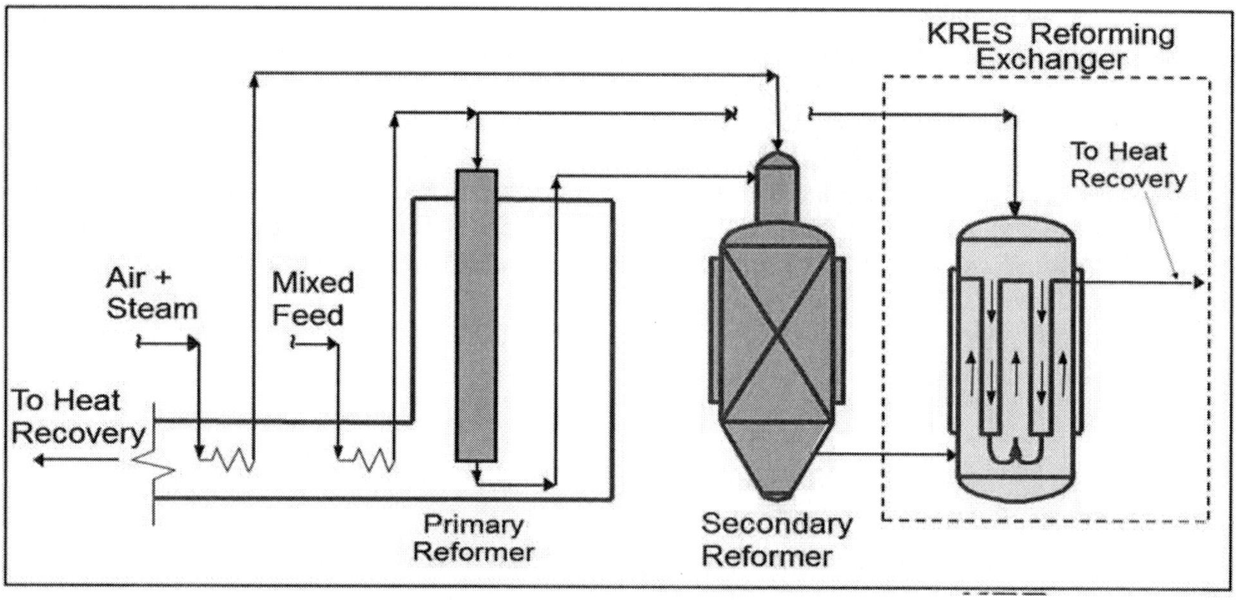

Figure 1: KRES Flow Scheme

In addition to the KRES exchanger, the secondary reformer and feed/effluent transfer lines are new equipment items. The new secondary reformer, based on KBR's combustion chamber technology, is a water jacketed type design with no metallic burner or internal metallic parts. Such design has proven to be highly reliable and robust. In this design the process air stream is mixed into the furnace effluent stream via a refractory port that is cast in place. This design is well proven, having been implemented in numerous KBR ammonia plants. The absence of a metallic burner increases operational reliability and avoids the routine maintenance that is required for conventional burner designs[5].

The feed transfer line carries primary reformer effluent to the secondary reformer and is a non-jacketed dual layer refractory design. This configuration matches the existing primary reformer effluent line. The KRES effluent transfer line carries gas to the RG WHBs and is a water jacketed dual layer refractory design.

KBR proposed this process scheme to PCS Nitrogen as the best solution for the desired revamp results. After some consideration and comparison with other licensors' offerings, PCS ultimately chose KBR to perform a process study. The results of the study were positive and the decision was made by PCS to proceed with the Engineering Design, Procurement, and Construction of the unit.

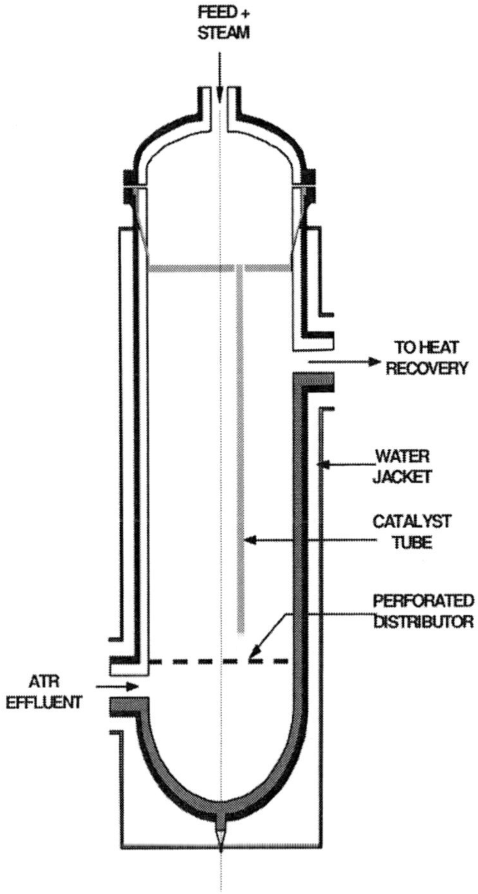

Figure 2: Sketch of KRES open tube arrangement

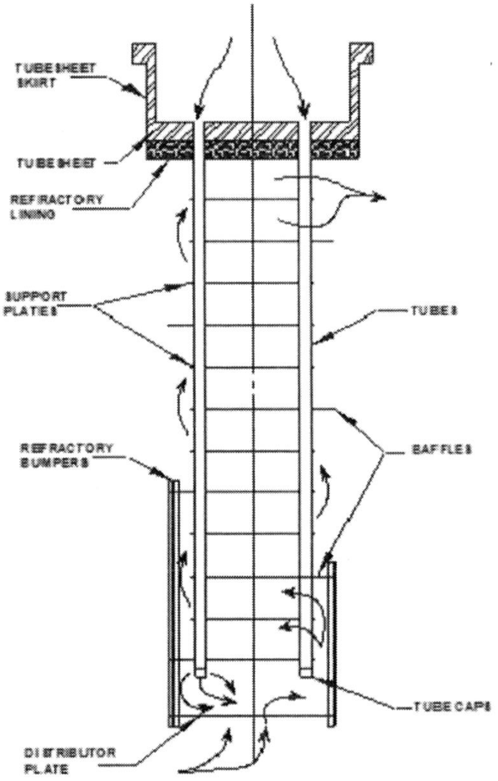

Figure 3: Sketch of KRES Tube Bundle

Study Phase

The process study was commissioned by PCS in May 2012. KBR collected base case data from the existing plant and began investigating options for capacity increase. KBR analyzed the existing system in conjunction with KRES and confirmed catalyst performance with Clariant. The study results confirmed that addition of the KRES was a good fit for this revamp. This conclusion was reviewed and PCS decided to proceed on this basis.

As with many revamps, physically locating the new equipment in the existing plot plan was a challenge. Various equipment arrangements were explored and vetted with PCS by KBR. The plot area was relatively tight, but just large enough to place the new secondary reformer and KRES exchanger.

One of the main decision points was how to utilize the existing secondary reformer and/or its foundation. There were basically two options; keep the existing secondary reformer or remove it. The option of removing the secondary reformer was eventually agreed as the most suitable arrangement (see Fig. 4 and 5 below).

A preliminary constructability study was conducted in this phase to make early determinations on the challenges to be faced. Performing excavation and foundation work including pile installation during plant operation as well as the demolition and extraction of the existing secondary reformer were significant concerns which had to be addressed.

Figure 4: KRES Installation

Figure 5: KRES Installation

Engineering Phase

The Basic Engineering Phase began in August 2012. During this phase the details of the agreed equipment arrangement from the study phase were solidified and confirmed. Process simulations were finalized with assistance from Clariant and systems, instrumentation, and equipment design work was commenced. Also included in KBR's scope were some components of detailed engineering as the scope split of the KRES area was vertical in nature.

KBR provided engineering design and technical requisition packages for the secondary reformer and transfer lines in addition to engineering design of the two steel structures (transfer line support structure and KRES support structure).

Design of the high temperature/high pressure air/steam and mixed feed piping was a collaborative effort between KBR and two of PCS's detailed engineering contractors. This made for a particularly challenging task. Operating temperatures of up to 732 °C (1350 °F) are much higher than typically seen in most other plants for these services. The higher operating temperatures required special consideration for pipe supports/guides to account for the extensive thermal growth at these conditions. KBR provided detailed pipe support designs for these two process lines (within KBR's scope boundary) as well as the feed/effluent transfer lines.

The KRES exchanger was supplied by KBR as a proprietary equipment item. KBR had full responsibility for engineering and design, supplier data review, shop inspection, fabrication and testing including installation of refractory, and packaging for ocean transport.

Along with supply of the KRES, KBR defined, with PCS approval, all the control logic for safe operation of the KRES including steam to carbon ratio, feed ratio split, etc. KBR also defined the critical shutdown logic associated with the KRES and worked closely with PCS to implement these functions into the existing trip logic. KBR participated in a number of review meetings with PCS and their contractor, including Process Review, P&ID Review, HAZOP, Detailed Engineering Review, Constructability Review, Control System Review, and Factory Acceptance Test (FAT).

Construction Phase

The KRES exchanger was fabricated by a KBR approved supplier with KBR inspection/supervision. The unit was shipped in August 2014. Upon reaching the Lima site, the KRES exchanger shell was installed on its foundation in December 2014. The bundle was shipped with the pressure shell, but separately in a shipping can under nitrogen purge and stored in the laydown yard at the Lima plant. Special care was taken to ensure that the support system for KRES was installed to meet all KBR requirements.

Figure 6: ReforMax® Catalyst

Catalyst loading was done in July 2015 by a specialty catalyst loading company recommended by KBR and contracted by PCS. The catalyst (ReforMax® - see Fig. 6) developed in KBR/Clariant alliance was loaded per proven loading pattern. KBR and Clariant representatives were present for the catalyst loading to assist and advise PCS. Prior to loading, the bundle was thoroughly checked for cleanliness with a borescope which revealed unexpected debris in the bottom of the tubes. Extra time was needed to clean the tubes of this debris. After

Figure 7: KRES Bundle Stored in Shipping Can

all of the tubes were cleaned and pressure drop checked, the loading activity was started. Catalyst loading was completed in three days after which pressure drop tests were confirmed and verified. The bundle was then placed back on nitrogen purge in the shipping can and stored in the vertical position (see Fig. 7).

Installation of the secondary reformer refractory lining and transfer lines was completed in July/August 2015. PCS elected to use a field joint detail that incorporated a slip fit detail instead of a field cast joint. Since this detail obviates the need for installation of new refractory at each joint, no additional dry-out was required. This saved significant installation time during the turnaround (TAR). Due to high work density, many activities would have been shut down during the dry-out process had it been required.

KBR prepared a series of checklists to aid PCS in addressing all of the important and critical items for the Construction phase. A checklist for the KRES exchanger, Secondary Reformer, and both Transfer Lines, including pipe supports, was developed and signed off by both PCS and KBR.

KBR technical representatives were present from mid-July to late October 2015 to support PCS in construction, commissioning, and startup activities. Clariant technical representatives were on-site to support PCS with other catalyst loadings as well as the startup of the KRES exchanger.

Commissioning and Startup Phase

As mentioned previously, KBR prepared checklists to be used for the construction phase. Similarly, checklists were prepared for the commissioning and startup phase. As part of the determination of plant readiness, these checklists were used to ensure that key items were addressed and in satisfactory condition for startup.

Commissioning of primary reformer/KRES system started with a Site Acceptance Test (SAT) of the DCS and interlock system in September 2015. All of the interlock and complex control logics were thoroughly checked by PCS and KBR teams. Punch lists were prepared for further action. Commissioning of equipment started at the end of September 2015. To verify mechanical completion of the systems and engineering intent, a complete check of the system was performed by teams from PCS and KBR comprising personnel from several different disciplines. Punch lists were prepared during such inspections for further action.

Figure 8: KRES 3D View

Operations meetings were held between PCS, KBR, Clariant, et al, to discuss details of the startup procedures including nitrogen circulation, steam lineup, and feed cut-in. The team also reviewed the operating procedures before commissioning. After mechanical completion of different systems, commissioning of the KRES system was started in the second week of October 2015. This included flushing of water jacket supply lines, water jackets of secondary reformer and KRES, and the instrument impulse lines. After lining up all the instruments including control valves and Emergency Shut Down (ESD) valves, detailed field checks were made to verify their operations including stroke checking and interlock actions.

Nitrogen circulation was started in mid-October 2015 and leak checks were done at all the flanges to verify tightness of the system. The jacket water connections were lined up and jacket water for the secondary reformer and KRES was placed in service. After confirming negligible oxygen levels in the circulating nitrogen, burners in the primary reformer were ignited.

The nitrogen circulation and heating of primary reformer catalyst, secondary reformer, and KRES continued until all of the bed temperatures were 50 °C (90 °F) above the dew point. Steam was introduced in both the primary Reformer and KRES in parallel, keeping the system pressure low to avoid any condensation. Nitrogen circulation was subsequently stopped while continuing on steam only. Firing in the Primary Reformer was continued to further increase the catalyst temperature. Feed gas from the desulfurizers was checked for sulfur slip and found to be less than 0.1 ppm. Feed gas to the primary reformer was introduced after achieving an outlet temperature above 677 °C (1250 °F). The feed gas to the primary reformer was increased to 40% of normal flow and outlet temperature was increased to 704 – 760 °C (1300 – 1400 °F).

After achieving steady operation of the primary reformer, process steam, KRES feed, and process air from the air compressor were introduced. Using the heat from the secondary reformer effluent, the KRES outlet temperature increased. The mixed feed to the KRES was then introduced. Process air to the secondary reformer and feed to the KRES were increased simultaneously. The KRES was initially operated at a higher steam to carbon ratio for reduction of the KRES catalyst. Methane slip was measured at the KRES outlet every four hours until it was stabilized. The KRES catalyst reduction took less than 8 hours. Afterwards the KRES feed and process air streams were increased to steady operation at approximately 60%. The primary reformer tripped once after achieving steady state operation of the KRES due to a positioner failure on the fuel gas control valve. After repairing the positioner, the KRES was again brought online. Overall start-up of the KRES was smooth and trouble free. Plant load was maintained at 60% for a few days to allow for downstream system startup.

Once the synthesis loop was lined up, the plant capacity was increased to design capacity without any limitation imposed by the KRES system. PCS has been operating the KRES since October 2015 and have not faced any operating issues with regards to the KRES.

Performance Test

As mentioned above, the KRES system operations have been very smooth and trouble free since initial startup. The 72 hour performance test was completed as planned in March 2016. Plant data was collected and reconciled daily by PCS and KBR against expected values. The KRES system performed very well and as expected, all performance guarantee parameters were met and exceeded. Guaranteed parameters were:

- Capacity based on quantity of H2 and CO
- Methane slip
- Pressure drop

Current Operations

The KRES system is operating better than anticipated. PCS has realized additional capacity near the future revamp capacity without implementing the total revamp scope outside the reforming section. Heat loss of the KRES system is also less than expected. The ReforMax® catalyst shows equilibrium performance from start of run with low pressure drop across the unit. There have been no operational issues or capacity limitations due to the KRES system.

Lessons Learned

With every installation there are areas that can be improved or done differently to enhance the final outcome. Below are a few items to note.

1. Tube cleanliness needs to be monitored more carefully. During catalyst loading the bottom of the tubes were checked with a borescope to ensure cleanliness. This was a worthwhile activity as there was debris from the fabrication shop still in the tubes; however, this should be taken care of prior to delivery. New shop inspection procedures are in place to prevent this occurrence in the future.
2. Construction, commissioning and startup check lists were highly useful. In the midst of fast paced construction activity with so many details that need to be addressed, it is very possible to miss critical items. Check lists were a useful tool to make sure nothing was overlooked.
3. Work space was limited. The physical area for installation of the KRES system was very congested resulting in high work density. Phasing and coordination of work is always key to scheduling, but even more so in this revamp. PCS managed the planning and executed very well to accomplish this challenging task.
4. Transfer line spools were assembled using slip fit joints. This type of refractory joint proved to be a time saver since dry-out of the field joints was no longer required.
5. Having licensor and catalyst field support services were beneficial to the construction, commissioning, and startup activities.

Conclusions

The KRES revamp was highly successful in increasing the capacity of the PCS Lima ammonia plant with minimal modifications to the primary reformer and also reduced operating severity for the RG WHBs. Heat loss in the KRES system was less than expected.

Installation of a KRES system into a congested plant layout is possible with the proper constructability studies and coordinated planning and execution of various activities in field.

Startup of KRES is relatively simple and catalyst reduction is very smooth.

References

1. A. Malhotra, P. Kramer, S. Singh, "Revamp of Liaohe's Ammonia Plant with KRES Technology to reduce Natural Gas Usage", AIChE Ammonia Safety Symposium 2004.
2. U. Singh, S. Singh, A. Malhotra, "Successful Operating Experience of CFCL Ammonia Plant Revamped with KBR KRES Technology", AIChE Ammonia Safety Symposium 2011.
3. S. Singh, J. Price, "KRES-ESTM Revamp makes more Ammonia from Less Natural Gas by Energy Substitution", Nitrogen & Syngas Conference 2011, Dusseldorf, Germany.
4. S. Singh, A. Jing, "KRES Technology Makes Ammonia Plant Revamping Viable", GazChem 2007, Trinidad.
5. S. Singh, D. Svec, "Secondary Reformer with Improved Combustion Chamber Technology proves highly reliable in CFCL Ammonia Plant", Nitrogen & Syngas Conference 2013

Failure of Level Bridle in Benfield Service

On 18th May, 2015 at approximately 11:39am, there was a release of Benfield Solution (Hot Potassium Carbonate) and Syngas from the level bridle of the CO_2 Absorber on the Tringen 2 Ammonia plant which resulted in a plant shutdown and subsequently a fire. Response action resulted in the fire being brought under control safely. There were no personnel injuries and damage was limited to the level bridle and the immediate vicinity of the bridle. During the event, the Site Emergency Response Plan was activated. Restorative work including demolition, material procurement, fabrication, NDE and insulation took seven days. This paper looks at the failure mechanism, shortcomings in the HAZOP for the site and difficulties experienced at the time of the incident when containment and isolation was key in reducing damage.

Kamla Balgobin
Yara Trinidad Ltd

Rodney Babwah
Yara Trinidad Ltd

Introduction

YARA TRINIDAD LIMITED operates three (3) ammonia plants each of a different design – Braun Conventional, Fluor and Brown and Root Purifier. The facility is situated south of the Point Lisas energy area which houses other ammonia and methanol facilities.

The incident occurred on the TRINGEN 2 Plant which is the Purifier design. The Carbon Dioxide (CO_2) removal system is Hot Potassium Carbonate solution with an activator and corrosion inhibitor – Benfield CO_2 Removal. The plant was commissioned in 1986.

Prior to the incident, the plant had been online and steady with normal operating conditions for a continuous 177 days when on May 18, 2015, at approximately 11:39 am, the acting Shift Supervisor on duty, was alerted by an abnormal venting sound in the vicinity of the CO_2 removal system. Closer observations indicated that Benfield solution, under pressure was blowing out, in the vicinity of the level bridle on the CO_2 absorber. The initial operational response by the acting Shift Supervisor and plant personnel included tripping the methanator, stopping production, stopping syngas flow to the CO_2 removal system, de-pressuring the CO_2 removal system and the stabilization of the safe operation of the front end of the plant up to the LTS outlet.

The release of Benfield solution and syngas continued for approximately eleven (11) minutes, after which the syngas ignited when the level in the absorber was below the rupture point of the level bridle. The situation was aggravated by the fact that the methanator trip valve was bypassing and isolation had to be done at the inlet of the cryogenic system (Coldbox), and also due to the fact that the preferential gas path was via the rupture. Nitrogen was introduced into the system upstream the CO_2 absorber to suppress the fire.

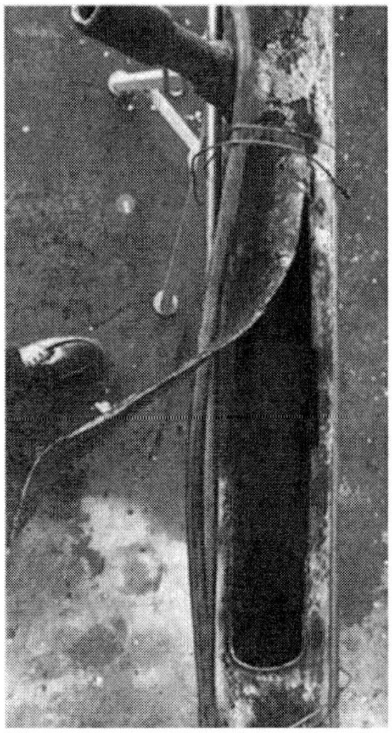

Figure 1 – Failed Level Bridle

The Site Emergency Response Plan was activated after the ignition and the emergency crew, together with plant personnel contained and extinguished it. The duration of the fire was approximately fourteen (14) minutes and the total duration of the event from the start of the initial liquid release to the system de-pressuring was approximately thirty (30) minutes. There were no personnel injuries. Adverse impact of the event included

- Loss of chemicals (Benfield Solution) inventory and release to plant drainage systems,
- Fire damage including limited damage to electrical wiring and instrumentation,
- Minor loss of installed insulation and,
- Production loss.

Given the circumstances of the situation, plant personnel handled the situation well and their actions to de-pressure the system and displace with nitrogen, prevented more widespread damage to the area.

Process Description

This CO2 removal system includes the absorption column T-3401, desorber column T-3402 and associated pumps and exchangers (see Figure 8). Lean Benfield solution composition contains approximately 29 % potassium carbonate, 3% Activator (LRS10) and 0.7% vanadium pentoxide.

Process gas enters the bottom of the absorption tower at 218°F (103°C) where it flows upwards through 4 packed beds of Raschig rings (SS) counter current to the lean Benfield solution (400psig/27.6 bar) that is tangentially introduced at the top of T-3401 at 145°F (63°C). The solution absorbs the CO_2 and leaves the bottom of the absorber as a "rich" solution. A relatively CO_2 free gas leaves the absorber at 136°F (58°C) at the top, and any liquid carryover is removed in the downstream knockout drum V-3411. The average potassium carbonate in solution is 27-30 vol%. Antifoaming agent UCON, is continuously dosed as an anti-foaming agent.

Damage Description

The bridle is made up of two nozzles attached to the tower (B1 & B2), four nipples attached to two separate level glasses (LG1, LG2, LG3 and LG4) and two further nozzles attached to a level transmitter LT-3038B, all of which provides the level of Benfield solution within the absorption tower. The normal level during operation is as indicated in Fig. 2 (marked "L.L") and operating data for the level bridle pipe is given in Table 1.

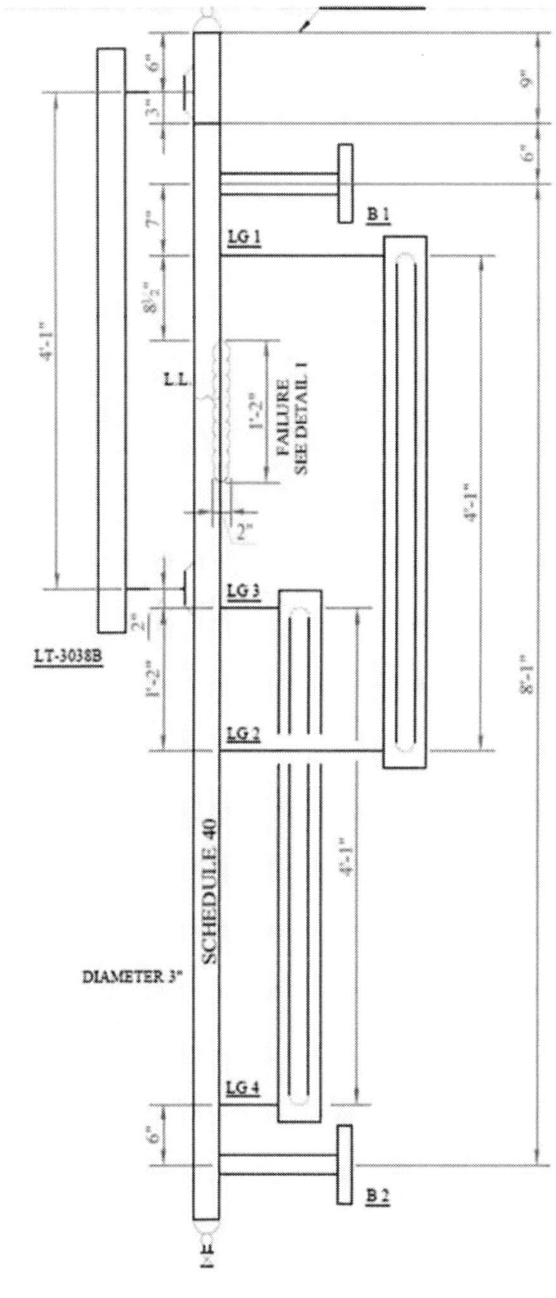

Figure 2 – Level Bridle Arrangement

Parameter	Value
Diameter	3"
Schedule	40
Nominal Thickness	5.5mm
Operating Temperature	240°F (116°C)
Operating Pressure	400psig (27.6 bar)
Material	SMLS API 5L (carbon steel)

Table 1. Design Data for T-3401 Level bridle pipe.

The failure was in the form of a sudden, total loss of containment with no previous identified leaks in the area. A section of the bridle was found to have peeled off. The failure took place directly in line with nozzle B1, approximately 1ft 3 inches below it. Upon inspection, immediately after the incident, vertical grooves emanating from the nozzle downwards were noted. There appears to have been some corrosive action due to condensation of the Benfield vapour at nozzle B1 which continuously ran down the sides of the bridle wall. The appearance of the transverse part of the fracture surface is shown in Figure 5.

Figure 3 – Failed level bridle

Figure 4 – Failed level bridle

Figure 5 – Traverse part of fracture surface (1st inspection at site).

Immediate Actions taken

Once the plant was secured and fire stopped the system was inspected.

- Detailed inspection was done on all areas affected by the fire damage, only insulation and instrumentation had to be repaired/replaced. First line of thought was that CO_2 Stress Corrosion Cracking was the culprit !!!!
- A new bridle (and adjacent), was rebuilt with carbon steel and all welds post weld heat treated for service – *investigation into materials used at other sites indicated Stainless Steel which was not readily available.*
- A higher schedule of thickness was used for new bridle assembly
- The new bridle assembly was pressure tested prior to installation
- UT thickness measurements were done on bridles on other vessels in CO_2 service (no wall loss found)
- Sent samples of failed bridle chamber to Yara Materials Engineering for analysis

Failure Analysis

Two independent investigations were done;

1. Failure analysis of bridle assembly by Yara Materials Engineering focusing on failure mode and,
2. Process safety and incident investigation focusing on root cause analysis and response to the incident.

Materials Analysis

The Yara International Materials Department was requested to do a thorough analysis on the failed bridle. Samples [Figure 6] were sent and the following is a summary of findings.

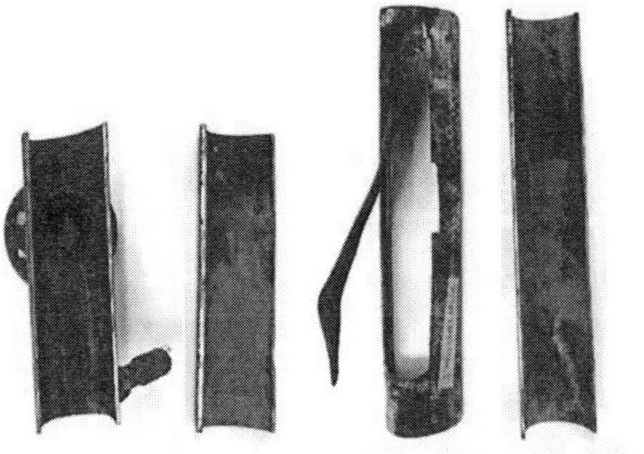

Figure 6 – Materials Lab samples

Sample 1a: Consists of a section of the upper corroded part of the level bridle, which has been split longitudinally, and includes nozzles B1 and LG1

Sample 1b: Consists of the opposite longitudinal half of tube Sample 1a

Sample 2a: Consists of the lower corroded part of the level bridle, which has been split longitudinally, and includes the ruptured part

Sample 2b: Consists of the opposite longitudinal half of tube Sample 2a

All four samples were individually analyzed both macroscopically and using fractography however most focus was placed on sample 2a as this was the ruptured part of bridle

The sample 2a was found to exhibit internal corrosion attacks and the rupture followed two of the deepest longitudinal corrosion grooves that were present on the internal surface (Figure 6 - right). The fracture surfaces along these grooves were very narrow, less than 1 mm in some areas and hard to identify with the naked eye. This suggested that the wall metal loss had almost been through-wall.

Figure 7 - Sample (2a) cut into smaller pieces for Fractographic analysis

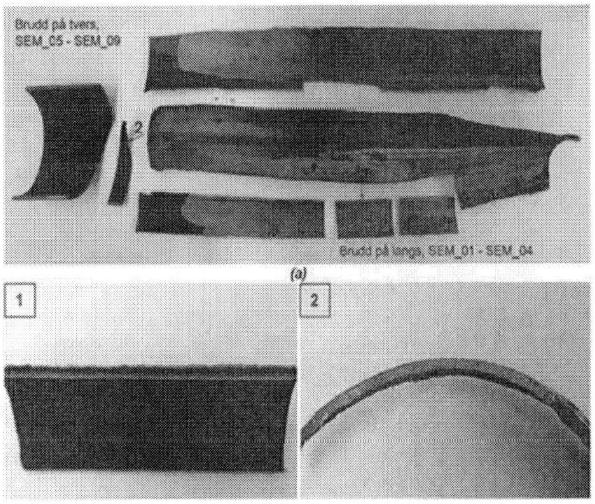

The longitudinal fracture surface exhibited a dimple rupture morphology, which confirmed that this part of the rupture was ductile. The longitudinal fracture surfaces were very narrow; the analysis revealed a wall thickness less than 0.5 mm in some areas.

Several smaller areas of undamaged fracture surface revealed that the transverse part also had ruptured in a ductile manner. There were no indications of brittle crack propagation and presence of secondary cracks or corrosion on any of the fracture surfaces.

The results of the analysis and observations made during the examination of the tube implies that the tube had ruptured due to ductile overload as a result of localized thinning in the form of longitudinal corrosion (grooving) on the internal tube surface. This localized thinning was caused by a corrosive liquid condensing within the bridle.

Materials Report Conclusion

- The tube ruptured due to **ductile overload** as a result of localised thinning from corrosion on the internal surface.
- The source of corrosive liquid appears to be condensation of Benfield vapours at nozzle B1, which had continuously run down the bridle wall. The depth of the corrosion grooves tapered off as the general Benfield liquid level was reached.
- The appearance of the rupture indicated that the tube first ruptured along two deep longitudinal corrosion grooves.
- The very narrow fracture surfaces on the longitudinal parts of the rupture showed that the bridle had suffered from severe local wall metal loss. The bridle wall thickness was less than 0.5 mm along parts of the deeper corrosion grooves. This thickness was below minimum required thickness to withhold the internal pressure of 400 psig (~28 bar) [3], and suggests that the rupture was initiated due to this reason (mechanical overload).
- ***The failure was not due to CO_2 Stress Corrosion Cracking as initially thought!***

The characteristics (shape and localization) of the internal thinning observed in the bridle supported the theory of a corrosive liquid phase condensing at the upper nozzle, going along the internal wall of the equipment down to the liquid level, most likely promoted by a deficient passivation layer on a carbon steel equipment.

Process Safety Analysis

A root cause analysis was completed by the process safety department (refer to Event Tree in appendices). The team focused on the process safety aspects which could have led to an early warning or a complete removal of the condition which resulted in the loss of containment. Had the gas not ignited sooner the possibility of more damage could have been a reality.

It was theorized that a broken light bulb close to the rupture was the source of ignition - possibly limiting the damage to the local area and not further to other nearby equipment.

Initially the horizontal crack was suspected having initiated the loss of containment, and several hypothesis was made:

1. Mechanical fatigue,
2. CO_2 Stress Corrosion Cracking,
3. Internal corrosion,
4. External corrosion.

CO_2 SCC was initially mainly discussed.

External corrosion was rapidly discarded due to the visual aspect of the crack and Corrosion Under Insulation (CUI) inspection was done in the last 2013 turnaround, showing no evidence of it. Thickness checks were not done in 2013 as this piping was not identified as requiring more than CUI.

The **direct cause** is a mechanical overload due to **localized internal corrosion** as per the Materials analysis. The Process Safety review however looked at the lack of the HAZOP to identify the presence of the corrosive liquid in the bridle.

There is no mention about this phenomena in the HAZOP done in 2007. The tracing system and its function is neither described, there is no mention about the risk of solution precipitation at low temperature that requires constant steam tracing.

In the original plant design, also no mention was made about this condensation phenomena and the tracing system designed to prevent the relatively stagnant Benfield solution precipitating/solidifying.

The failure, during HAZOP in plant operation, to identify such condensation phenomena can be explained by the HAZOP node definition that was not adequate (too large) to identify those local conditions in the level bridle.

Process Safety Report Conclusions

- The level bridle should have been handled as a separate node in the HAZOP. Reduce Node sizes during HAZOP.
- Process Hazard Analysis failure, first during design then during operation, can be considered as one of the root cause for this event.
- Designing/operating/maintaining an efficient tracing system including nozzles remains a challenge (electrical tracing is a viable option).
- Deficient localized passivation layer at the bridles and low circulation areas. Procedures to be developed/updated.

NOTE: This low circulation condition was not foreseen as an issue during the preparation of the passivation procedure, because the HAZOP made in the plant in 2007 did not further develop the analysis to this transient operating mode.

Analyzing risk in transient conditions requires a step by step detailed approach for which our actual PHA tool was not adequately set up. Training of individuals doing HAZOPS also requires attention. Recommendation is to improve guidance centrally to better develop HAZOP in transient conditions.

In the original design the level bridle came in carbon steel as for the main equipment. Design failure was considered by the team as one of the root cause for this event. Today ammonia plant design requires the level bridle to be in SS304. It is possible that we may have missed a lesson learned from ammonia plant licensors or ammonia industry in general. It is recommended to communicate this event via the network teams internally and externally to YARA in order to identify any similar situation.

FAILURE MECHANISM - Localized internal Corrosion

To explain this localized internal corrosion, the hypothesis of lack of control of the Benfield solution was rapidly discarded based on the regular solution chemistry follow up (showing no evidence of low corrosion inhibitor levels).

The characteristics (shape and localization) of the internal thinning observed in the bridle brings forward the hypothesis (supported by the metallurgical analysis) of a corrosive liquid phase condensing at the upper nozzle, going along the internal wall of the equipment down to the liquid level, most likely promoted by a deficient passivation layer, on a carbon steel equipment.

At this location the vapor phase is composed of syngas rich in CO_2 at 220°F (104°C) and water is present from the Benfield solution. If the temperature of this "mixture" is not maintained above the dew point, acid carbonic formation and condensation are inevitable.

Signs of long term exposure to acidic condensation in the upper nozzle and reflux demonstrate a lower temperature condition at that location, this can be associated to a weakness in the insulation/tracing system allowing this process condition (low temperature) to happen at the nozzle.

At the location of failure, in addition to the low turnover of the solution in the bottom part of the bridle, the upper part was most of the time in contact with a vapor phase. The conditions required to build up a proper passivation layer require constant renewal of the passivation solution at the surface of the metal.

Passivation is well performed for the tower and piping in the loop but it appears that little or no passivation are done on areas with low to no circulation of passivation fluid and in this case the bridle chamber.

Lessons Learned

1. Designing/operating/maintaining an efficient tracing system including nozzles that would reduce or prevent CO2 condensate corrosion remains a challenge. If this phenomenon was identified during original design or plant operation, the decision to upgrade this level bridle in stainless steel would have been taken.

2. Plant startup procedure in transient conditions or during passivation process was not detailed enough to include these type of bridles and also with such a complex design, it may not even be practical.

3. From our Risk Based Management Inspection program (RBMI), analysis at a loop/cluster level to a detailed inspection scope must be down to a component level and in this case, the level bridle should have been treated as a specific component. In this way, the bridle would have had defined degradation mechanisms. For this loop "Rich Benfield", the damage mechanisms were erosion/corrosion, CUI and CO_2 SCC.

4. Ultrasonic Testing measurements would have been the primary Non Destructive Examination method for erosion/corrosion.

5. CUI was part of the general inspection scope for small diameter connections to pressure equipment, however UT measurements were not done for the bridle.

Recommendations

1. Upgrade bridle chamber assembly at the next turnaround to stainless steel material, in this way avoiding the complexity of not knowing if passivation will take place or not.
2. In the corrosion study reviews, chambers and attachments to pressure equipment such as this should be classified as separate component for a better definition of the potential damage mechanisms according to the corrosion loop. As such it should also have its own detailed inspection plans.
3. Review the passivation procedure for potential dead leg areas in corrosive systems.
4. This recommendation is not only for Trinidad but plants with similar conditions.
5. Improving the process of draft inspection plan (outcome of RBMI) to the detailed equipment inspection plan to ensure nothing is taken for granted or not included.
6. Periodic inspection of carbon steel bridle chamber installed until the next turnaround cycle.
7. Review of papers on condensation corrosion in dead legs.

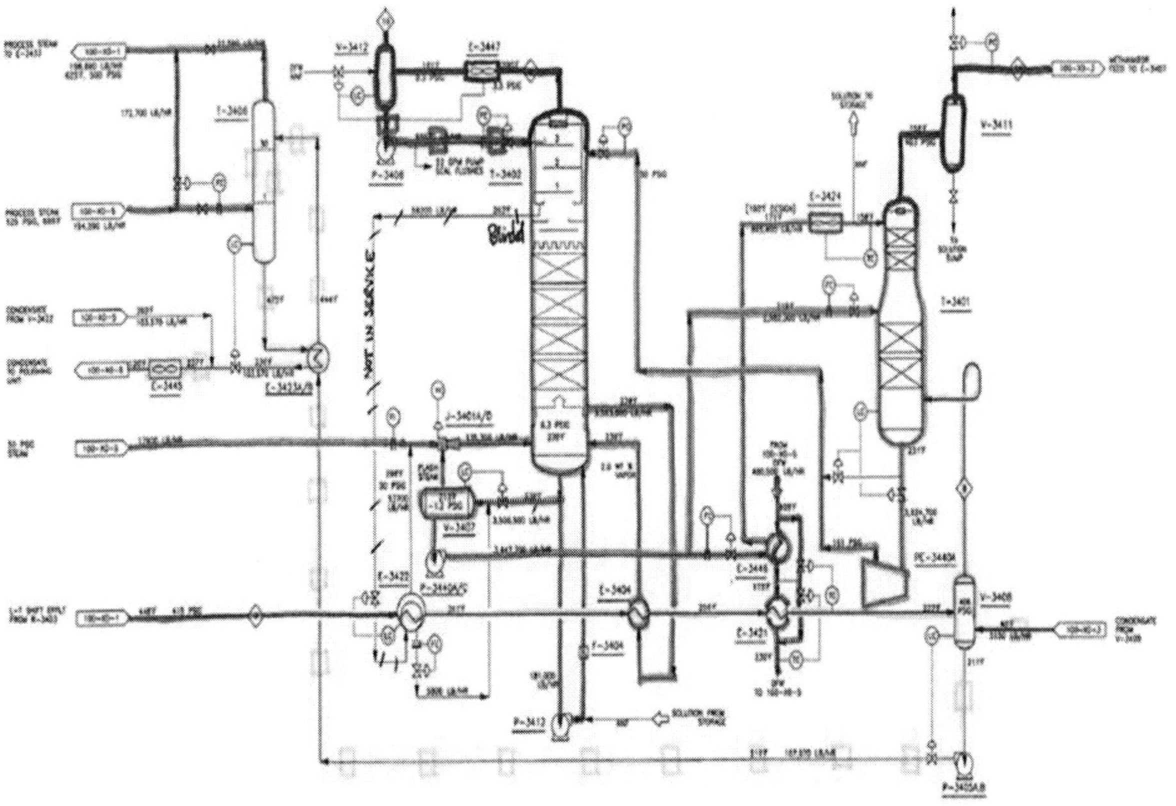

Figure 8 – Process Flow Diagram for CO_2 Removal system

EHS & Technology Risk Assessments for the Nitrogen Industry

Johnson Matthey's Process Technologies division supplies catalysts, absorbents, licensed technologies, plant equipment and services to the petrochemical, syngas, oil refining and gas processing industries. As such we design and operate our own catalyst manufacturing plants and we develop new products and technologies for our customers. This necessitates the use of a range of Environment, Health and Safety (EHS) assurance techniques to ensure the safety both of our own manufacturing facilities and of the products and technologies that we supply to third parties. The paper describes various EHS and technology risk assessments carried out by Johnson Matthey (JM) when developing new processes.

John Pach
Johnson Matthey Process Technologies

Introduction

Johnson Matthey's Process Technologies division supplies catalysts, absorbents, licensed technologies and services to the petrochemical, syngas, oil refining and gas processing industries. As such we design and operate our own catalyst manufacturing plants and we develop new products and technologies for our customers. Risk management is an essential prerequisite to the delivery of any project. There are many different types of risk – Environment, Health and Safety (EHS), commercial, technology, execution, market and reputation to name but a few.

This necessitates the use of a range of risk assessment techniques which are applied both to our own manufacturing facilities and the products, services and technologies that we supply to third parties. An example of this approach is described with reference to an innovative new process for Ammonia – Methanol - urea-formaldehyde concentrate (UFC) co-production recently launched by Johnson Matthey (Reference 1).

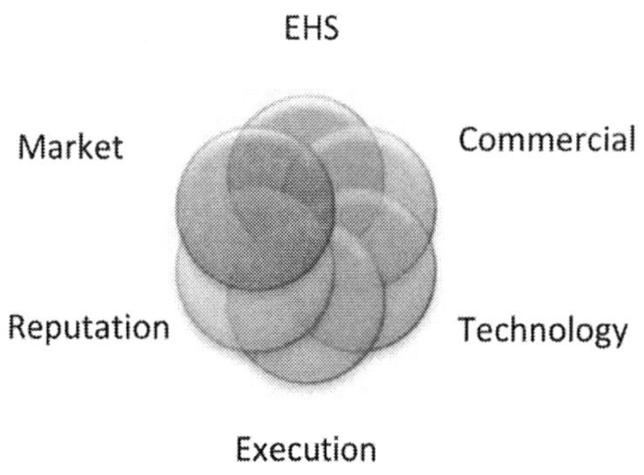

Figure 1: Various types of risk

Ammonia-Methanol-UFC Co-production

Carbon dioxide, a by-product of the synthesis gas generation process for ammonia manufacture, can be usefully combined with product ammonia to synthesize urea. Formaldehyde (in the form of UFC) is used to condition granular, and sometimes prilled urea. One of the most commonly used grades of UFC is UFC-85, a mixture of 60% formaldehyde, 25% urea and the balance water. Formaldehyde is produced from partial oxidation of methanol, which in turn is produced from synthesis gas.

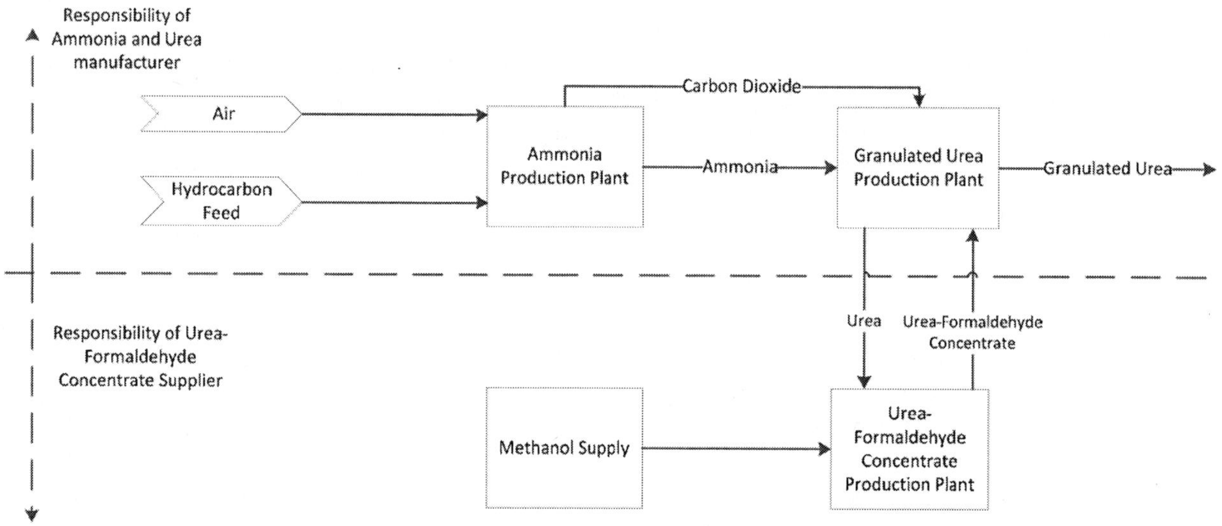

Figure 2: Standard arrangement of technologies

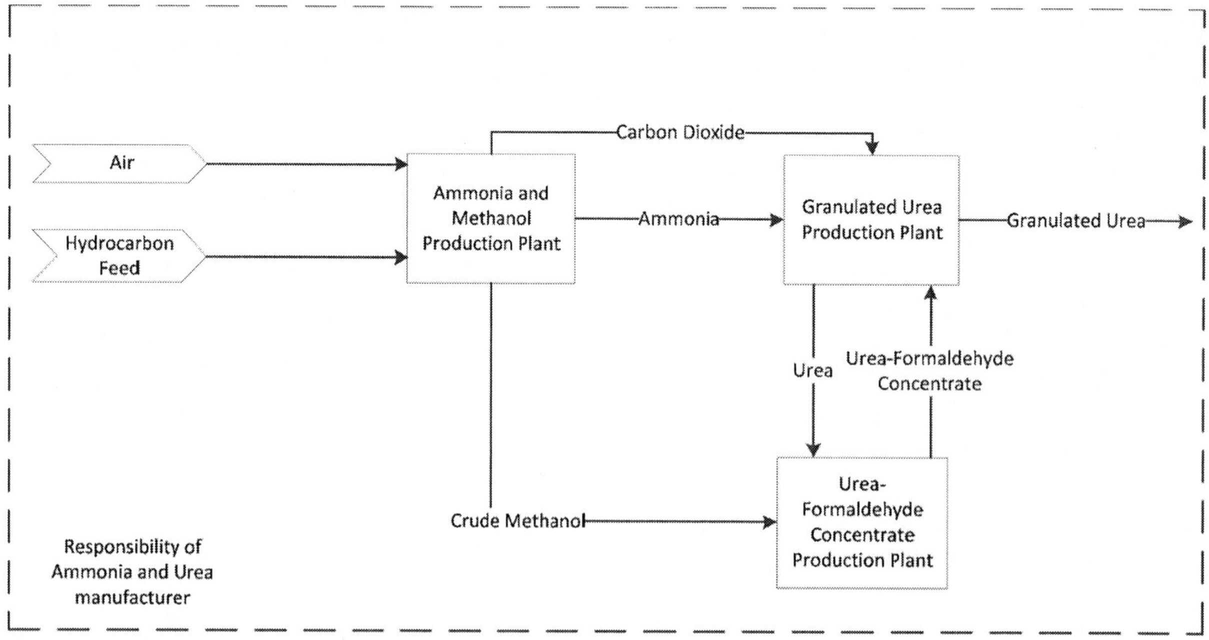

Figure 3: Arrangement of technologies when utilizing the Johnson Matthey integrated flowsheet

Generally, the formaldehyde additive required for urea conditioning is purchased from a third party as the quantities required for an individual ammonia/urea plant are too small to justify investment in a stand-alone formaldehyde plant along with an associated methanol plant to feed it.

As a result the UFC-85 is obtained from third parties whose cost structure is based the market price of methanol - which is normally at a considerable premium to the cost of the raw materials used to synthesize the methanol. The cost of the UFC-85 is further increased to cover transportation costs, the need to cover supplier overheads and return on investment criteria.

Johnson Matthey is a leader in catalyst and process technology development for the ammonia, methanol and formaldehyde industries and has developed the **FORMOX**™ integrated UFC process, a novel scheme which is the subject of patent applications, to integrate ammonia, methanol and UFC-85 production in a manner which ensures reliability of supply as well as capital and operating cost savings.

The scheme uses the residual carbon oxides downstream of the CO_2 removal system to produce methanol which is then used to produce UFC for the urea plant. The scheme can be applied as a retrofit as well as to new build projects.

Flowsheet Development

The first step in the integration of the three technologies is to produce methanol and ammonia on the same plant.

Although there are a multitude of flowsheet concepts for integrating ammonia and methanol production, some of which have been commercialized, the intent of most co-production schemes is to allow bulk production of both ammonia and methanol. For example, ICI's METHAMM flowsheet would have been capable of producing 1100 mtpd of ammonia along with 2200 mtpd of methanol.

For comparison, a typical 2200 tpd ammonia plant linked to a 3500 tpd granulated urea plant would need around 20-30 tpd UFC-85 which, in turn would only require 14-21 tpd of methanol. This is too small a volume to be produced economically using existing schemes.

Methanol Synthesis

The optimal location for small scale methanol production sufficient to satisfy the needs of an associated urea plant is downstream of the CO_2 removal system and upstream of the methanator. This allows methanol to be produced in concentrations high enough to allow UFC-85 to be produced without the need for distillation of methanol or formaldehyde which leads to significant operating and capital cost savings.

Urea Formaldehyde Concentrate (UFC-85)

Production of formaldehyde and UFC uses well established technologies. **FORMOX** technology from Johnson Matthey, employs mixed oxide catalyst technology due to its superior yield, high steam production and because it allows UFC-85 to be made within the same plant.

The main principle is to partially oxidize methanol in the presence of air to form formaldehyde and water. The gas mixture of formaldehyde, water and air is separated in an absorption column in which water is condensed and the formaldehyde is absorbed into the water. UFC is produced when formaldehyde is absorbed into a urea solution which is fed to the column. The oxidation process consumes oxygen; hence oxygen needs to be provided to the process and oxygen lean gas has to be removed. The oxygen lean gas contains traces of formaldehyde, methanol and carbon monoxide and is passed over a noble metal catalyst incinerator (emission control system or ECS) in which trace impurities are converted to carbon dioxide and water.

Ammonia and UFC Plant Integration

Standard Johnson Matthey **FORMOX** plants are designed with flexibility in mind and either UFC-85 or formaldehyde can be produced on the same plant. However, if the plant is only required to produce UFC-85, it is possible to reduce capital cost by replacing the second stage of the absorption tower with a gas cooler/condenser.

The plant can also be configured to produce UFC concentrations other than 85%.

The UFC plant can be designed to operate with crude methanol. Further integration is also possible and is offered on a case by case basis. An integrated UFC plant flowsheet is shown in Figure 4.

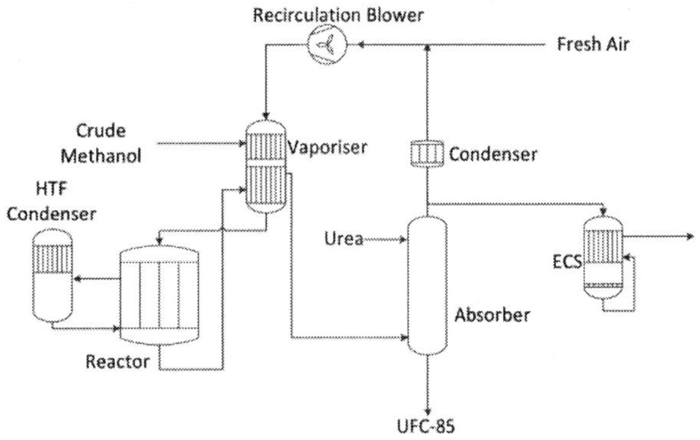

Figure 4: Integrated UFC flowsheet

Operating Costs

As crude methanol is synthesized at fuel cost (i.e.: 1 mole CH_4 per mole CH_3OH), there is no requirement for distillation, and the operating costs of an UFC plant are negligible, the result is a substantial saving in operating cost. Representative savings for a standard 2000 mtpd ammonia plant making around 3200 mtpd granulated urea are shown in Figure 5. The lower line represents sites that already have UFC production facilities on site that would save the cost of importing methanol to their UFC plant. The upper line represents sites that import UFC and so pay a premium for UFC. Further operational savings can be achieved and would be confirmed as part of a detailed project risk/feasibility study with a plant.

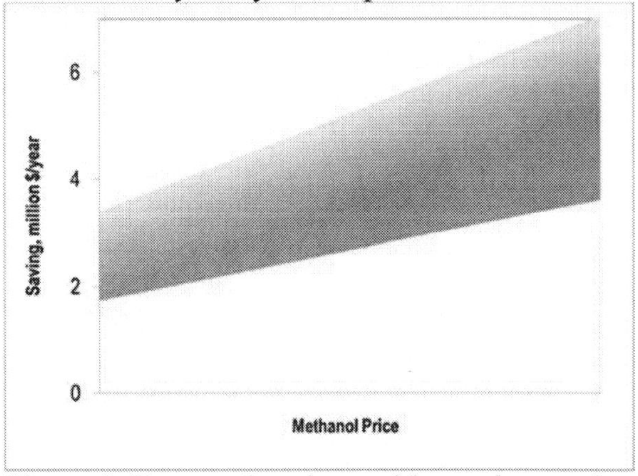

Figure 5: Expected savings with varying methanol price on a 2000 mtpd ammonia plant

Capital Cost

The capital cost of the process is competitive when compared to the alternative of purchasing UFC-85 at prices which reflect those of traded methanol.

Flexibility and Reliability

As the methanol synthesis unit can be bypassed, there is no impact on ammonia plant reliability. Instantaneous methanol/UFC unit capacity and intermediate storage tank size allows sufficient inventory to be built up to cover a UFC plant shutdown for formaldehyde catalyst change-out, without affecting ammonia or urea production.

Synthesis Gas and Ammonia Loop

The methanol synthesis reaction consumes less hydrogen than the methanation reaction, as shown for CO:

$CO + 2H_2 \rightarrow CH_3OH$ Methanol production
$CO + 3H_2 \rightarrow CH_4 + H_2O$ Methanation

As a result, for constant feed natural gas flow, more hydrogen is available to be fed to the loop. This means that the co-production scheme has the added benefit of allowing an increase in ammonia make or an improvement in efficiency of about 0.4%. In addition, less methane is added to the loop allowing the loop to run either with as lower inert concentration or a lower loop purge rate.

Crude Methanol Usage

The flowsheet allows the use of crude methanol as feed to the UFC-85 plant. This means that there is no need for a complex and energy intensive distillation system. It also reduces the capital cost of the plants compared to stand-alone units that could be supplied.

Process EHS Assessment

The EHS risk assessment techniques adopted by JM draw their inspiration from HAZOP, the multistage hazard study processes, and other techniques developed by ICI (Reference 2).

A multistage process is adopted in which each phase is linked to a particular project stage (figure 6), and as can be seen from the figure, other techniques can also be deployed (e.g.: fault and event trees, LOPA, FMEA, HAZCON).

Hazard Study 0 - Concept stage

The purpose of Hazard Study 0 is to identify hazard characteristics during conceptual design to allow them to be eliminated or reduced at an early stage of a project.

Hazard Study 1 - Concept stage

The purpose of Hazard Study 1 is to ensure that the understanding of the project, the process and the materials involved is sufficient to enable environmental, health and safety issues to be properly assessed.

Hazard Study 2 - Front-end engineering design and project definition

The purpose of Hazard Study 2 is to identify significant hazards and ensure that appropriate measures are implemented to eliminate risk or to reduce risk to tolerable levels. Identifying the hazards provides the opportunity for redesign to eliminate or significantly reduce risk. LOPA, fault trees or event tree are often used to augment this study.

Hazard Study 3 - Front-end engineering design and project definition

More commonly known as HAZOP, the purpose of Hazard Study 3 is to review the design and/or procedures to identify any hazards or obstacles to operability which could arise, particularly through deviations from the design intent. In the case of process plant, this will be based on the study of firm Engineering Line Diagrams and outline operating and commissioning procedures. The consequences of deviations are identified and, where necessary, appropriate corrective actions initiated.

The study also provides an opportunity to review potential maintenance and quality problems.

Hazard Study 4 - Construction/design verification

This review is performed at the end of the construction stage. The hardware is checked to ensure it has been built as intended and that there are no violations of the designer's intent. It also confirms that the actions from the detailed design hazard studies 1, 2 and 3 are incorporated, and operating and emergency procedures are checked.

<table>
<tr><th>Project Stage</th><th>Process Hazard Assessment Stage</th><th colspan="2">Study Method</th></tr>
<tr><td>Feasibility</td><td>PHA0</td><td colspan="2">Hazard Study 0</td></tr>
<tr><td>Development</td><td>PHA1</td><td colspan="2">Hazard Study 1</td></tr>
<tr><td rowspan="16">Definition</td><td rowspan="7">PHA2</td><td rowspan="4">Hazard Study 2</td><td>Process</td></tr>
<tr><td>Non-process</td></tr>
<tr><td>PES (Programmable Electronic Systems)</td></tr>
<tr><td>New Product</td></tr>
<tr><td colspan="2">LOPA (Layers of Protection Analysis)</td></tr>
<tr><td rowspan="2">Hazard Analysis</td><td>Fault Tree</td></tr>
<tr><td>Event Tree</td></tr>
<tr><td rowspan="9">PHA3</td><td rowspan="8">Hazard Study 3 (HAZOP)</td><td>Continuous Process</td></tr>
<tr><td>Batch Process</td></tr>
<tr><td>Maintenance & Operating Methods</td></tr>
<tr><td>PES (Programmable Electronic Systems)</td></tr>
<tr><td>Mechanical Handling</td></tr>
<tr><td>Electrical Systems</td></tr>
<tr><td>Buildings</td></tr>
<tr><td>New Products</td></tr>
<tr><td colspan="2">FMEA (Failure Modes Effect Analysis)</td></tr>
<tr><td rowspan="3">Execution</td><td>PHA4</td><td colspan="2">Hazard Study 4</td></tr>
<tr><td></td><td colspan="2">HazCon</td></tr>
<tr><td>PHA5</td><td colspan="2">Hazard Study 5</td></tr>
<tr><td>Completion</td><td>PHA6</td><td colspan="2">Hazard Study 6</td></tr>
<tr><td rowspan="2">Ongoing Operation</td><td rowspan="2">PHA7</td><td rowspan="2">Periodic Hazard Study</td><td>Process Hazard Review (PHR)</td></tr>
<tr><td>HazDem</td></tr>
<tr><td colspan="2" rowspan="3">Administration</td><td colspan="2">Auditing</td></tr>
<tr><td colspan="2">Hazard Study Leaders</td></tr>
<tr><td colspan="2">Action Management</td></tr>
</table>

Figure 6: Project Hazard Studies

Hazard Study 5 - Pre-commissioning safety review

The purpose of Hazard Study 5 is to provide an opportunity for those responsible for personal safety, employee health and environmental protection on the operating site to satisfy themselves that the detailed implementation of the project meets Statutory, Company and legislative requirements.

Hazard Study 6 - Project close-out/post start-up review

This study, carried out a few months into the production phase, confirms that all outstanding issues from the previous studies are complete and seeks any lessons that might give useful feedback to future design work.

Hazard Study 7 – On-going operation and EHS assurance

A Periodic Hazard Review (PHR) should be carried out on existing plant to review on-going operations. This can assess any changes that have occurred and support the continuing EHS assurance of the assets.

Technology Risk Assessment

When developing new processes, additional risk assessment techniques are required. This is facilitated by the use of a technology risk assessment (TRA) process in which the following is subject to critical review

- Experimental basis - Research data underpinning the project is reviewed to ensure that the results and conclusions can be extrapolated to a full scale plant. Particular attention is made considering the entire operating (both intentional and under fault conditions).
- Design Basis – technical and engineering assumptions and methodology are reviewed to ensure that they are appropriate.
- Operational and EHS learning relating to this or similar products/technologies – often facilitated by review of papers from events such as the AIChE Ammonia Safety Conference, Johnson Matthey's Formaldehyde Conferences and Johnson Matthey's International Methanol Technology Operators Forum (IMTOF).

The product lifecycle is also reviewed – in the case of a new catalyst; considerations would include aspects such as:

- Raw materials
- Catalyst manufacturing
- Charging
- Initial heating
- Reduction
- Start up
- Normal Operation
- Shut down
- In situ storage
- Restart
- Discharge
- Disposal
- Side reactions & byproducts
- Contaminants
- Ex situ storage
- Transportation

Aspects similar to those normally covered in a HAZOP are also addressed such as the consequences of deviations in

- Temperature
- Pressure
- Composition
- Flow velocity
- Impurities

The TRA meetings are most productive if used to review information that has already been circulated, hence the normal rule is that all participants apply their best endeavors to collate and share all pertinent information before the meeting. This means that the meeting is focused on interpreting and drawing conclusions from data rather than trying to quickly assimilate and understand something which has just been seen for the first time.

The composition of the review team is adjusted to reflect the needs of individual projects but would often include functional experts, product experts, research chemists, design engineers, project, technology and commercial managers.

The basic process is as follows:

STEP 1: Identify all technology "hazards",

STEP 2: Assess the likelihood and consequence of each hazard being realized, without any additional control measures and score against a word model. The results can conveniently be displayed in a standard risk matrix (figure 7).

	Likelihood					
Consequence	Remote	Highly Unlikely	Unlikely	Possible	Quite Likely	Likely
Catastrophic						SEVERE
Massive						
Major				HIGH		
Moderate			MEDIUM			
Minor						
Slight	LOW					

Figure 7: standard risk matrix

STEP 3: Identify the basic causes and means of reducing the risk factor, likely improvement, and the difficulty (manageability) of the technology hazard and, if appropriate, display on a balloon diagram (figure 8)

STEP 4: Assign clear actions and accountabilities.

STEP 5: Progress and complete the actions - the technology risk is only reduced when the mitigating actions have been implemented.

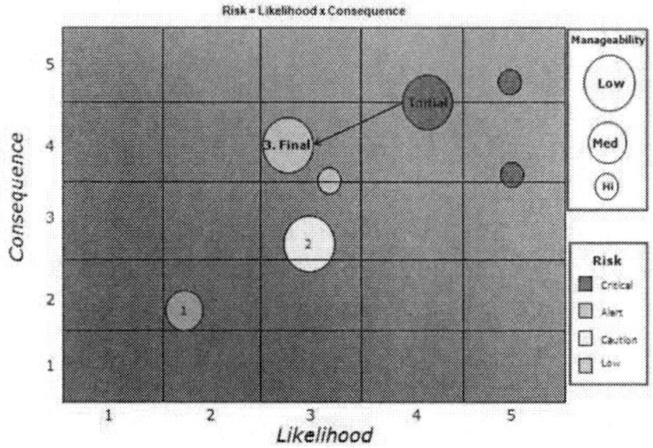

Figure 8: balloon diagram

Technology Risk Assessment – Ammonia-Methanol – Urea Co-production

The TRA team for this project consisted of

- Three product experts (one familiar with ammonia, one with methanol and one with UFC)
- Project Manager
- Research chemists
- Process design engineer
- Commercial manager
- Facilitator
- Independent expert (experienced engineer, familiar with developing and commercializing other processes but who had not previously been involved with this project)

Two distinct reviews were held – one focused on the modification of an ammonia plant to produce crude methanol which resulted in 66 actions. The other focused on the integration of the entire ammonia-urea-UFC scheme resulted in 72 actions

Most of the actions were points either of clarification (e.g.: what is the I/O count for a UFC plant), points of detail for detail design (e.g.: materials of construction), or project management items (e.g.: datasheets for the UFC plant should be in the same format as those for the methanol unit)

Follow up meetings were organized to ensure that items were closed out. Whilst these meetings may seem trivial, they are essential to ensure that the actions are carried out in a timely manner and in-line with the intent of the original risk assessment. As mentioned in Step 5 above, the risk has only been reduced once the actions are completed satisfactorily.

The major learning from the risk assessment was as follows:

- Each individual component of the scheme is well proven; the inherent risks are well understood and well managed.
- As with all similar projects, special attention needs to be paid to the interface between technologies. Although Johnson Matthey is fortunate in having extensive experience of all three technologies (ammonia, methanol, UFC), some minor

areas for improvement were highlighted (e.g.: common documentation standards)
- The only significant technological issue is the need to prevent ammonia from coming into contact either with methanol or with CO_2 as undesirable byproducts can be formed. As the flowsheet is configured in such a way as to prevent this from happening, this is not of concern.

Conclusion

Johnson Matthey employs a variety of risk management processes to manage risk both in our own catalyst manufacturing plants and in the products and technologies we offer to our customers.

Johnson Matthey's innovative ammonia-methanol-UFC co-production process has been subjected to detailed risk assessment which concluded that the inherent risks are both well understood and well managed.

References

1 "UFC Manufacture for Urea Production", A Magnusson, K McFarlane, J Pach, D Ward, Nitrogen and Syngas 2016

2 "Ammonia Production - A Century of Safety, Health & Environmental Improvements", J Brightling, J Pach, AIChE Ammonia Safety Symposium 2013

Integrated Safety Concepts for Rotating Machinery

For a long time the development of safety concepts for rotating equipment was seen as a job of rotating equipment vendors and specialists. Safety relevant scenarios were often seen more as part of "machine safety" than as a comprehensive task of inherent safe process design. With the development of modern Emergency Shut Down (ESD) systems and HAZOP studies more and more questions were raised, how the hazards, coming from rotating equipment should be treated in plant hazard studies to develop an integrated process safety approach. The aim of this paper is to review a number of safety relevant scenarios for centrifugal compressors and propose safety concepts, which are discussed with the learning from incidents. Major scenarios to be discussed are over speed on turbines in general and on turbo alternators in particular, axial shift of compressor internals, main lube oil and seal oil supply failures on syngas compressors.

Hans Juergen Essl, Leonhard Werner, Georg Zimmer

Borealis Group Austria

Introduction

With the development of M.W. Kellogg's single ammonia process in 1963 the basis for the design of most of the existing ammonia plants operating on natural gas feed stock was set. The main principle consists of the utilisation of the excess heat from steam reforming and synthesis section to drive non redundant main compressors with steam turbines. As this concept was allowing to build energetically self-sufficient large scale plants at competitive investment costs, this underlying principle was integrated into production processes from other licensors as well.

Typical arrangement of machinery in this type of plant setup consists of steam turbine driven centrifugal syngas, air and refrigeration compressor. In most cases a steam turbine driven boiler feed water pump and a steam turbine driven electrical generator (turbo alternator) to equalize the steam balance inside the plant complete the setup of heavy machinery [1].

Typical operational hazards of rotating equipment in ammonia plants

Depending on the exact process design turbine drives for syngas compressors already utilize a power of approximately 15 MW to drive a syngas compressor for a typical 1000 mtpd ammonia plant, whereas air compressors have a power consumptions of about 7 MW for the same size of production. Rotation speed of syngas compressors can go higher than 13,000 rotations per minute, whereas other machinery is operated in an area of below 10,000 rotations per minute.

As a consequence of the physical forces involved in this operation it is easy to imagine, what follows after a collision of rotating and static parts of machinery considering the additional hazards of fire, or explosion due to the media (hydrogen, natural gas, ammonia ...) present in the compressors and the compressor house.

Following main common causes of steam turbine failures are described in literature being responsible for the major portion of consequential losses [2]:

- Turbine over speed
- Liquid water introduction
- Loosening of parts inside
- Excessive vibration
- Lubrication problems including seal oil
- Fatigue and corrosion damages
- Misalignment of train

All these potential causes have to be considered in safety studies of the rotating equipment of the plant. In case of compressors also surging has to be considered as a main failure scenario.

Safety studies

For a long time safety concepts for rotating equipment were designed following a Standards-based, deterministic approach. These standards were developed by equipment manufacturers mainly. Consequently, the engineering of an overall safety concept of machinery was considered as exclusive know-how by the equipment vendor and seen as part of the deliverable of a whole machine package. Such an approach faces its limitations as not all process-specific aspects are covered appropriately.

Over time standardisation organisations started to introduce risk-based approaches into their codes and standards, allowing more flexibility but also assigning more responsibility to the performing party. One milestone was the EN 954-1:1997 (*Safety of machinery - Safety-related parts of control systems - Part 1: General principles for design*), where a risk graph for a semi-quantitative evaluation was introduced. Although the basic principles are still in use, the EN 954-1:1997 as such has been outdated by the new standards EN ISO 13849-1 and EN (IEC) 62061 to be compliant with the concept of functional safety and better aligned with the Machine Directive 2006/42/EC, and to address common-cause failures, etc. Obviously, re-evaluations of existing safety systems might end up in different results, not only because of different normative contents, but also depending on the team composition.

Process installations consist of a complex combination of different types of equipment, where only a rather limited number of them can be considered as machines. So additional to the evaluation of the machinery, a sound process hazard analysis (e.g. HAZOP) and risk assessments (e.g. Risk Graph or LOPA acc. to IEC 61508/11) needs to be performed in the course of the engineering.

However, as long as conforming to applicable Codes and Standards, machineries are considered as a "safe package" typically. So the consequences of attaching alternators, compressors or pumps to a turbine and the interaction of the whole set with the chemical process, are rarely analysed systematically.

Conducting Retrospective Hazard Reviews[1] for existing plants in a multidisciplinary team is a

---

[1] In order to maintain a high safety standard, Retrospective Hazard Reviews have to be carried out on a regular basis. For plants falling into the European Seveso III category classification parameters, this internal review of PS critical elements (e.g. conduct / or check accuracy and quality of existing HAZOPs, LOPAs, documentation of PED equipment,...) has to be done at least every 5 years or after major modification in the plant.

great chance of effectively evaluate and increase the overall safety standards. Process safety considerations need to involve a sound analysis of the hazards related to rotating equipment, especially the interaction between machinery and chemical process (Remark: the same principle applies for all units, where specific, international acceptable standards exists e.g. EN 746-2 for furnaces, EN 12952 for water tube boilers). Especially in retrofitting/revamping projects of existing plants it is absolutely key to deeply and consistently involve the client in the hazard and risk evaluation studies, as requested by different Codes.

Borealis Experience

Borealis was performing Retro Hazop studies on existing production plants in the years 2007 - 2012. In these studies a classification of safety critical interlocks according IEC 61508/11 took place by applying a Borealis internal *Layer of Protection Analysis (LOPA)*, focusing on Loss of Containment[2] scenarios. At that moment one of the ammonia plants was still operated on pneumatic and hard-wired *Emergency Shut Down (ESD)* systems. A project was started to migrate the ESD to a new state of the art PLC-based ESD system. The decision was taken to use these studies as a basis for the engineering of the new system. When reviewing the design, it initially seemed to be incomplete, especially when considering the potential hazards of rotating equipment. For over speed scenarios of steam turbines, for example, no LOPA was performed, because loss of containment was not considered as an immediate risk in these particular scenarios. As mentioned earlier, machineries are considered as safe units in process Hazops typically. In 2012 a study was started together with *Safewise* and the machine manufacturer *SULZER* to evaluate the hazards of machinery in interdisciplinary teams. The underlying philosophy of this evaluation was to install a state of the art machine protection system based on the state of the art functional safety philosophy (e.g. IEC 61511) integrated into the plant automation system to ensure a consistent approach in terms of design, validation and life-cycle management (e.g. testing, interlock override) of the safety instrument function. Moreover additional to the safety aspects the economic impact was also evaluated to cope with the risk of an excessive economic impact in case of heavy machine damage. As a conclusion of the team most scenarios were rated corresponding Safety Integrity Level (SIL) 1. Also the standards IEC 62061 and ISO 13849 are following a systematic approach to evaluate the quality for the safety systems as well. However, these standards were not applied, because they are not suitable for the so called "high demand" mode with a continuous or high frequency presence of hazards.

Independently from the ongoing evaluation process during the course of the Borealis investment project, a vendor cross functional working group comprising several international, reputable machine manufactures like *MAN, Siemens, Voith, Alstom*... was established to work on a solution to close the expressed gap in terms of safety codes and standards for machinery as integral part of chemical processes. The overall objective was to define an applicable code and standard in line with the nowadays functional safety principles according IEC 61508. The final outcome of this high qualified work group has led to the development of a dedicated German standard for turbomachinery called VDMA 4315 "Turbomachinery and generators – Application of the principles of functional safety". It was published by the German association for machinery and plant engineering in February 2013. At that time the Borealis customized

[2] In the Borealis LOPA methodology LOC is the defined cut-off-criteria. As the further development of the scenario is depending on highly unpredictable parameters (presence of people, probability of ignition, meteorological data, ...), this approach is providing more consistent results, but is limited in applicability. Remark: A release of toxic substances to atmosphere via PSV is also seen as LOC; LOPA to be conducted.

approach was already in the execution and pre-commissioning phase.

Practical example syngas compressor

The syngas compressor, which was investigated in 2012 consists of a Siemens 120 bar steam extraction turbine and 40 bar MP extraction turbine with low pressure condensation stage attached. The compressor (*Nouvo Pignone*) consists of 4 stages in 4 barrels [Figure 2]. In the study also the general scenarios creating hazards and incidents of the past were reviewed. The considerations resulted in recommendations for safety and business critical interlocks in SIL 1 quality [see Table 1]. As a general philosophy a proper selection of availability (e.g. 2oo2), reliability (e.g. 1oo2) and combined architecture (e.g. 2oo3) for sensor and signal processing was chosen. 2oo3 voting was applied on speed sensors for control and trip purposes [Figure 1]. 2oo2 voting was applied on the axial displacement on each stage of the turbines and the compressors. Like often published in literature the exchange of mechanical based over-speed protection devices to electronic systems is strongly recommended [2, 3, 4].

***Figure 1** – Turbine equipped with 2oo3 speed, bearing temperature, and axial displacement probes.*

After a full study of the performance and maintenance history (proven for more than 40 years) of existing shut off valves including closing times, it was decided to keep those shut off valves and verify them according the so called prior in use in accordance to IEC61511. Furthermore the shut-off valves of a steam turbine are seen as integral part of the turbine in order to minimise the enclosed steam volume in case of a trip. Therefore an exchange of the valves is always linked to unreasonable investment costs.

In general, it needs to be noted, that the concept of the IEC 61508/11 is based on large numbers of components, which are installed in comparable applications/environments, which is typical for components of instrumentation (transmitters, etc.). For mechanical parts of safety functions (e.g. valves) exposed to process media, this is only very limited applicable. Statistic-based approaches face their limitations there, which is another strong argument to aim for proven in use design in case of mechanical parts of safety functions.

The start-up of emergency oil pumps was also indicated as a business critical reaction of the ESD system.

Following the philosophy to apply the most appropriate methodology (LOPA, Risk graph, company internal risk matrix) to evaluate the different hazard scenarios in a cross functional team the following incidents were reviewed and SIL-classified.

Table 1 summarises the safety functions, which were identified and their quality requirements determined. The first column is the unique number for the Safety Instrumented Function (SIF); which is indicated on PIDs and SRSs. Next the Safety Function as such is described, followed by a description of the development of the unmitigated consequences. Whenever LOC is considered as main risk, Borealis LOPA was applied to determine the SIL requirement for the specific safety function. For other hazardous scenarios, not resulting into a LOC, a Risk Graph evaluation acc. to IEC 61511-3/Appendix E was performed. The selected methodology for each SIF is indicated in the table as well. Scenarios resulting into severe machine damages and consequently into significant business interruption, were evaluated by applying a simplified, two-dimensional risk matrix, simply taking probability and costs into account. According to an internal Borealis procedure these functions needs to be designed (and operated and maintained) equivalent to min. SIL1. However, a precise calculation of the required response time is not mandatory (in most cases this is not even possible for machine protection systems).

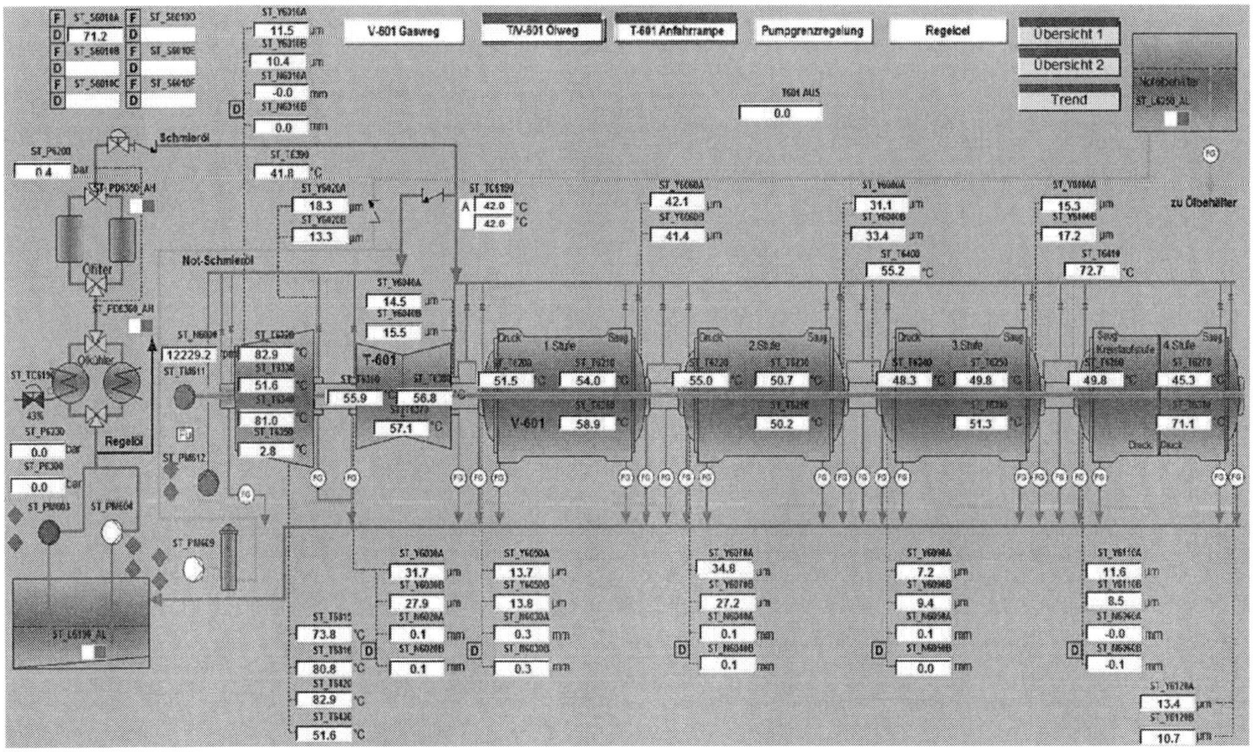

Figure 2 *– Syngas compressor setup*

SIF-No	Description of Safety Function	Consequence(s)	Result	Method for Determination	Required Response Time
SIF 14	Lubrication oil pressure low initiates a start of emergency lubrication oil pump	Loss of seal -, lubrication and control oil pressure -> Failure of control system of steam turbine -> Severe damage of machine but LOC improbable	**SIL 1** (Business Critical Interlock)	Borealis internal risk matrix for Business Interruption	n.a.
SIF 15	Lubrication oil pressure low initiates a trip of the steam turbine and synthesis gas compressor	Loss of seal -, lubrication and control oil pressure -> Failure of control system of steam turbine -> Severe damage of machine but LOC improbable	**SIL 1** (Business Critical Interlock)	Borealis internal risk matrix for Business Interruption	n.a.
SIF 16	High differential pressure between pressure and suction side 4. stage initiates opening of bypass valve and trip of steam turbine and synthesis gas compressor	Reverse flow from pressure side to suction side of the 4.stage of the compressor -> damage of synthesis gas compressor (damage of separation between recycle - and 4th compressor wheel) but LOC improbable	**SIL 1** (Business Critical Interlock)	Borealis internal risk matrix for Business Interruption	n.a.
SIF 17	Low differential pressure between pressure and suction side 4. stage initiates opening of bypass valve and trip of steam turbine and synthesis gas compressor	Reverse flow from pressure side to suction side of the 4.stage of the compressor -> damage of synthesis gas compressor (damage of separation between recycle - and 4th compressor wheel) but LOC improbable	**SIL 1** (Business Critical Interlock)	Borealis internal risk matrix for Business Interruption	n.a.
SIF 18	Low water level in waste heat boiler synthesis unit initiates a trip of the steam turbine and synthesis gas compressor	Low water level in waste heat boiler synthesis unit -> overheating of waste heat boiler -> above design temperature -> release of process gas	**SIL2**	LOPA	Max. response time total 190 sec.
SIF 19	High level of ammonia liquid initiates a trip of the steam turbine and synthesis gas compressor	High level of ammonia liquid in converter -> activation of safety valves -> release of ammonia within process area on safe location	**SIL1**	LOPA	Max. response time total 97 sec.
SIF 180	High speed initiates a trip of the steam turbine and syngas compressor	High speed of turbine ->damage of turbine and compressor -> risk for people due to flying objects	**SIL1**	Risk Graph IEC 61511-3/Appendix E	Max. response time total 300 msec.
SIF	Axial displacement of the steam turbine and synthesis gas compressor initiates a trip of the steam turbine and syngas compressor	High axial displacement of the turbine -> damage of bearings and static parts of the turbine and compressor but LOC improbable	**SIL 1** (Business Critical Interlock)	Borealis internal risk matrix for Business Interruption	n.a.
SIF 57 – 061	Limit switches on the compressor isolation valves	Damage of the compressor but LOC improbable	**SIL 1** (Business Critical Interlock)	Borealis internal risk matrix for Business Interruption	n.a.
SIF 62 - 63	Limit switches of the shut-off valves (+ permissive interlock)	Damage of the compressor but LOC improbable	**SIL 1** (Business Critical Interlock)	Borealis internal risk matrix for Business Interruption	n.a.

Table 1: *Result of safety concept review on syngas compressor*

Incident review 1: Trip of syngas compressor after external power grid failure

In 2011 the ammonia plant in Linz tripped as a consequence of a disconnection of power supply (decoupling of grid occurred as a consequence of a power dip). The syngas compressor shut down was initiated automatically as a consequence of another failure in the electric supply provided by a turbo alternator integrated in the plant. The compressor was isolated from gas supply and the main steam shut-off valves of the HP and MP turbine were closing automatically, but the compressor did not stop and continued to turn at about 1000 – 2000 rotations per minute. Shift operators started to close manual valves immediately. When the manual valve on the extraction line to the 40 bar steam grid was closed the machine finally stopped. Damage on the extraction steam jack valve was assumed, allowing steam flowing backward into the extraction turbine.

The turbine was equipped with an overhead emergency oil tank, whereas the emergency lubrication of the compressor was consisting of two redundant electrical driven main oil pumps and one additional electrical emergency oil pump with backup supply from a diesel generator. Due to an auxiliary contact on a fuse, the emergency oil pump did not start up automatically as foreseen in the safety concept of the train. As a consequence the bearings of the compressor were melting due to the missing oil lubrication [Figure 3].

The axial bearing of the second stage was completely welded with white metal, which was partly solidifying after passing the oil lubrication holes [Figure 4].

Due to damage of the bearings slight axial displacement of the first and second stage took place resulting in a damage of the labyrinth seals. The incident finally had led to a production loss of 18 days.

Figure 3: *Main bearing 1 stage with molten metal layer.*

Figure 4: *Welded axial bearing second stage.*

As a consequence the following **corrective actions** were implemented:

- Overhaul of extraction steam jack valves
- Overhaul of manual steam shut off valves and attachment of connection pieces for pneumatic operating device to accelerate closing time.
- Full redundant emergency oil supply, by connection of both main oil pumps to emergency diesel.
- Full redundant setup of sensors and signal processing in SIL 1 quality to start pumps in case of limited oil supply.

- Test of interlock according SIL 1 requirements with a proof-test interval of every second year.

It has to be mentioned at this point, that the reliability of the interlock may be improve less with the installation of SIL-certified components, but the proper lifecycle management. Maintenance and testing procedures and frequent testing create an essential contribution to increase reliability of business critical interlocks.

Incident review 2: Malfunction of high pressure steam shut off valve syngas turbine

The steam shut off valves of the same compressor were overhauled in 2013 by a well-known international contractor. As mentioned above these valves were qualified to be part of an interlock, which was classified as SIL 1. During recommissioning of the turbine an over speed trip test was performed with decoupled compressor. The machine stopped after activation of the trip sequence in the expected time frame. After the turnaround several plant trips have occurred. In some trips the machine shut offs did not occur in an acceptable time frame or the machine even had to be stopped manually by closing the manual steam valves. The main steam XV valve was investigated in a small turnaround together with the OEM Siemens. It turned out, that the shut off valve was wrongly adjusted by the contractor performing the overhaul of the machine. After readjusting the connection of the drive and the valve stem the turbine was properly shut off by the trip system again. The effect of the wrongly adjusted connection between drive and valve stem is demonstrated on Figures 5 and 6.

Learnings and corrective actions:

After detecting the default on the shut off valve it could not be explained, why the over speed test could be passed. In further analysis it was found, that the regulation valve, which was also overhauled in the turnaround and which is also closed in case the trip sequence was shutting off the steam during the over speed trip test and the first trips of the plant. During further operation as regulation valve the full tightness of the valve was lost. As a consequence the machine did not stop automatically any more, when the trip sequence was activated, because steam could leak through regulation and shut-off valve.

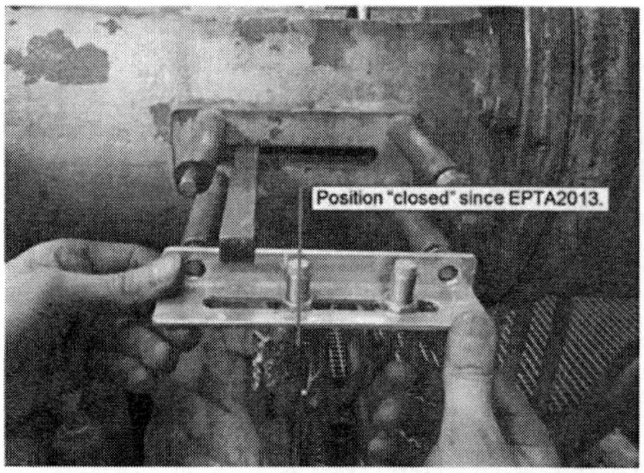

Figure 5 and 6: *Position indicator of wrongly adjusted shut-off valve*

OEMs do not share all details of regulation and ESD systems of rotating equipment in general. Although these details are not available for contractors, they are accepting maintenance contracts including work on critical interlocks, which is increasing the probability of quality defaults. In general it is extremely difficult to reveal quality problems prior to restart of the plant as these are mostly revealed only under process condition with all hazards of a running plant present.

So in the recommissioning after big overhauls *full functional tests including tightness control* on single valves should be considered. For the same reason *partial stroke tests* should be more seen as an exception to perform interlock tests, while the plant is in operation, but not be applied as a general rule for testing of critical interlocks.

Incident review 3: Over speed damage on turbo alternator

Incident 3 was occurring on a turbo alternator consisting of 36 bar MP turbine with extraction possibility to 7 bar grid and LP condensation turbine with a total power of 10 MW in France. After start-up of the machine no steam extraction to the 7 bar grid could be produced. In first consideration of likely causes it was concluded that one of the XV valves in the extraction line was blocked in closed position. That was why each of the valves was forced "closed" first and then the other into "open" position by an E&I technician. In the second possible configuration the machine suddenly accelerated due to backflow of steam from the 7 bar grid.

***Figure 7:** Broken coupling of turbo alternator*

As a consequence the over speed trip of the machine was activated leading to closure of the main steam supply valve and detachment of the alternator from the grid. Due to the strongly reduced resistance of the alternator the steam backflow to the LP turbine accelerated the machine further until the coupling failed and the rotor crashed with the blade carrier attached to the casing of the turbine [Figure 7].

In the later investigation it was found out that one of the *shut-off valves (XV valves)* was constructed as a jack valve with a disk adjustable in open position. Due to unknown cause the disk got detached from the lever [Figure 8] and had blocked the steam extraction line after start-up of the turbine. When the first XV was forced closed the disc fell back into the casing because of missing counter pressure.

***Figure 8:** Disk detached from the lever in XV valve was leading to steam backflow into turbine.*

As a consequence of forcing the damaged XV valve "closed" and the intact XV valve "open" steam could pass back to the turbine from the 7 bar grid and led to the uncontrolled acceleration of the LP steam turbine.

The investigation of this incident revealed that this interlock was not seen as a safety critical interlock, but treated as a kind of "normal" process interlock. That was why the override of the XV valves was taking place at night time without following a procedure necessary in case of a safety critical interlock. Testing and maintenance of this interlock was seen as a task of the contractor when performing overhauls of the machine. Of course, there is no guarantee that appropriate actions are taken, when overriding/bypassing an interlock, which is designed as SIF according to latest applicable codes and standards. But if an organization is able to identify and classify all relevant safety function and ensures proper documentation by keeping the total amount (and administrative burden) to a reasonable minimum, these safety critical functions will get the required focus during maintenance and whenever they need to be bypassed.

Discussion

In a chemical process plant a lot of different types of equipment are typically installed. All these equipment are introducing certain risks into the plant operation. Especially the rotating machinery like compressor and steam turbines comprises different risk elements caused by mechanically driven parts and interlink to the process via the pressurized part. The above described difficulties during the course of an investment project with the overall aim to lift-up the plant safety level to current standards, outlines immediately a certain gap in the available codes and standards in terms of turbomachinery. However this gap is seen and identified over the last year in the branch and the German association for machinery and plant engineering reacted with the development of the VDMA 4315 standard.

Moreover considering the holistic view on plant safety based on the example of the turbomachinery in an ammonia plant gives an impression how versatile the topic of functional safety can be. Only an integral and customized approach will ensure ultimately a safety level within an acceptable residual border.

Acceptable safety function is not achieved with SIL-certificates of single components but by identifying all safety relevant functions following an integrated approach and consequently applying a proper Life Cycle Management (e.g. acc. to the IEC 61508/11).

Conclusion

The application of IEC 61511 for machine protection system of rotating equipment offers a frame to develop an integrated safety concept for ammonia plants. Nevertheless for individual equipment it is necessary to evaluate the safety risk based on their dedicated codes and standards without losing the focus on the process interconnection. Moreover the design, installation and validation of safety instrumented function is equally important compared to proper life cycle management (proof-testing, bypassing of safety interlocks) during the operational phase until it is decommissioned. Also the right marking and tagging in the field ensure full functionality of the safety related interlocks and represents an important protection barrier for an uncontrolled malfunction during operations and maintenance works.

Finally the IEC 61511 gives the right basis to establish an appropriate safety concept and philosophy, but the example of the turbomachinery incidents identifies that for high complex equipment the collaborative approach within a specific industry to develop dedicated safety standards is required. In fact it provides a standardized approach for the end user, which can be adapted to the individual needs of their assets. This finally provides the best solution in terms of safety and costs.

References

1) Appl, M., "Ammonia: Principles and Industrial Practice", Wiley-Vch, Weinheim, p. 140 (1999)

2) Clark Edward, E.; "Rotating Equipment Loss Prevention – An Insurer's Standpoint" in Proceedings of the Twenty-Fith-Turbomachinery Symposium (1996)

3) Fisher L. L., Feeney R.L., "How Important is Turbine control", in Safety in Ammonia Plants & Related Facilities, Vol. 24, American Institute of Chemical Engineers, New York (1983)

4) Bailey T.R., Harclerode C.S., "Retrofitting the Ammonia Plant with Integrated Turbine Compressor Control" in Safety in Ammonia Plants & Related Facilities, Vol. 37, American Institute of Chemical Engineers, New York (1996)

Syngas Compressor hydrodynamic thrust bearing failure

The Syngas Centrifugal compressor of the ammonia plant at the Misr site consisting of High Pressure and Low Pressure stages had a thrust bearing failure at HP stage due to malfunction in controlling instrumentations which protect the train. This paper reviews the incident and the investigation analysis which identified thrust bearing failure as the root cause.

Ibrahim AbdAl-Wahab
Misr Fertilizer Production Company, Egypt
Hatem Abdel-Rahman
Misr Fertilizer Production Company, Egypt

Introduction

The synthesis gas (gas mix) compressor train located in Misr Fertilizer Production Company is part of the biggest urea ammonia plant in Egypt. (2 million tons of urea per year, Uhde design) The compressor experienced thrust bearing failure due to control instrumentation failures. We will discuss how we identified the problem and how it has been solved. We will also discuss the learned lessons and the modified actions which have been done to avoid repeating it in the future

Compressor Train Description

The train has HP and LP barrel compressors 2BCL 408 and 2BCL 458 manufactured by General Electric (GE) which are driven by a 19MW SIEMENS extraction steam turbine at 10,600 rpm. [Table 1.]

The HP barrel is divided into two stages, one is the make-up gas compressor which has 7 impellers, the second is the recycle stage which has 1 impeller, all on the same shaft. It has floating type barrier shaft end seals and the journal and thrust bearings are tilting pad type (figure 1, 2).

The HP compressor sections are protected by a single anti-surge controller FIC with anti-surge recycle flow FV. Additional valve HV is balancing the pressure between make-up gas and recycle gas during start up or trip in order to protect the compressor from axial forces which could be established. The HV valve has no controller during normal operation.

The speed governor is directed by CPC (Current to pressure converter) (I/H).

Technical Specifications

stages	Make up stages			Recycle
	1st stage	2nd stage	3rd stage	4th
Model	LP 2BCL458		HP 2BCL408	
Gas handled	Syngas (H2+N2)	Syngas (H2+N2)	Syngas (H2+N2)	Recycle Gas
No of impellers	4	4	7	1
Capacity NM3/Hr	142235	142235	142235	515053
Suction pressure bar g	26	48.5	81.7	160.2
Suction temperature C	8.5	34.7	31.2	29.7
Discharge pressure barg	48.5	81.7	162.5	168
Discharge temperature C	81	97	126	37

Table 1, compressor technical specifications

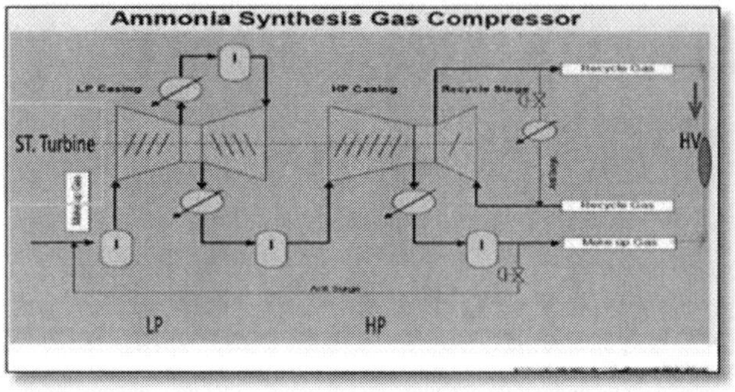

Figure 1. Ammonia synthesis gas compressor diagram

Equipment History

The compressor train has been in operation since 2008 and it worked well to this date since all parameters and required maintenance activity have been executed. The failure occurred in September 2013

Compressor Pro-Active Maintenance – Engineering Program in normal operating condition

In normal operating condition, by applying the MOPCO Pro-Active Maintenance program, we can prevent and predict failures of equipment parts. Through different action with different intervals, 1Day , 1Week , 2W, 1Month , 2M, 6M, 1year, 2Y, 3Y, for instance, check and act if necessary for leakage, vibration analysis, tightness, lubricant and control oil condition, loose parts, misalignment, fracture, cracks, cooling system condition, wear and tear, intercooling system, compressor flow and pressure, steam turbine performance, external dirties and deposits, lubrication system devices, oil analysis, record for different operating indicator devices.

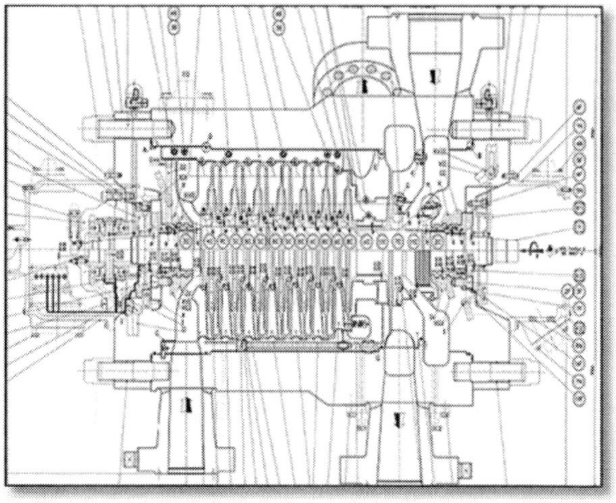

Figure 2. HP compressor sectional drawing

Catastrophic failure in 9-2013

During the normal operation with all parameters of the turbine and the compressor in normal values, suddenly the logic and governor of the steam turbine deviated and hesitated. The governor closed but apparently the trip signal wasn't sensed as the driver power was lost. The anti-surge valves started to operate due to the deviations in flow, and finally 12 seconds later the machine tripped due to high axial movement (thrust) signal, which reaches to 1mm as the trip setting was at 0.7mm. The total axial clearance (end play) of the bearing was 0.37mm.

Damage Description

High axial movement of the rotor led to severe damage for many parts. See below Figures 3 - 7.

Figure 3. Thrust bearing active side, damage pads

Figure 4. Thrust collar scratch

Figure 5. Non Active side

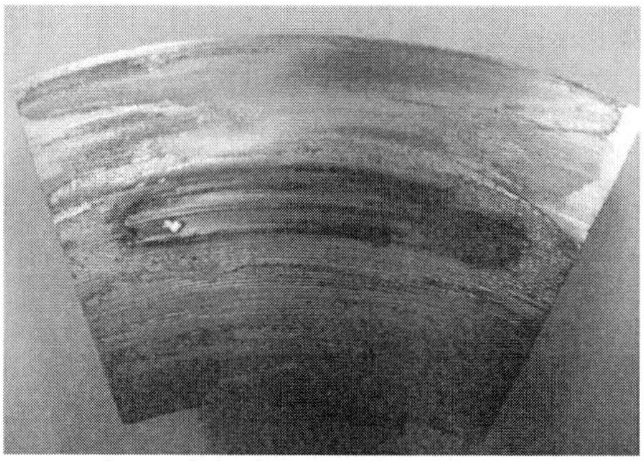

Figure 6. Pad thickness reduced by 1 mm

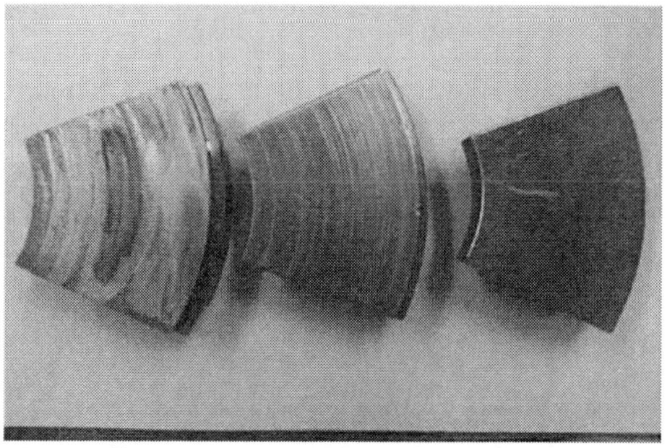

Figure 7. Non Active side pad –active side- new pad

Root cause analysis

An investigation into the incident determined the following:

The hand valve operated through DCS (HV) is balancing the pressure between make-up gas and recycle gas during compressor start–up or trip condition

The HV valve should be fully open during machine trip or speed slowdown to equalize the pressure at compressor rotor. The HV valve is to fail open at shutdown or trip.

The CPC of the driver power which controlling the steam inlet to the driver had failed but this failure was not sensed immediately by the trip system (ESD type is ABB-Safety guard 400 series)

Since to the trip signal had not occurred, this led to the HV valve not opening.

The late of opening HV valve caused an unequalized pressure arrangement at the compressor stages (sections) which result in a high axial load on the compressor rotor i.e. high axial movement.

The normal differential pressure between the compressor stages make-up gas and recycle gas normally 3 to 6 bar. This measuring point alarm is set at 15 bar. The differential pressure between compressor stages during the problem reached to 80 bar. (Figure 8, 9) which resulted in a high axial load on the compressor and high axial movement.

Figure 8. DCS screen shot for HP compressor operating condition

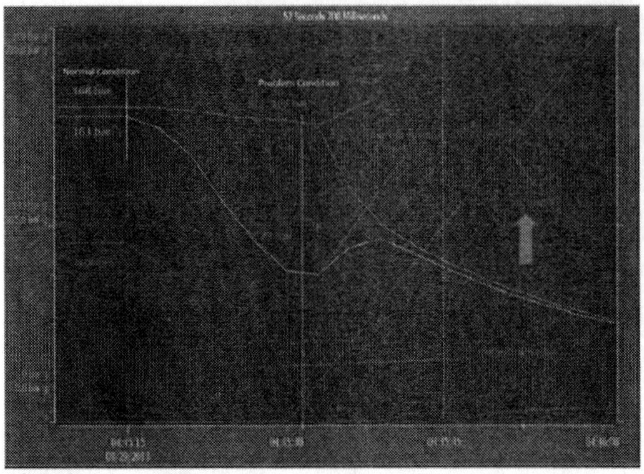

Figure 9. Relation between pressure difference and axial movement during the accident

Causes of thrust

As a result of the pressure ratio created by the impeller, a thrust force is generated which acts from the hub side (backside) of the impeller towards the suction side. Additional changes in gas composition, inlet temperature, speed and adjustable guide vane angle will cause compressor ratio and force to vary.

Thrust calculation basics

The average thrust of each impeller is calculated and summed to determine the net thrust of rotor, and a balance drum is designed in an opposed configuration to reduce the rotor thrust.

The thrust bearing carries the net residual thrust between the difference of net impeller and balance drum, with the thrust bearing normally carrying a small load compared to that of the balance drum. (Figures 10 - 18)

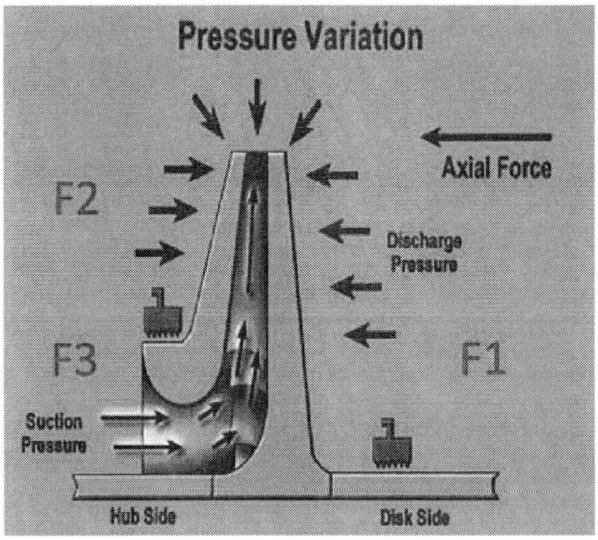

Figure 10. Impeller axial pressure forces arrangement

- Total impeller thrust (LB) = $\sum$ individual impeller thrust
- Balance drum thrust (LB) = ΔP x Balance drum area
- Thrust bearing load (LB) = Total impeller thrust – Balance drum thrust
- Force = Pressure x Area, pressure psig, area in IN^2, force in Lbs.
- Area= π (impeller radius r) ^{2}

Calculation of the net axial forces established at the thrust bearing in 2BCL408

The rated load for the thrust bearing carrying capacity is 75 kilo newton.

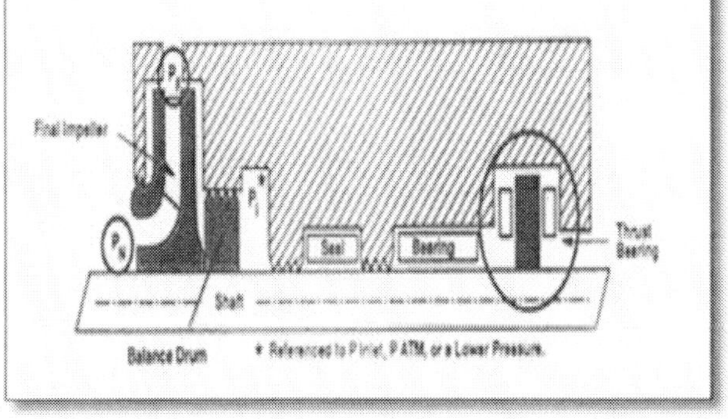

Figure 11 Rotor thrust force

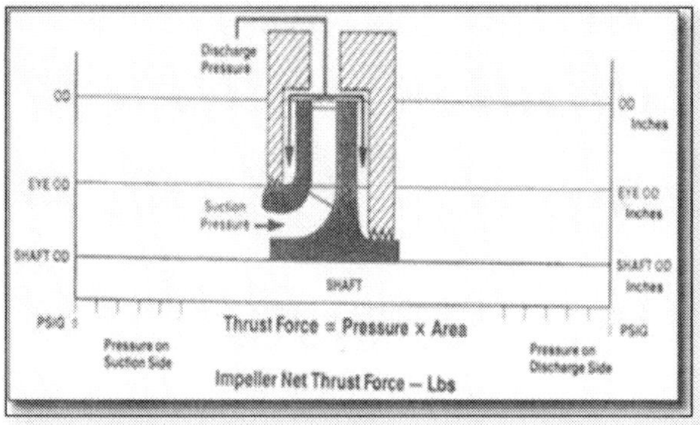

Figure 12. Impeller net thrust force

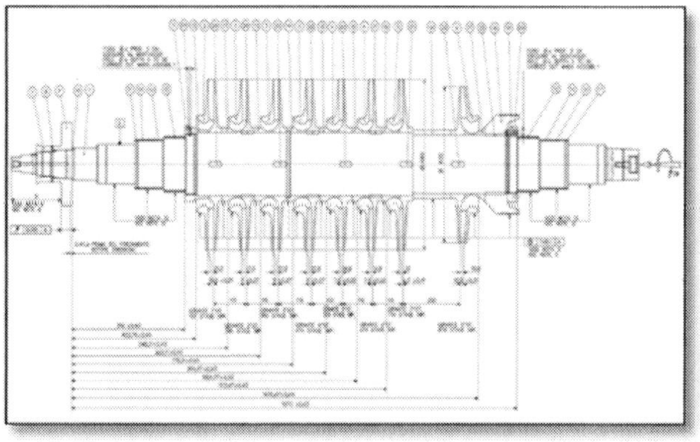
Figure 13. Rotor arrangement drawing

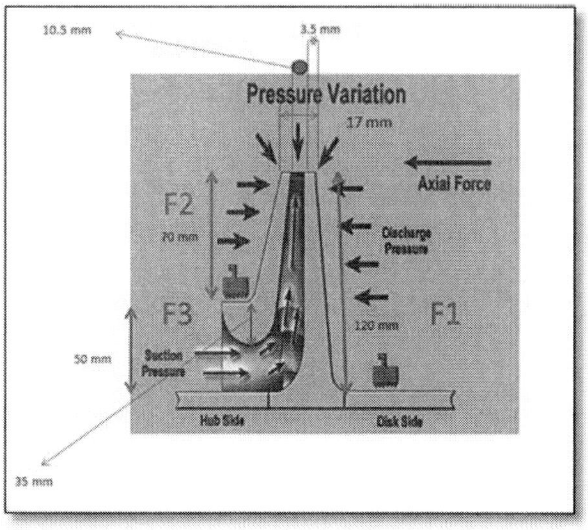

Figure 14. Thrust force at impeller

Calculations during the compressor problem

1- **Balancing Force calculations for Recycle impeller figure 15**

Impeller outer Radius R outer
Impeller eye Radius R eye
Shaft Radius R shaft

The Equations used in below calculations as per description at Thrust calculation basics paragraph

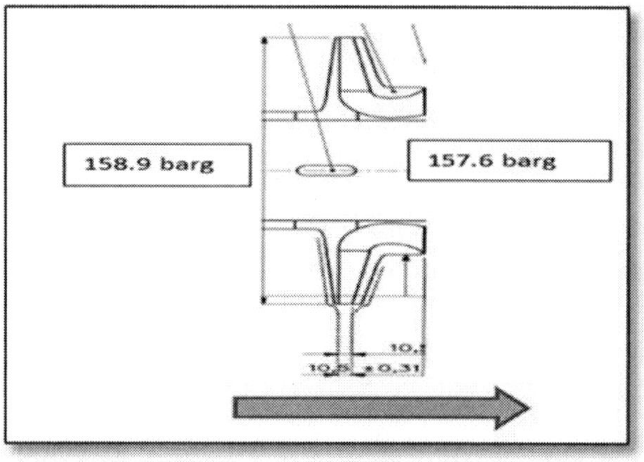

Figure 15. Thrust force at Recycle impeller

Dimensions are (R outer 8.49 in, R eye outer 5.58 in, R shaft 3.79 in).

- - 158.9 is gas pressure at impeller left side by barg
- - 157.6 is gas pressure at impeller opposite side by barg
- 14.5 is to convert the pressure to be Psig
- F1, 2,3 is the established forces on impeller body due to gas pressure AS related in figure 14

$F1=158.9 \times 14.5 \times \pi (8.49^2 - 3.79^2)=417770.7$ lbs.
$F2=158.9 \times 14.5 \times \pi (8.49^2 - 5.85^2)=274027.9$ lbs.
$F3=157.6 \times 14.5 \times \pi (5.85^2 - 3.79^2)=142566.7$ lbs.

Resultant force towards right side of the impeller i.e. it pushing the rotor towards right direction is 1176.1 lbs = F1-(F2+F3)

2- **Thrust Force calculations for balance drum (figure 16)**

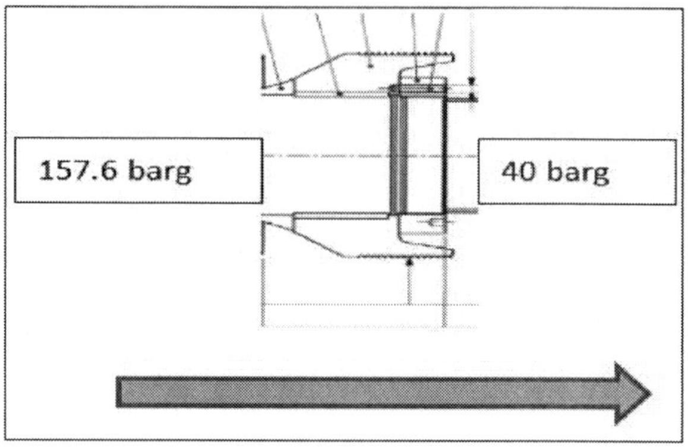

Figure 16-Balancing Force arrangement at balancing drum

Dimensions are (Ro 5.5 in & R shaft 3.79 in).
$Fth = (157.6-40) \times \pi \times 14.5 (5.5^2 - 3.79^2) = 85691.27$ lbs.
Balancing drum resultant force in right direction is 85691.27 lbs.

3- **Balancing Force calculations for make-up (figure 17)**

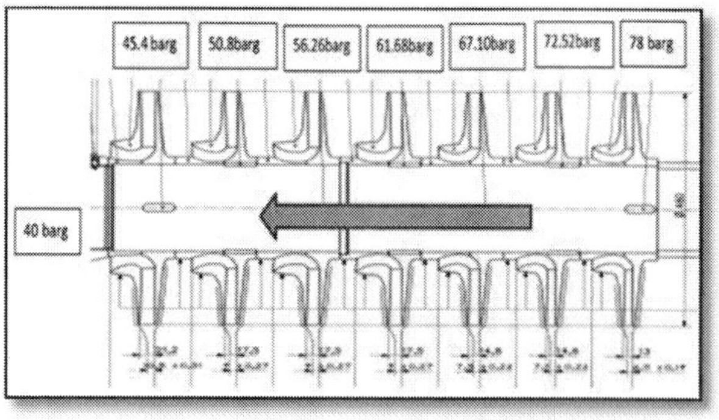

Figure 17. Balancing Force arrangement at make-up stage

All impeller diameters are the same (only width is different)

Dimensions is (R outer 9.46 in, R eye outer 5.32 in, R shaft 3.79 in)

$F1 = 45.4 X 14.5 X \pi (9.46^2 - 3.79^2) = 155371.96$ lbs.
$F2 = 45.4 X 14.5 X \pi (9.46^2 - 5.32^2) = 126546$ lbs.
$F3 = 40 X 14.5 X \pi (5.32^2 - 3.79^2) = 25397.3$ lbs.

Resultant force impeller 1 is 3428.6 lbs.
Resultant force impeller 2 is 3428.5 lbs.
Resultant force impeller 3 is 3466.6 lbs.
Resultant force impeller 4 is 3441.2 lbs.
Resultant force impeller 5 is 3441.32 lbs.
Resultant force impeller 6 is 3441.32 lbs.
Resultant force impeller 7 is 3479.3 lbs.

Total make-up Resultant force in left direction is 24126.8 lbs.

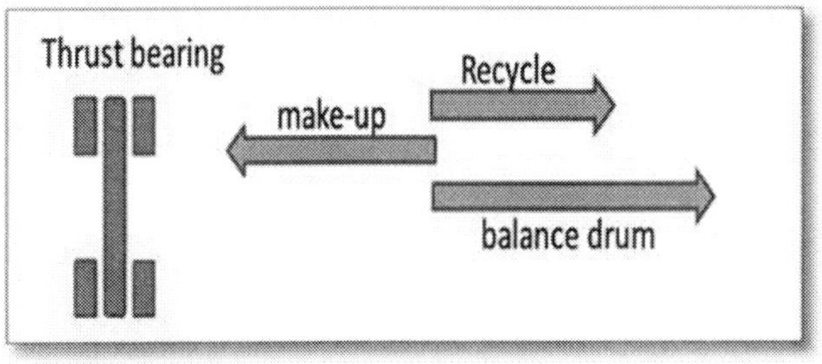

Figure 18. Rotor forces diagram

The thrust force on the bearing for the failure case (figure 18)

The resultant force on bearing = axial force by balance drum + axial force by recycle stage–Σ axial forces by make-up stage = 62740.5 Ibs. = 279 KN.

Rotor axial force analysis Calculations during the compressor normal operating condition

Calculation the forces by the same way and equations at problem condition above

Pressure arrangement

ΔP at Make-up is 77 bar g

ΔP at Recycle 7.6 bar g
ΔP at Balance drum is 73.9bar g

-Recycle stage resultant force in left direction is 6875 lbs.
-Balancing drum resultant force in right direction is 53848.5 lbs.
-Make up stages resultant force in left direction is 48890.1 lbs.

The resultant force on bearing = axial force by balance drum + axial force by recycle stage– Σ axial forces by make-up stage = 11833.5 Ibs. = 52KN

Calculations analysis

The net axial load during the failure reached 279 kilo newton.

The net axial load during the normal condition is calculated at 52 kilo newton.

The rated load for the thrust bearing carrying capacity is 75 kilo newton. (Figure 19)

The axial load on the thrust bearing was more than three times the rated load, which led to the high axial movement and severe damage of tilting pad thrust bearing active side. This caused the failure of the bearing.

Hydrodynamic thrust bearing failure occurred due to high axial load: the supporting hydrodynamic oil film had been lost its ability to support the rotor thrust load.

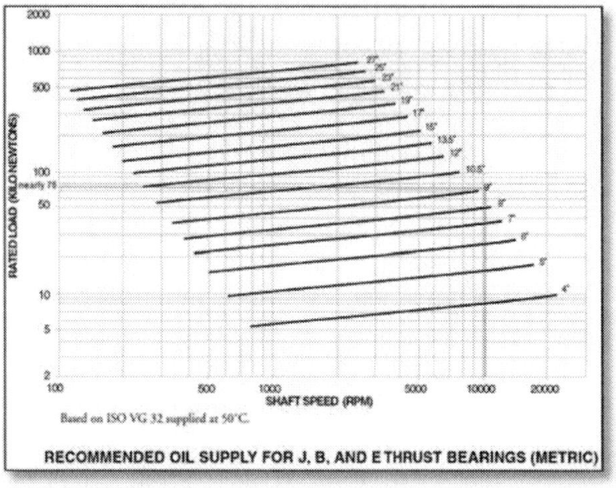

Figure 19. Thrust bearing chart

Recycle Impeller Stress Analysis

A structure analysis was performed for the recycle impeller as shown in figure 20, and all the calculations for the stresses at impeller by solid work program (V 9.0)

SolidWorks is a solid modeler, and utilizes a parametric feature-based approach to create models and assemblies. Parameters refer to constraints whose values determine the shape or geometry of the model or assembly. Numeric parameters can be associated with each other through the use of a relation, which allows them to capture design intent.

The results are shown in figure 20. The results show the modulating pressure did not have a huge effect on the net forces.

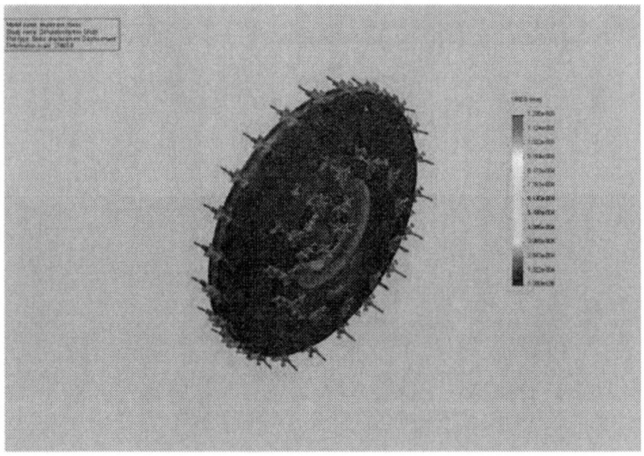

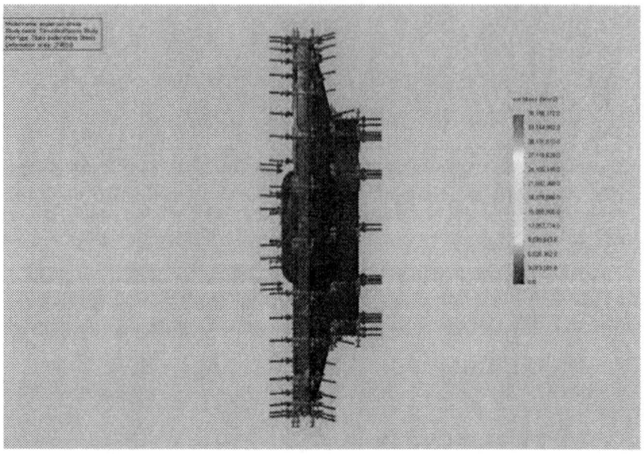

Figure 20. Stress analysis of recycle impeller

The same analysis applied to the thrust collar of thrust bearing figure 21, which when applied at the normal and the problem condition, shows the deformation occurred due to the huge net force at the problem condition

At 50 KN (normal condition)

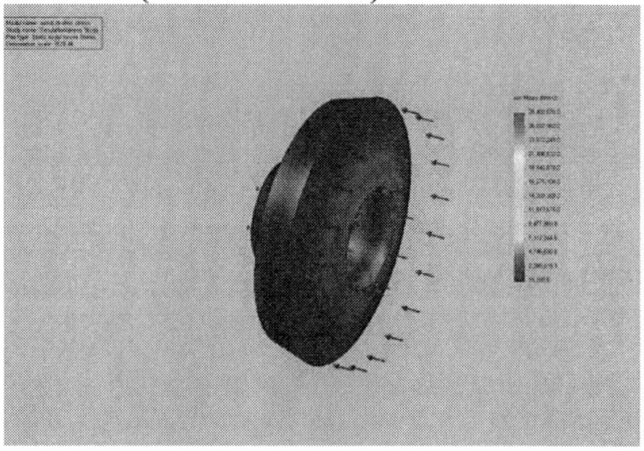

At 279 KN (at problem condition)

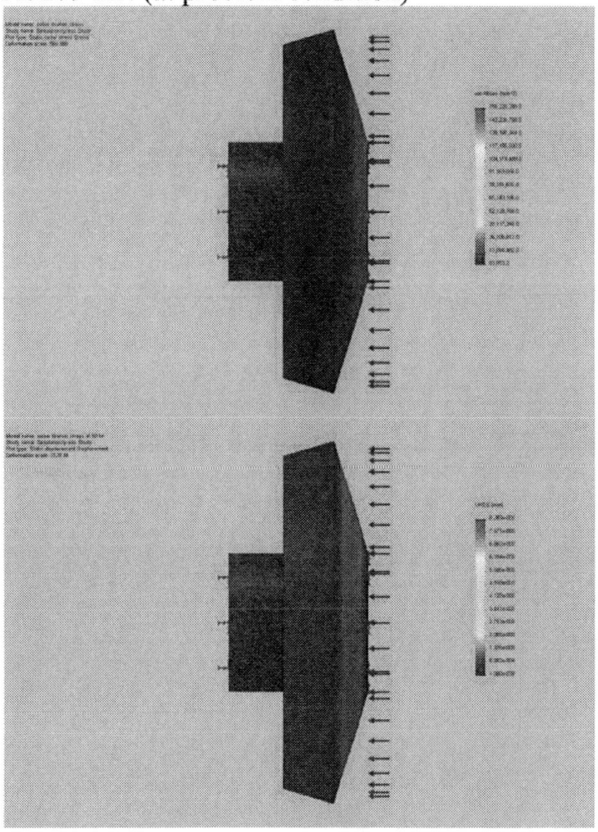

Figure 21. Stress analysis of Thrust Collar

Also the same analysis is applied at the active thrust pad. The stress analysis at the problem condition shows the deformation occurring to the pad due to the huge established net force. (figure 22)

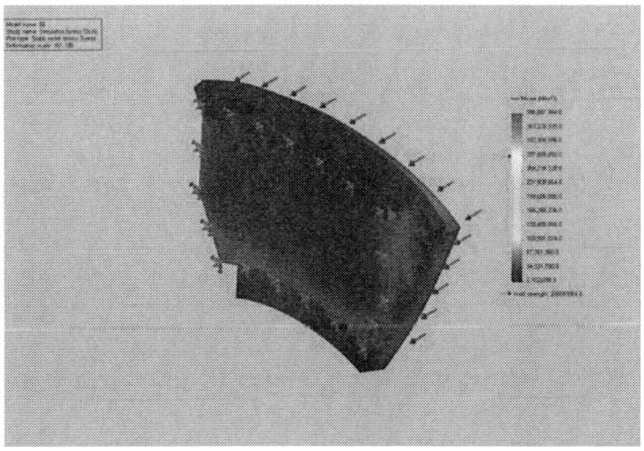

At 279KN

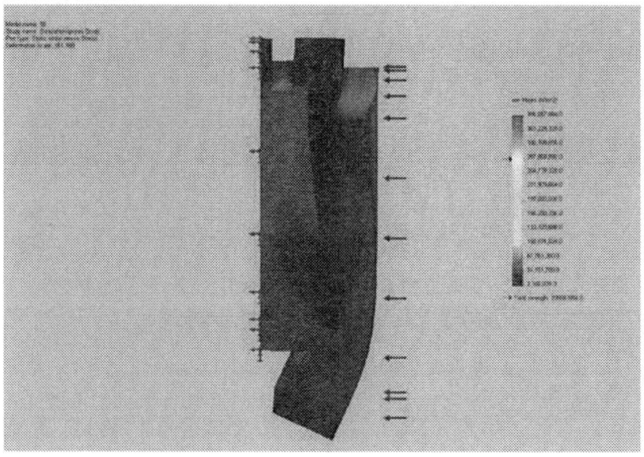

Figure 22. Stress analysis of Thrust pad

Action and assembly activity

All compressor minor parts were inspected. (seal oil rings- end labyrinth seal- journal bearing – active/non active thrust bearing side) All clearances were measured and adjusted to tolerance range. The thrust bearing pads and thrust collar were replaced. (Figure 23)

Figure 23. Inspected parts

Figure 24. HP compressor

Test of compressor at service

The compressor (figure 24) was tested in operation and running normally, keeping the rotor axial position in the center emphasized well balancing and stability at compressor stages.

Technical assistance

The case was under investigation from a high technical committee includes the top managers of MOPCO company

Conclusion

In the majority of multi-section machines, not only the failure of the balance drum and thrust bearing but also the absence of aerodynamic balancing between stages must be in consideration

Alarming for balancing valves between stages is not enough as you should put the human fault and instrument fault in consideration to avoid catastrophic failure, so in critical valves as our HV we designed and installed a trip signal at the pressure difference between stages.

Lessons learned

Like in our case, we added a new protection setting to trip the machine when the differential pressure between the make-up and recycle reaches 20 bar g.

References

{1} Machinery Message Sept.1993 Orbit, by W.E. Forsthoffer

Ammonia Converter Reaction Loss followed by Fire Incident

This paper describes reaction loss in ammonia synthesis converter followed by a fire incident. Reaction loss occurred during emergency handling after tripping of the urea plant which directly impacts ammonia recovery/refrigeration circuit loads. As a result of rapid pressure increase and temperature decrease, Synthesis gas leakage occurred from ammonia converter outlet flange which turned into fire. Following paper will highlight the sequence of events, key findings, root cause, contributory causes and corrective actions to avoid such kind of events at ammonia plants.

Rizwan Majeed and Shahbaz Ali
Fatima Fertilizer Company Limited.

Introduction

Fatima fertilizer ammonia plant is designed by CF Braun & Co. Alhambra California built in 1967. This Plant was initially operated Exxon. Then plant was bought by Kemira in 1985 and kept the same operational philosophy with exemplary service factor and production record of 1767 MTPD. Due to some business reason this plant was shut down in 2000. Kemira initial plan was to scrap this plant therefore it was never mothballed. Fatima moved this plant from the Netherlands to Pakistan in 2007 and started successful production on 15th March 2010 with a design capacity 1500 MTPD. The plant was successfully revamped to 1650 MTPD design capacity in November-2015

Back Ground

On 10th of June-2015 at 01:15 hrs, the ammonia plant was running smooth at 106% load. The utility steam generator tripped and the urea plant was stopped to manage steam load. Ammonia plant load was reduced from 106% to 88%. During turndown of the synthesis converter, synthesis loop pressure increased rapidly from 168 kg/cm^2 to 203 kg/cm^2 (2390 to 2890 psig).

At 01:43hrs a fire erupted at the converter outlet flange with a flame length of 7 feet. The plant backend was shut down immediately, synthesis loop depressurised and fire extinguished within 5 minutes. The fire emergency was handled in efficient and timely manner by the emergency response squad.

Synthesis Converter-Introduction

The Fatima ammonia synthesis converter is a Topsoe vertical design with S-300 removable basket contained within a pressure shell. The reactor is adiabatic featuring three radial flow catalyst beds with two inter-bed heat exchangers. Synthesis converter pressure shell was installed when the plant was originally built. However it was revamped from S-200 to S-300 after relocation from Netherlands to Pakistan. (see figure 1)

The converter feed gas is divided into three inlet streams (see Figure 1):

1. Main feed (~38% of total flow)
2. Quench (~41% of the total flow).
3. Cold shot (~21% of the total flow).

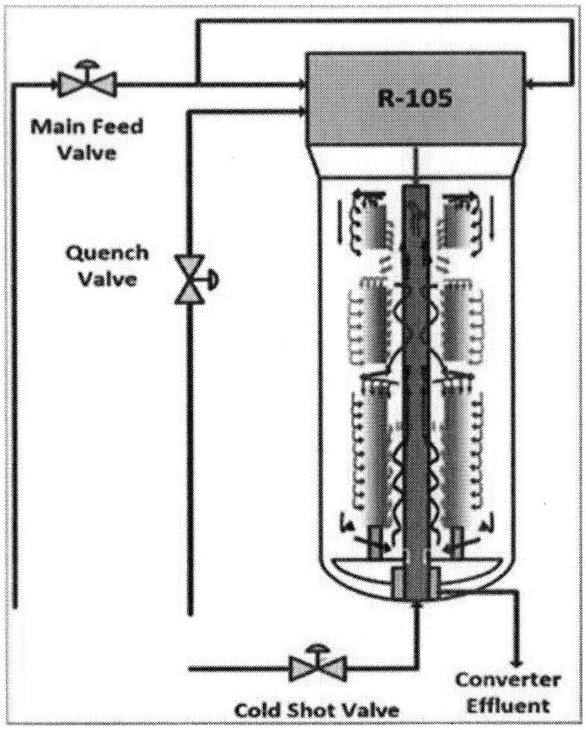

Figure 1. Synthesis Converter Sketch

Sequence Of events

The following table summarizes the sequence of events:

1. 1:15hrs Utility steam generator (USG) Tripped
2. 1:16hrs Urea plant Stopped
3. 1:15hrs synthesis compressor unloading started and pressure decreased from 211 kg/cm^2 to 169.2 kg/cm^2 (3001 psig to 2406.7 psig) within 18 minutes.
4. 1:20hrs Cold shot valve opened from 70% to 76%
5. 1:32hrs Cold shot valve closing started.
6. 1:35hrs Synthesis pressure started rising rapidly from 169 kg/cm^2(2400 psig) to 203 kg/cm^2 (2890 psig) within 10 minutes.
7. 01:43hrs Fire erupted at ammonia converter outlet flange.
8. 01:46hrs Synthesis gas compressor stopped and loop depressurization started.
9. 01:48hrs Fire extinguished

Key Observations

Salient observations of the incident are described below:

1. During plant load reduction, synthesis loop pressure was reduced at relatively fast rate i.e. pressure reduced from 211 kg/cm^2 to 169 kg/cm2 (3000 to 2400 psig) in 17minutes. (Figure 6, Curve #4)The converter cold shot valve was opened from 70 to 76% initially and remained same for next 13 minutes. (Figure 6, Curve #1)
2. Afterwards, the cold shot valve was quickly closed to help arrest the converter 1st bed inlet temperature but was ineffective and caused 1st bed cooling. (cold shot valve behaviour was confirmed normal in later stroke checking). During normal operation hot ammonia product is supplied directly to the urea plant. In the event of a urea plant trip, load on refrigeration circuit increases a large amount and is managed by ammonia plant load reduction. Due to increased load on refrigeration circuit, the ammonia content at inlet of converter increased from 3.6%. to 6.3 % (Figure 7, Curve #1 and 2)
3. The synthesis loop pressure increased from 169 kg/cm^2 to 202 kg/cm^2 (2400 to 2870 psig) within 10 minutes even as the speed of synthesis machine was constant (9800 RPM). This rapid rise in loop pressure was caused by loss of reaction and constant addition of makeup gas by the synthesis compressor. At the same time, outlet temperature of the converter

dropped by 143°C (257 °F) or in actual temperatures, 405°C to 262°C (760 to 500°F) over 20 minutes as a result of reaction loss in all three beds of converter.

4. A liner, with insulation material, is installed at outlet flange of ammonia converter. (see figures 4 and 5) The purpose of this liner is to protect the flange and gasket from direct impact of temperature changes (hot gasses). During inspection, the liner was found damaged and brittle. Moreover the insulation was also not located during inspection in TA-2015. Operating temperature of converter effluent gases is ~470°C
5. In the metallurgical report it was concluded that the liner was nitrided on both sides and a middle zone of only 0.3mm was not nitrided. Original thickness of the liner was 1mm. There were several cracks observed along the surface that formed during service (prior to detachment). It is most likely that the rapid pressure changes that took place during the fire incident could have caused the nitrided liner to fail and detach. Liner pieces were found from downstream steam generators during inspection in TA-2015. (see figure 2)

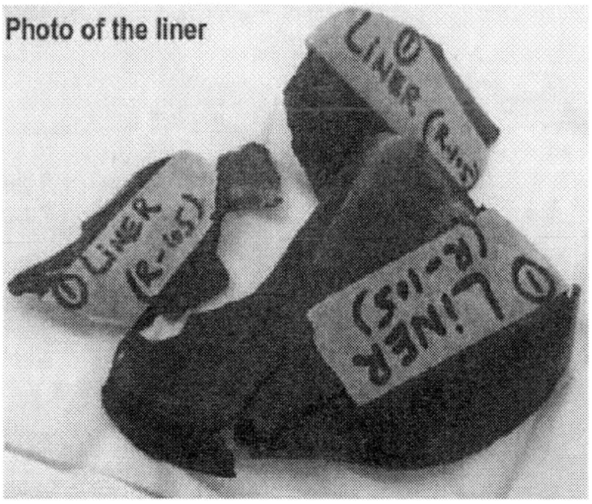

Figure 2. Detached Liner

6. The hardness test of converter outlet flange and associated bolts indicated no significant impact of ageing. (see figure 3)

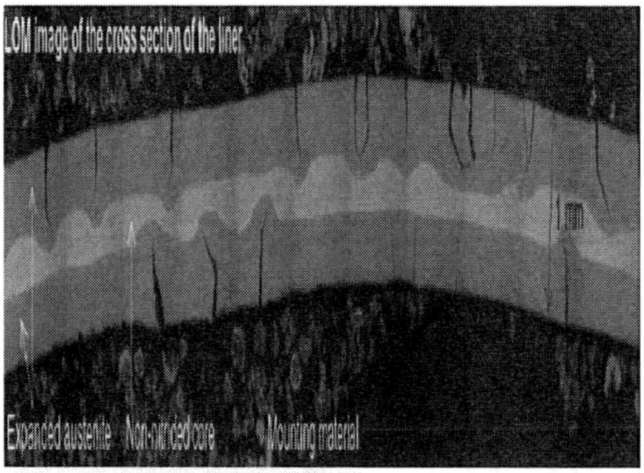

Figure 3. Light Optical Microscopy

Conclusion

Synthesis gas leaked from converter outlet flange due to slightly faster contraction of forging than the bolting. This caused less gasket seating pressure, leading to the leakage when the pressure started increasing while the catalyst temperatures were falling. The leaking gas was close to its auto-ignition temperature and caught fire immediately when mixed with air.

Forging contraction was caused by;

1. Loss of reaction in converter and
2. Damaged liner/insulation

•

Loss of reaction - was caused by (a) rapid cooling of 1st catalyst bed, (b) fast decrease in loop pressure. This decrease in pressure was caused by sudden unloading (relatively faster) of synthesis compressor in the event of urea plant trip to avoid excessive load on refrigeration compressor and (c) increase in ammonia contents at converter inlet. The ammonia reaction is reversible and decrease in loop pressure, converter inlet temperature and increase in product (ammonia) concentration at inlet of converter has adversely affected the equilibrium. All these factors caused rate of backward reaction to increase and point for loss of ammonia reaction reached. As governed by Le-Chatelier Principle (When any system at equilibrium is subjected to change in concentration, temperature, volume, or pressure, then the system readjusts itself to (partially) counteract the effect of the applied change and a new equilibrium is established), combined effect of above three factors caused the equilibrium to adversely shift to left and led to complete loss of reaction.

Damaged liner/insulation - was responsible for relatively higher heat transfer directly on the forging which caused rapid contraction, whereas bolting was cooled at a lesser rate due to obvious difference in convection and conduction heat transfer rates.

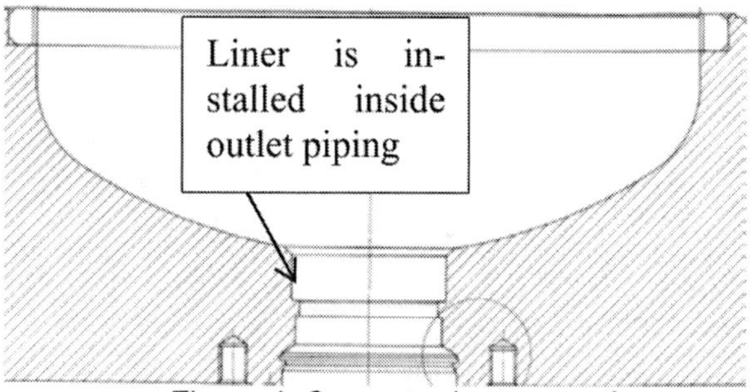

Figure 4. Converter bottom portion

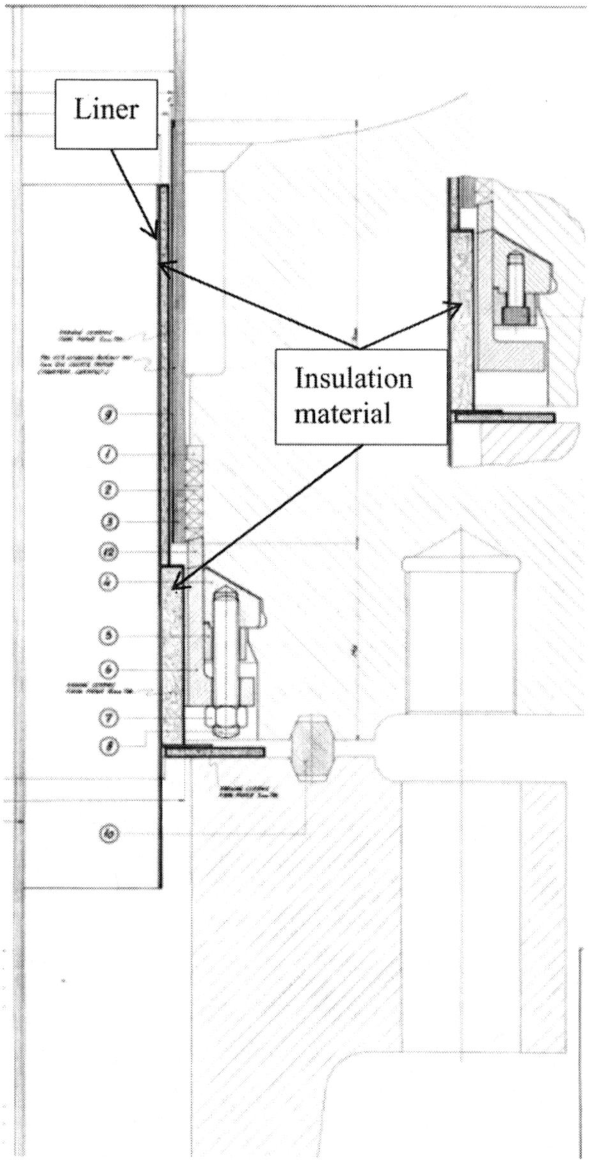

Figure 5. Cross sectional view of liner

Sharp increase in loop pressure - was caused by loss of reaction and continued addition of fresh make-up gas into the synloop by the compressor.

Key Learnings

The loss of ammonia converter reaction followed by a fire at outlet flange was a tragic incident which not only caused a production loss but also had potential for catastrophic failure.

Some of important learnings from this incident are;

1. Manipulate any given parameters in small steps and wait until the system has stabilized before next step
2. Plant emergency handling procedures should be revised by incorporating actions considering all scenarios.
3. Periodic inspection of the internal liner(s) of the converter flanges is to be ensured. Repair/replace when found damaged is necessary to avoid a similar incident at your site.
4. To avoid brittle damage, we upgraded liner material (SS304 to SS321). You may consider replacement if similar damage is witnessed.

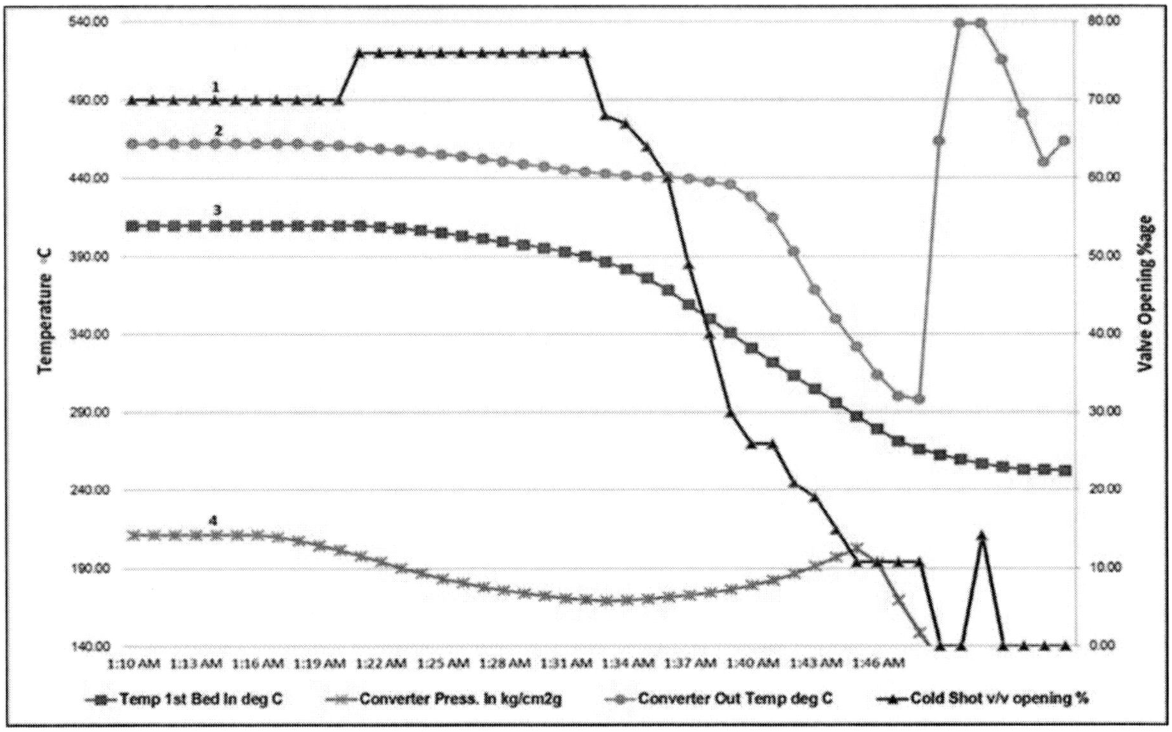

Figure 6. Converter Parameters Analysis

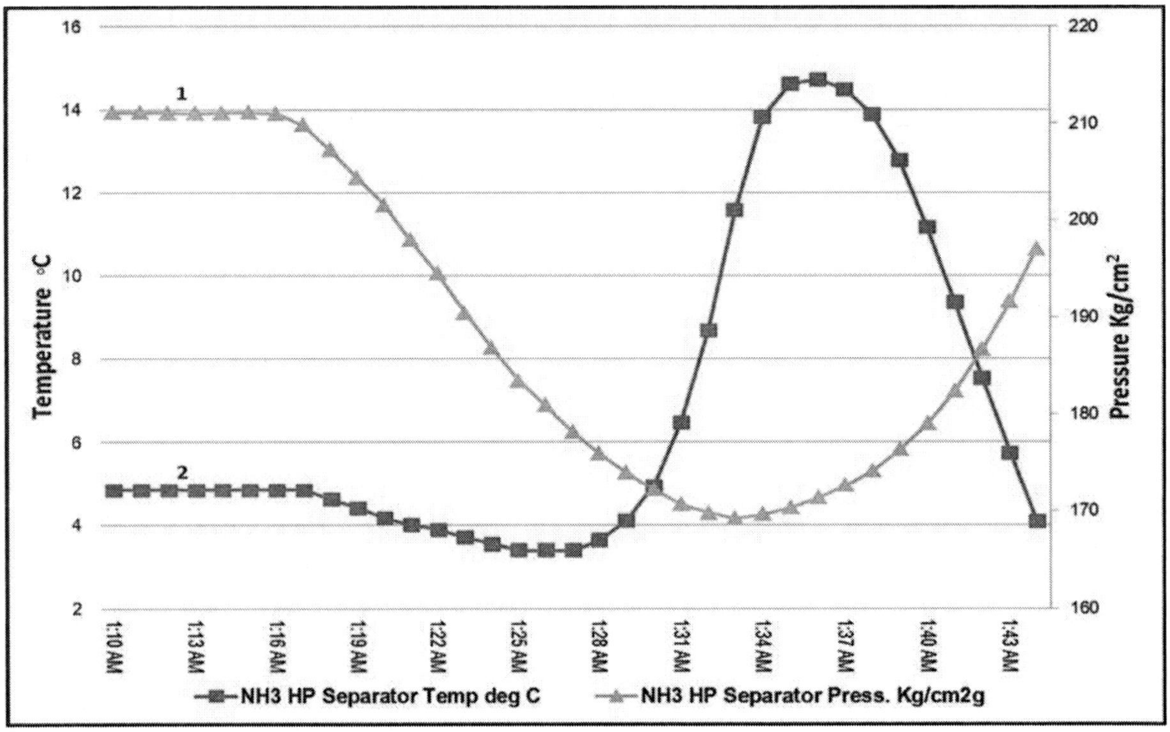

Figure 7. HP Ammonia Separator Temperature/Pressure Profile

Ammonia Synthesis Loop Water Exposure: Identification and Re-reduction of AmoMax®-10

After the 2015 outage at the Koch Fertilizer Dodge City Ammonia Plant, the site found their ammonia synthesis converters had been exposed to water and water vapor. Koch Fertilizer and Clariant Technical Services worked together to avoid significant loss of conversion in the ammonia synthesis converters and implement procedures to attempt to salvage the charge of AmoMax®-10 ammonia synthesis catalyst. This paper will discuss the troubleshooting that was completed, safety precautions taken to ensure the ammonia synthesis converters were started back up without incident, actions taken to re-reduce the catalyst, and lessons learned.

Kevin Turner
Koch Fertilizer, Dodge City, KS, USA

Carrie Schaffer
Clariant, Louisville, KY, USA

Introduction

Koch Fertilizer operates a Kellogg design ammonia plant in Dodge City, Kansas. The plant has two parallel ammonia synthesis converters that are of Casale axial-radial 3-bed retrofit design and are loaded with a total of approximately 1,400 ft^3 (40 m^3) of AmoMax®-10 catalyst from Clariant, one of the world's leading specialty chemical companies based in Muttenz, Switzerland. This charge of catalyst was started up in May, 2007 and had been on-stream for 8 years with excellent performance.

Following a shutdown, it was found that the ammonia synthesis catalyst had been exposed to water and water vapor.

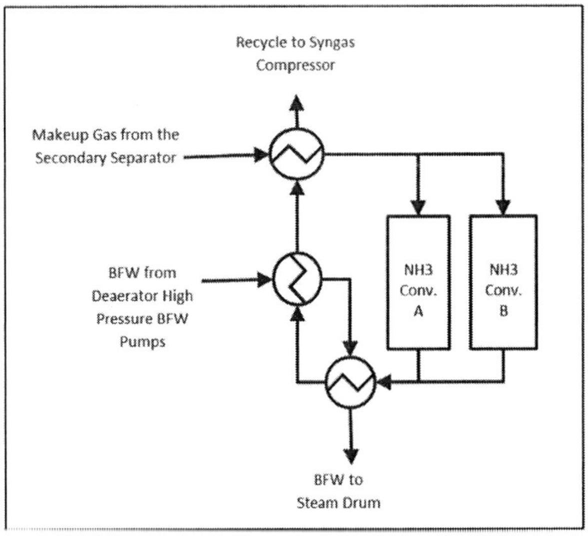

Figure 1. Simplified flow diagram

Troubleshooting

The Koch Fertilizer Dodge City ammonia plant was down for an extended shutdown. As the work was nearing completion, the steam system startup was initiated in order to coincide with the work completion, therefore reducing the overall startup time. Unfortunately, the compressor work that was wrapping up required additional work to meet quality standards. As a result, the high pressure steam system was pressurized to approximately 1,200 psig (8,275 kPa) for one week, while the synthesis loop was under low pressure.

During the subsequent startup, operators noticed that they could not obtain flow from the 103-J synthesis gas compressor through the loop. Initial troubleshooting discovered the refrigeration system and exchangers had formed ice in the process piping which restricted flow. For safety reasons, the startup was aborted at that time.

As troubleshooting continued, drains were checked and a sizeable amount of water was found. Koch Fertilizer reached out to Clariant for assistance and got immediate response from Clariant, and both companies began to troubleshoot the incident jointly.

As the source of the water became apparent, Koch and Clariant estimated that near ambient temperature catalyst was exposed to approximately 1,000-2,000 gallons (3800-7600 liters) of water and water vapor. This created concern for re-oxidation and/or wetting of the catalyst in both the A and B converters.

Safety Precautions

Koch Fertilizer and Clariant collaborated to determine the fastest way to safely restart the plant without damaging the catalyst or the equipment. Safety was the single most important factor of re-reducing the ammonia synthesis catalyst.

The following precautions were taken to minimize any potential safety risks during the reduction:

- Process water collected during the dry-out procedure was analyzed for contaminants before being properly disposed
- Reduction piping and sample setup were walked down and reviewed
- Clariant and Koch Engineers provided 24-hour support of the re-reduction
- Single points of communication were setup between Koch Operations, Koch Engineering, and Clariant Engineering

Ammonia Synthesis Catalyst

The ammonia synthesis catalyst installed in the Koch Fertilizer Dodge City ammonia plant converters is Clariant AmoMax®-10. AmoMax®-10 is based on $Fe_{(1-x)}O$ (wustite) instead of conventional Fe_3O_4 (magnetite) and offers the following advantages when compared to other iron-based ammonia synthesis catalysts, including those containing cobalt:

- Higher activity
- Higher mechanical stability
- Better thermal stability
- Higher resistance to deactivation
- Lower light-off temperature due to improved low temperature activity

The following AmoMax®-10 data was compiled through cooperation between Clariant and the University of Queensland [1].

Activity

As shown in the Arrhenius plot in Figure 2, AmoMax®-10 provides higher activity at lower temperatures. This property allows for operation at lower inlet temperatures while providing more conversion per pass due to a more favorable thermodynamic equilibrium at lower temperatures.

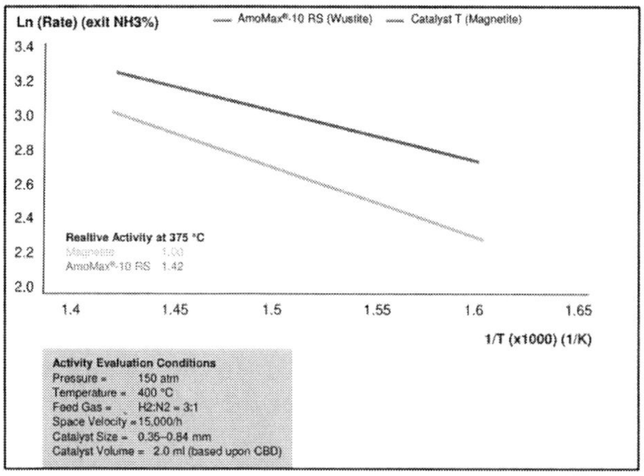

Figure 2. Activity of AmoMax-10 vs. magnetite-based catalysts

Resistance to Poisoning

Oxygenates in the ammonia synthesis gas such as CO, CO_2, and H_2O affect catalyst activity and are considered temporary poisons to ammonia synthesis catalysts by oxidizing the active α-Fe sites. The data shown in Figure 3 illustrate the outstanding resistance of AmoMax®-10RS against poisoning by oxygenates compared to magnetite with CO_2 used as the model molecule.

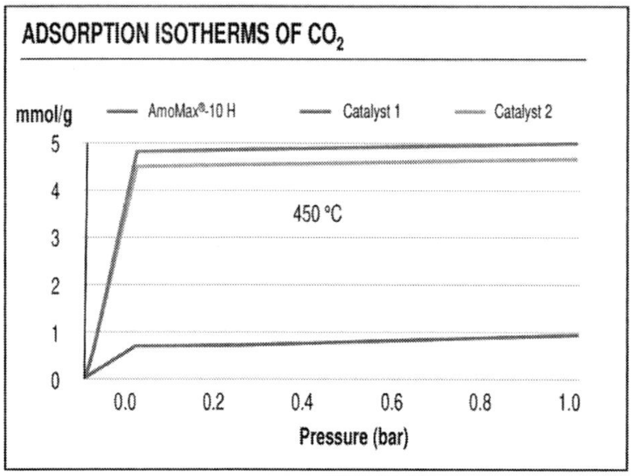

Figure 3. Superior resistance to poisoning by oxygenates of AmoMax-10RS.

Promoter Stability

Based on Clariant laboratory and commercial experience as well as knowledge of ammonia synthesis products, most of the AmoMax®-10's proprietary promoters are very stable against leaching. However, it is possible for certain potassium compounds to become mobile and to be leached out of both magnetite and wustite catalysts if the converter is flooded with water.

The activity of both the magnetite and wustite catalysts have been tested after re-reducing severely wetted catalysts in Clariant's laboratory. Both catalyst types showed an activity loss of approximately 10% compared to the fresh catalysts. However, even after some of the potassium compounds were leached, the AmoMax-10 still provided more activity than the fresh magnetite. Also, a proper drying procedure of the wetted catalyst may result in a deposition of the mobile potassium on the catalyst again.

Re-reduction of Ammonia Synthesis Catalyst

When first notified of the issue, Clariant began jointly troubleshooting with Koch Fertilizer personnel. Within 48 hours, two Clariant Engineers drove to the Dodge City ammonia plant to assist with the re-reduction of the ammonia synthesis catalyst, taking a detour to stop at another Koch plant to pick up an Ascarite apparatus, which is used as part of the moisture testing method described in Clariant's AmoMax®-10 Operating Manual [2] to measure moisture content during an ammonia converter catalyst reduction.

The plant immediately took steps to provide nitrogen purging to clear out the remaining liquid water and the water vapor from the loop. This was done in a manner to push the water away from the AmoMax-10 catalyst, in order to prevent further exposure.

Clariant provided Koch with guidelines for the dry-out and re-reduction of the catalyst based on the procedure for a normal AmoMax®-10 oxide catalyst reduction, while taking additional precaution to monitor water levels at the exit of both the A and B converter at 'dry-out' temperatures.

The Ascarite apparatus was initially not giving reliable results. During this time, the bed temperatures were slowly increased at less than 10°F (5.6°C) per hour until it reached 500°F (260°C) and then it was held between 500°F (260°C) and 520°F (271°C) until reliable water analysis results could be measured. To resolve the Ascarite measurement issue, the sample line length was reduced and the pressure regulator was removed.

Once the Ascarite apparatus was repaired, the moisture content was measured every hour alternating between the exit of the A and B converters. While monitoring the moisture values, the bed heat-up continued at 20°F (11.2°C) per hour to slowly drive off the moisture. Exposing wet catalyst to high temperature gas has been shown to cause physical fracture of the catalyst. If heated too quickly, the water adsorbed into the pores of the catalyst will be released quickly causing physical damage to the catalyst and leading to increased pressure drop across the catalyst.

After the re-reduction of the ammonia synthesis catalyst, the plant was able to return to normal operation with similar temperature profiles and conversion rates with production higher than pre-event rate. Current performance results do not indicate any long-term damage to the AmoMax®-10 catalyst.

Lessons Learned

A number of lessons were taken away from this event, as detailed below.

Consider deeper potential consequences when attempting to partially start up systems in a plant before the whole work scope has been completed in an outage.

Working closely with the catalyst provider with deep expertise allows for the ammonia synthesis catalyst to be safely re-reduced if water exposure is suspected.

Having an Ascarite apparatus and procedure available that can give reliable moisture content during the reduction of ammonia synthesis catalyst is critical to successfully starting up an ammonia converter as quickly as possible with minimal negative impact to the catalyst.

Having a catalyst that has higher activity and is more resistance to poisoning can increase the chances of surviving a severe event with minimal long term impact even after the catalyst has been on-stream for over eight years.

Conclusion

Following exposure to water, the re-reduction of the AmoMax®-10 catalysts at the Koch Fertilizer Dodge City ammonia plant went well with no major incidents or concerns. The ammonia synthesis converters are back in service with minimal long-term damage.

References

[1] V. Pattabathula, A. Grace, Incitec Pivot, S. Bhatia, University of Queensland, J. Richardson, Süd-Chemie, Nitrogen + Syngas Meeting 2008.

[2] AmoMax®-10 Operating Manual, 3rd ed., Clariant Intl., 2013.

Severe cracks on a new duplex syngas loop cooler

Within 2 weeks after taking into service a new duplex heat exchanger, severe cracking was discovered in the tube to tube sheet welds. The duplex tube side is exposed to brackish canal water. Welding and hydraulic expansion is investigated to determine the cause of the cracking.

Richard Bosselaar
Yara Sluiskil, The Netherlands

Introduction

This paper describes the inspections and investigations executed on cracks in the tube to tube sheet welds of a newly installed duplex heat exchanger. These cracks were discovered within 2 weeks after taking the exchanger into service.

Syngas loop cooler description

Yara Sluiskil, The Netherlands, has 3 ammonia plants. The oldest ammonia plant, ammonia C, originally constructed in 1969, is the subject of this paper. The cooling system of this plant is once through canal water and is brackish. Almost all the canal water coolers were originally constructed of carbon steel tubes with a heat cured coating on the inside of the tubes and the channels. This kind of coating is sensitive for mechanical damages and other disadvantages. Due to the brackish water service, the coolers are sensitive for fouling and need to be cleaned every turnaround. The cleaning is done by means of high pressure washing. In time the high pressure washing will damage the internal coating of the tubes, is difficult to inspect and is practically impossible to repair. In case of a damage of the coating, severe corrosion of the tube will occur and lead to leakage.

In recent years, many of the canal water coolers have been replaced by coolers with standard duplex tubes and channels (SAF 2205). The choice of duplex is based on the chloride content of the cooling water. Chlorine is dosed to avoid mussel growth and other biological fouling. In normal conditions, the chloride content is low enough to use normal stainless steel materials like 316L, but dosing upset conditions can lead to too high chloride concentrations. Also chloride can concentrate locally in dead zones and damage the standard stainless steel.

Based on past good experience, it was decided in 2013 to replace the syngas effluent condenser, E317. This condenser exists of 2 stacked exchangers with syngas on the shell side and cooling water at the tube side. (Figure 1) The overall length of each exchanger is 11300 mm (37 ft.) with a diameter of 1200 mm (47 inches).

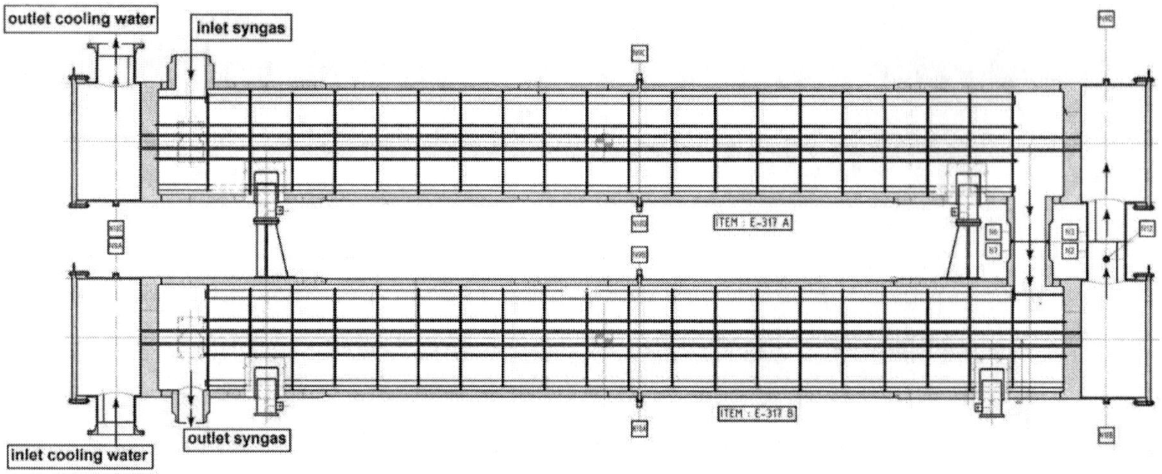

Figure 1 Syngas effluent condenser

The shell side has a design pressure of 190 barg and 120 °C (2755 psig & 248 °F), the tube side 7 barg and 120 °C. (101 psig & 248 °F). The operating inlet temperature of the syngas (shell side) of the first exchanger is approximately 75 °C (167 °F), the outlet temperature of the second exchanger is around 30 °C (86 °F). The cooling water inlet temperature is ± 15 °C (59 °F), with an outlet temperature of ± 25 °C (77 °F). The exchangers are each mechanically designed for a temperature difference of 65 °C (117 °F), the actual operating temperature difference over the stacked exchangers is only 45 °C (81 °F). The shell side has the highest operating temperature. This will introduce a tensile stress in the condenser tubes. There is no need for an expansion bellow, because the tube stresses are within the allowable limits. The tube sheets are designed with tubes acting as stays. This allows designing a relative thinner tube sheet. Even with this stayed tube sheet design, the required thickness is 180 mm. (7 inch) The tube sheet is constructed of carbon steel (thickness 180 mm / 7 inch) with a 10 mm (0.4 inch) duplex weld overlay. The duplex tubes are strength welded to the weld overlay and hydraulically expanded in the tube sheet. The tube holes are drilled according to TEMA special close fit.

The replacement exchanger was successfully installed during the planned turnaround of 2013.

Detected failures

Shortly after the startup of the plant, a leak into the canal water system was detected. The pH measurement at the exit of the cooling water system of the plant indicated a pH increase. Normally this is caused by an ammonia leakage. Laboratory analyses confirmed the higher ammonia content in the cooling water. There are several coolers on the cooling water system which can leak ammonia into the canal water system and it took some time to determine the origin of the leakage. Of course, the new exchangers were not the first suspect, after a deep investigation with several trials of cooling water sampling and resampling, the conclusion was that the new exchangers were the source. Meanwhile, a temperature increase of the synloop was detected. This temperature increase starts slowly, but increased rapidly in time. Trying to increasing the water flow over the exchangers by opening the cooling water valves didn't stop the temperature increase of the synloop. It was suspected there was no water flow anymore over the synloop exchangers. Opening a vent valve on the water side resulted into a gas escape. H2 was measured during this escape.

Approximately 2 weeks after the startup, the plant was shut down for an inspection of the exchangers. The channels were opened and immediately a severe deposit of calcium carbonate was detected on the tube sheets. (Figure 2 and 3) The canal water contains naturally calcium bicarbonate which is under normal conditions completely in solution. The solubility of calcium carbonate depends strongly on pH. The leakage of ammonia into the cooling water increased the pH locally. Due to this pH increase, the calcium bicarbonate deposits as calcium carbonate which is visible on the tube sheets. The most of the deposits were at the cooling water inlet side. The exchangers are in a counter flow service. The inlet tube sheet is the coldest tube sheet of the exchanger with the lowest delta T.

Inspections executed

Before a proper investigation of the tube sheets was possible, the exchangers need to be cleaned. The first inspection, to determine the leakage, was a tightness test with low pressure (0,5 barg/ 7 psig) nitrogen. This test was prepared parallel to the cleaning activities. The exchangers was pressurized and tested before the cleaning was started. During this test, severe leakage of the tube sheet was detected, even with the carbonate deposits on the tube sheets. With some soap, a lot of bubbles were visible, but due to the fouling, the exact location of the leakage was not yet visible.

Figure 2

Figure 3

After cleaning the tube sheets, the soap tightness was repeated. Between 50 and 100 leaks were detected in the tube to tube sheet welds. At some locations, cracks were visible in the welds. (Figure 4)

The leakage was felt in front of the tube sheet. This is strange because the tubes are expanded into the tube sheet according to the drawing.
Almost all the leakages were detected at the cooling water inlet tube sheet. At the outlet tube sheet of the lower exchanger only 5 leaks were detected.

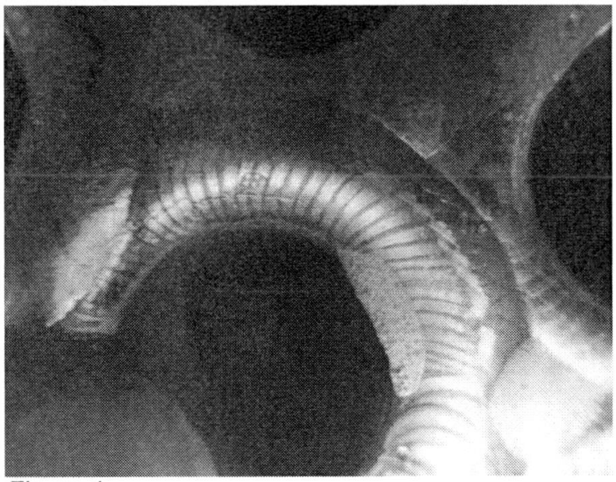

Figure 4

The upper exchanger, with the higher gas inlet temperature and therefore a higher delta T over the tube sheets, didn't show any leaks. The exchangers were manufactured together in parallel at the same manufacturer.

The severe damage of the lower exchanger was a big surprise. The first decision to be taken was the ability to repair the exchanger. Before a repair can be done, the cause of the damage must be clear. To determine the cause of the damage, several inspections were executed on the exchanger.

Material investigations

First, a PMI examination (positive material identification) was executed on the tube to tube sheet welds. This PMI confirmed the chemical composition of the duplex welds and the duplex weld overlay. The possibility of wrong welding consumable was excluded.

To get more information about the cracks in the tube sheet welds, field metallographic replications were taken. Different locations were prepared for the replicas. The weld overlay, some cracked and whole welds were inspected. Already after preparation the locations for the replicas, cracking was clearly visible. The cracks followed circumferential the welds, but also stepped over to the adjacent welds.

The replicas indicated cracks which were relatively straight and ran step wise and transgranular through the weld structure. The crack tips looked very sharp. Also no deformation was observed. The structure looks like a common duplex structure. Furthermore, the direction of the crack growth seems to be from the shell side to the tube side. This finding was based on the results of grinding the cracks - after grinding, even more cracks were visible. Some additional indicative ferrite measurements were performed by means of a ferritescope. (Figure 5) The results of a ferritescope are not very reliable, but give an idea of the ferrite content of the materials. The results of the ferrite measurements showed common values for duplex weld material:

Results of the ferrite measurements.

Position no.	measurements (% ferrite)
position 4, weld overlay	53-51-52
randomly chosen positions of the welds tube/tubesheet E317 B	43-51-66-56-61

Figure 5

Based on the information of the replicas, the welds seem to have a brittle failure behavior.

Parallel to the material investigations, the process conditions were reviewed. No deviations in the process conditions were found, neither were any indications for heavy pressure fluctuations. Only the continuous increase of the outlet temperature was detected due to failing cooling capacity as a result of blocking the tubes with carbonate deposits.

The Yara materials department was consulted to verify the original material choice. It was confirmed by this department that the materials are suitable for the process conditions. The brackish canal water should not introduce any cracking in the duplex material. From the shell side, the ammonia and the hydrogen should also not introduce any cracking under these process conditions.

Because the real cause of the cracking was still not determined, repairing was deemed not feasible. The decision was taken to replace the both new exchangers with the old exchangers, which fortunately still available on site.

For a deeper investigation, a material sample with a crack must be available. The intention was still to repair the new exchangers, and therefore it was requested not to damage the tube sheets and the weld overlay. Onsite machining tools were used to take a sample from the tube sheet and examined by the Yara materials department. (Figure 6)

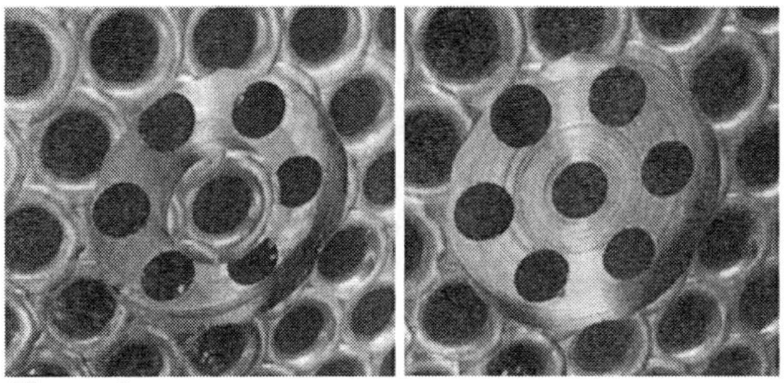

Figure 6

The results of the examination are:

- The EDS analysis confirms the chemistry of the welds as duplex 2205.
- The cracks seem to have been started from the end of the tube sheet to tube crevice, so at the root of the tube to tube sheet weld.
- The fracture is brittle, transgranular and the crack does not contain any corrosion products. This indicates that environmental cracking is not the damage mechanism.
- The structure of the weld overlay and the tube material appear to be a sound duplex structure.
- No signs of sigma phases were detected.
- The structure of the tube to tube sheet weld with the cracking appears to be ferritic with very low austenite content. This structure can be formed when the weld is cooled too fast during welding. The duplex material has ferrite solidification. The austenite is formed during the further cooling down of the weld. (Figure 7)

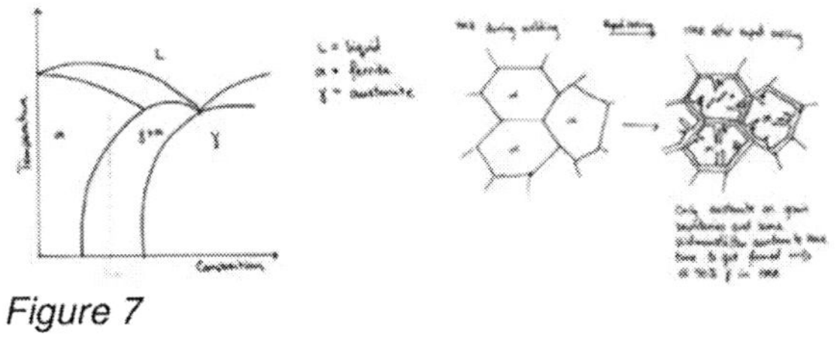

Figure 7

After the exchanger was send back to the original manufacturer, there were more samples taken. (Figure 8)

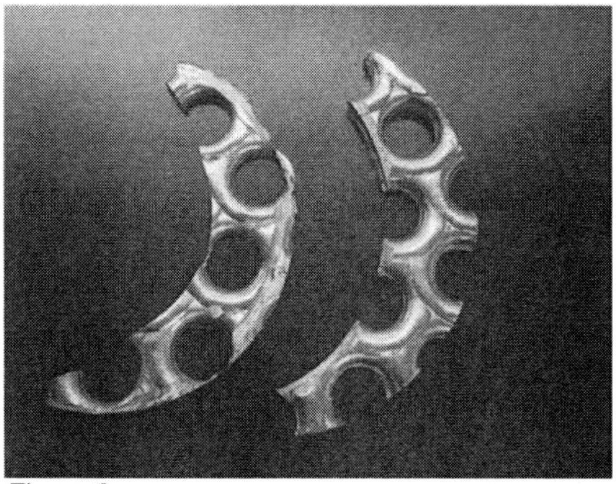

Figure 8

These samples were examined by the manufacturer. Their conclusions confirmed the high ferrite findings of the Yara investigations:

After ASTM E 562 tests, the average ferrite value is:

Sample	Location	Ferritic content [%]
Second Coupon	Weld Metal - Near fracture	74 %
Third Coupon	Weld Metal - No presence of fracture	66 %

Figure 9

Intermediate findings/conclusions

Intermediate findings:

- Cracking is not the result of the process environment. The initial material choice is correct.
- The weld material structure is not as expected:
  - The ferrite content is to high
  - This is only possible due to a rapid cooling of the weld.
  - The high ferrite content gives a brittle structure
- This brittle structure is caused by a manufacturing error.

Reviewing welding documentation

The original manufacturing data book is reviewed to check the original welding procedures. The welding qualification fulfilled all the requirements. The measured ferrite content on the mock up was normal (55.7 % according to ASTM E562-08 (point count)). The qualification was made on a 70 mm (2.75 inch) thick block and has limited dimensions. The actual tube sheet thickness is 180 mm (7 inch) and large in diameter. The question can be raised if the mockup is representative enough for the actual tube sheet dimensions.

Other duplex exchangers in comparable services didn't have problems with cracking in the tube to tube sheet welds. The welding qualifications of these exchangers, build by different manufacturers, was compared. (Figure 10) A difference was found in the heat input used by the different manufacturers. The heat input used for the cracked exchanger was significantly lower.

	Heat input (kJ/mm [kJ/inch])		
	min	max	average
E317 (Manufacturer A)	0.42 [10.6]	0.95 [24]	0.66 [16.6]
Manufacturer B	0.84 [21.2]	1.51 [38.1]	1.14 [28.7]
Manufacturer C	0.82 [20.6]	1.26 [31.8]	1.03 [26]

Figure 2

This lower heat input can be an explanation for the high ferrite content as a result of the rapid cooling. If a higher heat input is used, the weld will cool down slower and more austenite can be formed.

Other observations

A brittle material only, doesn't conduct into a rupture. An impact or forces are required to rupture a brittle material. No evidence was found in the process data for pressure and /or temperature impacts on the exchangers. As mentioned earlier, during the first soap test, the leakage was clearly felt. This was a strange observation, because the tubes were hydraulically expanded in to the tube sheet according to the drawing. When an expansion is properly executed, this should be more or less gas tight at a tightness pressure test of 1 barg (14.5 psig), even without a weld. So another suspicious item was the execution of the hydraulic expansion. This is further investigated.

First, internal diameter measurements of the tubes were randomly executed. These measurements confirmed the execution of the expansion. The drilled holes in the tube sheet are always a little bit bigger as the external diameter of the tubes. The TEMA (Tubular Exchanger Manufacturers Association) prescribe the dimensions of these holes. The special close fit drilling was used during the manufacturing. The internal diameter of the tube will be greater after expansion. In the tube sheet also an expansion groove was applied. This groove was also found back in the measurements. The difference in internal diameter indicates that a kind of an expansion was executed.

The manufacturing data book contained only a statement about the executed hydraulic expansion. The hydraulic expansion was executed by a sub-vendor of the manufacturer. On request of the manufacturer, the calculations for determining the hydraulic pressure were issued. These calculations are based on the work of Krips and Podhorsky. By reviewing these calculations, several mistakes were discovered.

The yield values used in the calculations for the tubes were below the minimum values according to the material standard. The material certificates of the tubes indicated much higher yield strength values than minimum required by the material standard:

	used in calculation	material standard	material certificates	
Tubes	397 [57]	450 [65]	647 [94] - 662 [96]	Mpa [ksi]
Tube sheet	314,5 [45]	295 [43]	300 [44] - 329 [48]	Mpa [ksi]

Figure 11

Furthermore, there were some doubts on the calculation results of the software used. There was a big difference between the supplier results and the Yara calculations. To check the results, manual calculations were requested. A first set of calculations was made based on the minimum material standard values. The results are in the table below:

	original software	manual	Yara calc.	
Lower limiting pressure	1186 [17197]	2461 [35685]	2378 [34481]	bar [psi]
Expansion pressure	2681 [38875]	2840 [41180]	2840 [41180]	bar [psi]
Adhesion pressure	1495 [21678]	142 [2059]	177 [2567]	bar [psi]
Calculated pull out force	24194 [53338]	552 [1150]		kg [lb]

Figure 3

The manual calculations and the Yara calculations are comparable, but it is clear there is a mistake in the supplier's software.

After these findings, new calculations were requested to determine the required hydraulic expansion pressure based on the actual material certificate values. Unfortunately the sub-supplier used again the wrong software program:

	Software	Yara calc.	
Lower limiting pressure	1858 [26941]	3725 [54012]	bar [psi]
Choosen expansion pressure	3353 [48618]	3353 [48618]	bar [psi]
Adhesion pressure	1495 [21678]	-143 [-2073]	bar [psi]
Calculated pull out force	28462 [62747]		kg [lb]

Figure 4

It was clear the expansion sub-contractor didn't understand the problem completely and was not very willing to help to solve the problems. The Yara calculations indicated a negative adhesion pressure. This means there is actual no adhesion between the tube and tube hole.

The principle of hydraulic expansion is illustrated in the figure 14.

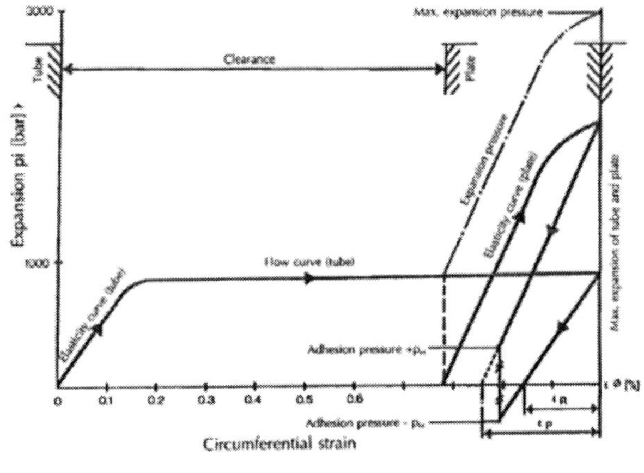

Fig. 5.10 Principle of hydraulic expansion.

Figure 5

The tube is pressurized until its starts to flow (plastic deformation) and the clearance between the tube and borehole is closed. When the tube is in contact with the tube hole, the expansion pressure will increase further. The tube sheet is subjected to elastic stress and sometimes also to partial plastic stress. Care should be taken by the locally plastic deformation of the tube sheet. The expansion success can be impaired, especially for the adjacent tubes. If the expansion pressure is removed, reverse deformation occurs to differing degrees because, in most cases, the tube and tube sheet have different material properties and geometries. The difference between the reverse deformation of the tube and tube sheet gives the adhesion pressure. This adhesion pressure occurs only when the elastic reverse deformation of the tubes is less than the elastic deformation of the tube sheet. The reverse deformation (elastic deformation) of a material depends on the Young's modulus of a material. The Young's modulus is determined by the angle of a tensile curve. If materials have the same Young's modulus, the maximum amount of reverse elastic deformation is prescribed by the yield value. If the yield value is higher, more reverse plastic deformation can occur. (Figure 15)

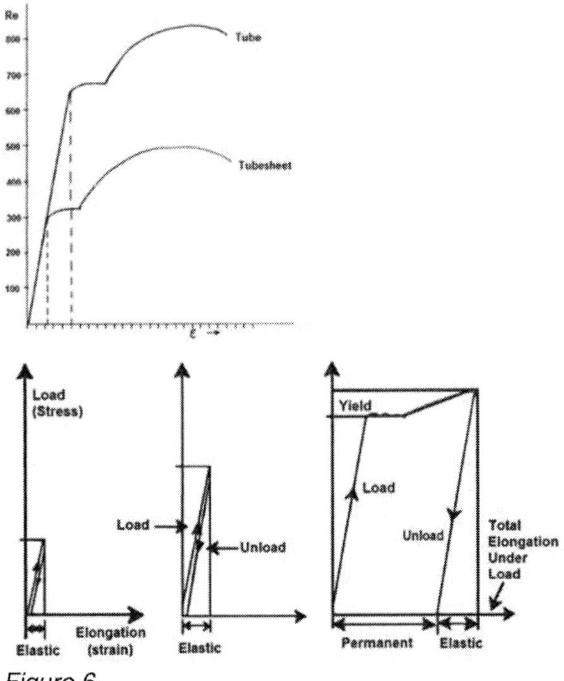

Figure 6

In this case, the yield value of the tubes was approximately 1.6 times higher than the tube sheet. Based purely on deformation properties, the reverse behavior of the tubes will always be higher than the tube holes, so no adhesion pressure will occur. Geometric differences between the tube and tube hole are influencing the actual reverse deformation, but when the yield strength of the tube is very much higher than the tube sheet, it's almost certain that hydraulic expansion is not possible. This behavior is indicated in figure 16, εr >εp, so no remaining adhesion pressure.

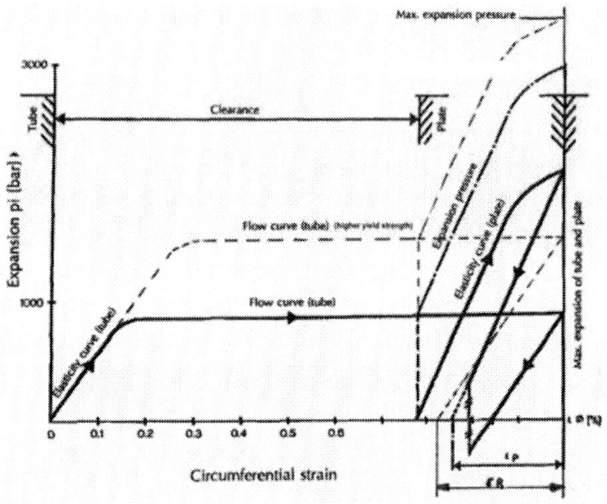

Figure 7

Based on the Yara calculations, it was decided to make a mockup for testing the hydraulic expansions for the repair job. (Figure 17) A hydraulic expansion was executed on the original expansion pressure 2700 barg and then 2900, 3100, 3300 and 3500 barg. (39160 psig, then 42050, 44960, 47860, and 50760 psig) According to the expanding company, 3300 barg (47860 psig) should be sufficient. For the mockup, the same tube materials and comparable tube sheet material (comparable yield strength) is used.

Figure 8

After the hydraulic expansion, all the tubes were quite easy to move with slight hammering, no pull out tests were executed. The mockup test confirmed the Yara calculations with no adhesion as a result. It was confirmed by the mockup test that hydraulic expansion was not possible for this material combination.

A mechanical expansion (5 - 7%) was executed on a tube. This resulted into a strong connection between the tube and tube sheet hole which was not movable with a hammer. The difference between a mechanical and hydraulic expansion is the local plastic deformation of the tube during a mechanical expansion. During a mechanical expansion the plastic deformation of the tube will occur under the contact pressure of the rollers and is very locally. The locally deformation has a limited length and therefore the absolute reverse elastic deformation is limited. The tube hole is, due to the diameter increase of the tube, stressed around the complete circumference and has, due to the longer length, more absolute reverse elastic deformation and gives the required adhesion pressure.

The absence of the adhesion force between the tube and tube sheet gives direct forces on the tube to tube sheet welds. The combination of a brittle weld and forces on the weld has introduced the severe cracking of the tube to tube sheet welds.

During the repair of the heat exchangers the loose connection of the tubes was further confirmed. Almost all the tubes were more or less easy removable from the tube sheet.

Executed modifications for the repair

The following modifications are made for making a reliable repair possible.

Expansion:
Based on the expansion trials and calculations, it was decided to execute a mechanical expansion instead of the hydraulic expansion.

Welding:
The welding qualification for the tube sheet welds is adjusted. The first step is to increase the heat input. To make it possible to weld with a higher heat input, the welding is executed "under hand". For the execution of these welds on the equipment, the exchanger was placed vertical to make this possible. The material supplier Sandvik advised to increase the heat input to around 1 kJ/mm (25.4 kJ/inch) for the tube to tube sheet welding. This is also in line with the values used by the other companies as indicated earlier.

Another important modification is to use a shielding and backing gas with approximately 2% nitrogen. Nitrogen is an austenite former so it makes the weld less ferritic. In orbital TIG welding, the weakest spot will be the "slope down" area where a very low energy weld is produced. This low energy area will have a rapid cooling with a ferritic structure.

Conclusions

The failure of the tube to tube sheet welds was a combination of factors:

- When a proper expansion is executed, practically no forces are acting on the tube to tube sheet weld when the exchanger is in service. Even a brittle weld should not fail during service.
- A proper weld (even without expansion) should not fail in these operating conditions. The strength of the weld is sufficient to withstand the tube forces due to pressure and thermal expansion.
- Mockup tests blocks are not 100% representative for production welds due to geometric differences and therefore different thermal behavior during welding.

The cause of the leakages was as a combination of brittle welds due to wrong welding parameters and an insufficient hydraulic expansion of the tubes into the tube sheet which conduct into a brittle facture of the welds.

It's possible there are more duplex exchangers in service with brittle tube to tube sheet welds, but as long as there is a sufficient expansion of the tubes into the tube sheet this will not lead into a failure.

References

Examinations into the nature of cracking of tube sheet E317B north at the production location of Yara, Sluiskil.
Element report: ERO002100, 31/01/2014
By: A. Dhont

Heat exchangers: A practical approach to mechanical construction, design and calculations
By: M. Podhorsky and H. Krips
1998

Laboratory report SMT RP-56276 and RP-56275
By: L. Giroletti, G. Zambarbieri
March 2014

Welding practice for the Sandvik duplex stainless steels SAF 2304, SAF 2205 and SAF 2507.
By: Claes-Ove Pettersson and Sven-Ake Fager
August 1995

Yara incident investigation report: Cracked tube welding in duplex material, E317A/B
By: P. Bleijenbergh
April 2014

TEMA
Tubular Exchanger Manufacturers Association.
9th edition

Ammonia Plants Compressor Room Fire; Prevention and Remedies

On May 6th, 2014 the disconnection of a 1/2 inch instrumental tube caused a fire in the compressor room of Pardis ammonia plant No.2. The failing of this connection, which was connected to the PDT-3027 pressure differential transmitter and installed between the inlet and outlet of the fourth stage in the synthesis compressor, caused leakage of high pressure gas at the disconnection point. This created a jet fire with a flame length of approximately 10 meters. The plant fire station with the cooperation of fire departments of other neighboring petrochemical companies in the Pars Special Economic Energy Zone controlled the fire in less than an hour. The cost of the reconstruction was more than two million USD and a shutdown of 60 days.

Alireza Orooji
Sajjad Hosseininia
Ehsan Askari Mahvelati

Process Engineering Department, Pardis Petrochemical Company, Assaluyeh Port, Bushehr, Iran

Introduction

Pardis Petrochemical Company is located on a site with an area of 64 hectares, 270 kilometers southeast of the Bushehr port in the Pars Energy Special Economic Zone in the Assalouyeh Port, Iran. Pardis has two Ammonia plants each with a capacity of 2050 metric tons per day and two urea plants each with a capacity of 3250 metric tons per day in operation. The Phase 3 is under construction and a feasibility study is ongoing for phase 4. The production in ammonia plants No.1 and 2 have been started since 2007 and 2009, respectively.

This paper covers an incident case study which occurred in Pardis Petrochemical Company, ammonia phase 2 compressor room.

At 14:54 on May 6th, 2014 the disconnection of the 1/2 inch instrument tube caused a fire in Pardis ammonia plant compressor room. The failing in this tube connection to the PDT-3027 pressure differential transmitter between the inlet and the outlet of the fourth stage in the synthesis compressor caused the high pressure gas leak in the disconnection point of the instrument tube tapping.

This incident created a jet fire with a flame length of approximately 10 meters. In this incident ammonia synthesis and refrigeration compressors, piping insulations, instrument tools, electrical cables and part of the structure of the compressor room incurred losses.

Because of the damages to the lubricating oil piping flanges and spreading of oil, both of ammonia synthesis and refrigeration compressors were engulfed in fire and severely damaged.

The incident was caused by a collection of factors such as the improper fitting of the 1/2 inch tubes threads, poor arrangement and securing of the tube according to the standard guidelines, possible human error, and physical damage to the instrument tube during unfastening the scaffolds near the tubing.

The incident led to severe damage to both synthesis and refrigeration compressors. The ammonia plant was out of service for 60 days to repair and overhaul the compressors.

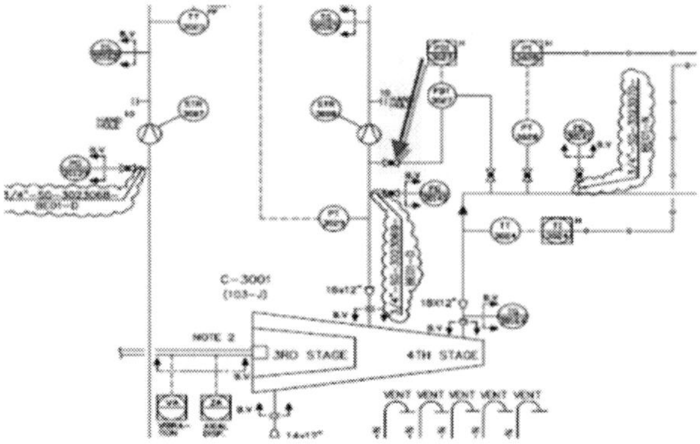

Fig. 1: Instrument Tube Disconnection Point

Synthesis Compressors & Refrigeration System in Pardis Ammonia Plants

The ammonia synthesis compressor, with a power of 22 MW (29500 hp), is a 4-stage design which increases the pressure of the synthesis gas from 32 to 176 bars (464 to 2552 psig). The volume of gas at the output of the final stage of this compressor is approximately 880,000 kNm^3 /hr (32.845 MM SCFH).

Approximately 25% of this gas is fresh gas and the remaining 75% is non converted gas in the ammonia synthesis loop which is returned again to the interior of the reactor. This gas, which is known as recycle gas is returned to the third stage of the synthesis compressor at a pressure of approximately 170 bar (2465 psig) and following an increase in pressure to 176 bars (2552 psig) in the fourth stage of the synthesis compressor, it is once again injected back to the inlet of the ammonia synthesis reactor.

PDT – 3027 shows the pressure difference between the recycle gas from ammonia synthesis section and fourth stages of the synthetic compressors. At the inlet of the third stage of the compressor, a strainer is installed and PDT – 3027 is an indicator for the level of clogging in this strainer.

The Fire Survey; Optimization Scenario Which Lead to an Incident

In the weeks prior to the incident, the plant operators done optimizing operation of the low pressure steam system used by the ejectors in the surface condenser system. The main suspect in loss of vacuum was the steam duct, outlet of the synthesis and air compressor turbines. In order to access, examine and seal the suspect air leak points, scaffolding was required below the compressors.

Fig.2: The PDT-3027 instrument tube low pressure leg tapping location. Fire jet from here hit the compressor room. The hanging damage tube also is seen in the image.

At the time of the incident, scaffolding removal was underway very close to the location in which the incident occurred. Following the leakage of hydrogen gas, due to disconnection of the high leg of the instrument tube related to PDT–3027, the scaffolding workers quickly left the scene.

Later, after some examinations and discussions with the workers, no reliable information was obtained with respect to contact with or pressure exerted on the instrument tube when the scaffolds were removed.

Fig.3: The view of the fire rising from the roof of the compressor room

Fig.4: Ammonia synthesis and refrigeration compressors, both on fire.

Yet, considering the fact that the instrument tubes were unprotected and that repeated scaffolding operations were performed for a period of over one month, exertion of pressure on the instrument tube was one of the factors which likely played a role in the incident.

Scope of Damages Incurred

Damages included the loss of almost all the lubrication oil piping and fittings, instruments cables and transmitters. In addition there was severe damage to the compressor governor and actuators system, temperature shock to the turbines and deformation of the turbine casings, and the turbine. In addition, there was loss of axial alignment of the ammonia refrigeration and synthesis compressors. The total financial cost of the incident was estimated to be approximately 2 million USD.

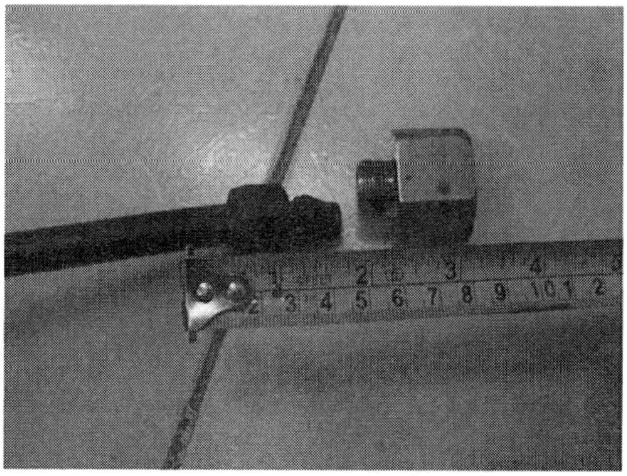

Fig.5: A small size tube and connection which caused a huge incident. Thread attrition of the instrument nut is clear in the image.

Plant Operation Modality Analyzing During Incident Occurrence

Performance of Control Room Operators and Plant Operation Management

The performance of the operation unit in finding the problem and the approach in dealing with it had some positive and blind points.

Positive Points

1. Timely identification of the location of leakage of synthesis gas following the alarm issued by DCS control system related to PDT-3027 by the operators in the control room.

2. The operators in the ammonia site quickly found the jet gas location under the ammonia synthesis compressor room structure and announced it to the control room.

Fig.6: Synthesis compressor steam turbine damages

Blind Points

Considering that the jet of hydrogen gas and then the flame jet were directed from the leakage point toward the interior of the compressor room and that in the initial moments no visual proof of the amount and intensity of the hydrogen gas leakage and flame was available, the assumption of the shift supervisor, the ammonia unit manager and also the operation manager was that the situation might be controllable and that there would be no need to shut down the plant.

This resulted in the ammonia unit staying in service followed by attempts of the operators to find the exact leakage point and determine the reasons behind it. This led to invaluable time being lost that could have been used for decision making and corrective actions.

Even after observing the widespread fire and smoke, the order for shut down the compressors oil lubricating system was never issued.

Fig.7: Refrigeration compressor damages. Severe damage on steam turbine and HP case of compressor occurred.

The ammonia synthesis compressor was shut down after 2 minutes from the time of the incident, but the total plant shut down command was not given. Much time was spent in the control room to keep the situation in other sections of the unit stay stable.

Although the incident did not occur in the urea unit, the experienced urea plant manager considered it in an emergency situation and issued the command for total shut down of the urea plant and evacuation of the control room. This action was proven to be correct and a preventative measure to be used in future situations.

Fig. 8: Fire blazes below the compressor room

Performance of Safety Department and Fire Station Team

Positive Points

1. Speediness in arriving at the incident location and activating the stationary and mobile monitors.
2. Quick announcement the occurrence of the incident to the neighboring petrochemical complexes and requesting the dispatch of fire-fighting crews.

Negative Points

1. Low pressure in the fire water loop during the initial minutes and non-existence of independent fire water loop inside the complex. It was needed to use the common fire water loop present in the petrochemical zone.
2. Weakness of the management system and lack of timely decision making with regard to accurate identification of the possible dimensions of the incident. The order of appropriate commands to the operations and safety units for taking the necessary actions was too slow.
3. Inadequacy of the fire-fighting crew in the complex (approximately 10 persons in each shift) in comparison of the numerous operational plants. There was no possibility of forming several groups in order to attack the fire in several fronts and bring it under control.
4. Entry of individuals and unrelated personnel to the scene. There was an issuance of non-technical commands leading to the strategy of dealing with the situation that led to it getting out of control.
5. Spreading water directly over the compressors' turbines and the parts burning. This led to a quick drop in temperature and resulted in the intensification of the temperature shock and damage to the casing, even the twisting of pedestals supports.
6. Entry of fire-fighting teams from other companies of the petrochemical zone to the scene of the incident. There was no coordinated program to put out the fire and there operational interference with each other.

Fig.9: Empty oil console. Near 16500 liters oil was consumed and intensifying the fire on synthesis and refrigeration compressors.

Gas Detection System

The gas detector of synthesis compressor, GD-5802 alarmed was recorded in the control system at 14:55:24. Because of high number of alarms in the control system at that moment, and also bad record of performance of gas detectors in the plant due to numerous errors, the GD alarm was not taken seriously.

Causes which Intensified the Incident

1. Slow reaction in shutting down the synthesis and refrigeration compressors lost approximately 2 minutes. The special location of the leakage between the third and fourth stages of the ammonia synthesis compressor and took too long (approximately 40 minutes) for the pressure to drop to the atmospheric level and resulted in spreading fire during that time.
2. Lack of attention to the lubrication system being in service. Considering the fact that the lubrication system of the compressor must remain in service following the

compressor being offline in order to decrease the temperature of compressor bearings, after the fire incident, there was no appropriate reaction in the stopping the oil lubrication system.

Fig.10: lubrication and steam seal piping damages around synthesis compressor steam turbines.

Approximately 30 minutes after the incident, steps were taken to put the turbine oil pump out of service. Due to the destruction of the related switch this became impossible and closing the low pressure steam valve was then done. Considering the lengthy time taken to do this task actually, all the existing oil, in the oil console was lost over the oil lubrication piping fittings which had failed due to fire.

Problems in Instrumental Tube Systems of Pressure & Temperature Transmitters in This Case

The problems contrary to normal standards in this section were:

1. The tubes not having safety guards.
2. Not performing routine inspection of the tubes and instrumental connections.
3. Not installing piping instead of instrument tubing in many segments.

Failing in the Plant Incident Command System (ICS)

In the examination of the compressor room fire incident, it was determined that there was a serious deficiency in the plant incident command system. Slow reaction in recognizing the occurrence of fire and its spread, not stopping the oil lubricating system fast enough, improper management of the fire-fighting team of the complex and also the other fire-fighting crews of petrochemical zone, spreading water over the turbines and the lines containing high temperature steam led to severe damages, especially to the casing of the turbines and compressors, in addition to temperature shock to the steam piping were some of major faults during the incident occurrence.

Fig.11: The compressor room roof damages by fire.

Lessons Learned

1. The importance of the ammonia plants compressor room's safety due to the high price equipment and importance of compressors in production operation when placed in an enclosed area.
2. Training the operation and safety personnel in the methods of operation in case of gas leakage or fire inside the compressor room.
3. Providing an action plan specifically for incidents inside the compressor room and describing the duties of the various

personnel and groups, separately and in detail.
4. Creating and updating incident command system (ICS).
5. HAZOP & HAZID study specific to compressor rooms at the time of constructing the plant and reviewing in the operation.
6. Performing incident and fire drills inside the compressor room according to the programming scenario.
7. Keeping areas free unnecessary scaffolding under the compressor rooms.

Fig.12: Synthesis compressor rotor. No major damages seen.

Measures Taken for Preventing the Recurrence of the Problem

1. Installing closed circuit camera inside the compressor room for online monitoring and recording.
2. Installing mist spray system inside the compressor room.
3. Replacing high pressure instrument tubes by piping.
4. Increasing the number of foam fire extinguishers and placing several high capacity foam capsules inside the compressor room.
5. Creating low pressure steam spray nozzles inside the compressor room for the purpose of extinguishing small fires with steam prior to its spread
6. Providing an action plan specific to fire incidents inside the ammonia plant compressor rooms.
7. Revised the incident command management system (ICS) by training all of the responsible personnel and managers, providing the detailed job descriptions and installing on the work desks.
8. Installing a solenoid valve in the steam line to the oil turbine pump in order to immediate shut down it in case of fire.
9. Change in the interlocks related to the emergency AC and DC electric oil pumps for the purpose of quickly shut down during fire case.

Fig.13: scaffolding work very close to the incident point remaining as a suspected reason of the incident.

10. Training the site operators and the control room personnel with respect to the importance of the shut down the circulation of compressor lubrication system at the times of fire for avoidance of damage to the equipment.
11. No direct spraying of cold water on the turbines and steam lines due to their high temperature to avoid thermal shock.

Actions which must be taken in the Compressor Rooms of Ammonia Plants in Case of Fire

1. Avoid spraying water directly on the steam turbines and the steam lines.
2. Use low pressure steam or foam for extinguishing small fires.
3. Shut down oil lubricating system immediately.
4. Reduce the synthesis section pressure quickly as much as possible by using all existing vents.

Fig.14: Gasket rupturing due to fire jet and oil spreading from lubrication piping around the steam turbines caused fire.

Practical Suggestions for Preventing Fire Incidents in Compressor Rooms

1. Installing closed circuit cameras in different directions and constant monitoring of the conditions inside the compressor room.
2. Installing and increasing leak detectors in different locations close to the synthesis and refrigerating compressors and mechanical and instrument connections.
3. Installing infrared cameras and sensors for the purpose of quick detection of any type of fire.
4. Installing smoke sensors inside the compressor room (F & G systems).
5. Installing firewater lines and creating connections for installation of hydrants and spray of water to the walls and certain cold connections and equipment for prevention of spread of fire.
6. Developing hose stations and low pressure steam hoses for extinguishing minor fires by steam.
7. Creating a system of mist spray over the compressors for putting out fires quickly and preventing its spread.
8. Creating instrumental interlocks and solenoid valves necessary for shutting down the compressors and also lubrication system in case of ascertainment of oil leakage and spread of fire resulting from it.

References

1. American Petroleum Institute, API 614: Lubrication, Shaft-Sealing, and Control-Oil Systems and Auxiliaries for Petroleum, Chemical and Gas Industry Services.

2. Instrument Tube Fitting Installation Manual, Bulletin 4200-B4, Parker Hannifin Corporation, Instrumentation Products Division, Huntsville, AL,USA, May 2014.

Construction and Commissioning of Kaltim-5 2500 mtpd, World's Largest Ammonia Purifier™ Plant

Towards the end of 2014 Pupuk Kaltim completed construction of the *Kaltim-5 project and started up its 2500 MTPD Ammonia Plant the world's largest, single train Purifier based ammonia plant. The plant is licensed by Kellogg Brown & Root (KBR) with Purifier Technology incorporated in the flow sheet and is located in Bontang, Borneo Island in East Kalimantan province of Indonesia.*

The world's largest Purifier based ammonia plant was commissioned successfully in January 2015 with the supply of ammoia to its new 3500 MTPD Urea plant. Performance test was successfully completed with net energy compsumption within the guaranteed number, and also the production capacity exeeded the guarantee value. The plant was also proven to successfully produce 2700 MTPD ammonia. This paper presents an overview of the project and its journey through the various stages of project development, design, engineering, construction and commissioning as well as challenges encountered throughout the project.

Luther Sirait
PT. Pupuk Kaltim

Jakub Tarigan
PT. Pupuk Kaltim

George Colman
Kellogg Brown & Root

Introduction

Pupuk Kaltim commenced commercial production from it's newly constructed Kaltim 5 ammonia urea complex located in Bontang, East Kalimantan, Indonesia in October 2015. The Pupuk Kaltim site comprises of seven ammonia and six urea plants. Four of the ammonia plants utilize technology supplied by Haldor Topsoe, 2 utilize technology supplied by KBR, and 1 utilizes technology supplied by Lurgi. Ammonia production can be routed to 2 x 26000 mt and 1 x 50000 mt ammonia storage tanks.

The new Kaltim 5 ammonia plant utilizes KBR Purifier ammonia technology. The plant is the world's largest KBR Purifier plant having a nominal capacity of 2500 MTPD and a name-plate capacity of 2700 MTPD. A 3500 MTPD urea plant, utilizing Toyo ACES 21 process technology with granulated product, was also constructed.

Upon successful commissioning of the Kaltim 5 ammonia plant the Kaltim 1 plant which was constructed in 1979 was shutdown.

Ammonia Process Design Features

The ammonia plant is based on the KBR Purifier flowsheet as shown in figure-A1 and utilizes the BASF licensed OASE White (activated MDEA) solvent for the CO_2 removal process.

The plant design is based upon the natural gas composition listed in Table 1.

Components	Mole Fraction,%
CH_4	83.72
C_2H_6	5.40
C_3H_8	2.89
C_4H_{10}	1.17
C_5H_{12}	0.35
C_6H14	0.12
C_7H16	0.15
N_2	0.09
CO_2	6.13

Table 1. Natural Gas Feed Composition

The unique design features of the Kaltim 5 Purifier plant are:

- Mild conditions for primary reforming via excess air to the secondary reformer
- Primary reformer having 2 steam turbine driven forced draft fans and 2 steam turbine steam turbine driven induced draft fans (all 4 fan operating at full plant load)
- Fire tube waste heat boiler design – supplied by Borsig
- KBR's nonmetallic mixing chamber (no metallic mixer or burner) in the secondary reformer
- Low temperature shift guard bed
- High pressure flash gas compression and re-injection upstream of the CO_2 absorber
- KBR Purifier for adjustment of synthesis gas hydrogen: nitrogen ratio and removal of all methane and the majority of the argon present.
- KBR horizontal ammonia synthesis convertor with 3 catalyst beds and 2 interchangers
- KBR unitized chiller
- Front end hot vent and a back end flare
- Use of KBR's operator training simulator

For plant start-up ,the feed gas compressor is utilized for front end nitrogen recirculation.

Plant Performance

The plant performance guarantee test was conducted over a 15 days period in October 2015 with all guarantees being successfully met:

Guarantee item	Guarantee item	Measured Value
Ammonia production,mt/d	2500 min.	2545
Ammonia product quality: NH_3,% H_2O,% Oil,ppm	 99.9 min. 0.1 max. 5 max.	 99.99 0.028 0.0476
Net energy consumption,Gcal/mt	7.15 max.	7.05
Carbon dioxide production,mt/d	2633 min.	3885
Carbon dioxide purity: CO_2,%	 99.0 min.	 99.46

Table 2. Guarantee & Measured Numbers

Additionally, the plant capacity test was conducted over a 5 day period in September

2015 with an average ammonia production of 2720 MTPD being achieved for the duration of the test.

Project Execution

The project kick-off was in September 2011 with an Engineering Procurement Construction scheme (EPC Contract) as a lump sum project. The EPC contract was awarded to Inti Karya Persada Tehnik (IKPT) as consortium leader and Toyo Engineering Corporation (TEC) as partner. Beside the ammonia and urea plants, the project also included a new utility unit. Ground breaking occurred in November 2011 with the plant mechanical acceptance being achieved in December 2014, and commercial operation declared in October 2015.

The main project milestones are listed in Table 3.

Major Events	Date
Effective date of contract	14 Sept. 2011
Basic engineering completion	Jun. 2012
Detail engineering completion	Sept. 2013
Ground breaking	Nov. 2011
Mechanical acceptance	29 Dec. 2014
Precommissioning phase	Mar.-Nov. 2014
Natural gas receiving	26 Nov. 2014
Commissioning start (Reformer 1st firing)	28 Nov. 2014
Feed-in to primary reformer	5 Dec. 2014
Syngas compressor start	7 Jan. 2015
Catalyst reduction of ammonia synthesis converter start	9 Jan. 2015
First ammonia produced	12 Jan. 2015
Performance testing completed	19 Oct. 2015
Plant acceptance	9 Nov. 2015

Table 3. Key Milestones of Kaltim-5 Project

First ammonia production was successfully accomplished 38 days after the initial introduction of natural gas feed.

Commissioning and Start Up

Unlike a conventional turn key project in which the contractor provides a complete team for start-up, Pupuk Kaltim used its own operation team to start the plant. Kaltim's operation team training consisted of in-house training in the contractor's office, licensor's classroom training, and simulator training using KBR's dynamic process simulator. Problems that arose during plant commissioning were jointly resolved by the contractor and owner electrical, instrument, and mechanical teams.

Problems Faced During Plant Initial Start-Up and Operation

Below is a summary of some of the main issues encountered during the plant initial start-up and operation.

1. Front End Vent Silencer Internal Damage

The front end vent header collects gases being vented front the front end of the plant and routes them to a vent silencer located at an elevation of 55 meters (180 ft.) above grade. (Figure 1) On the 4th December 2014 while steaming the primary reformer with 65 mt/h of steam the wire mesh flame arrestor contained within the front end vent silencer was emitted from the silencer housing. The root cause of this event was the vent silencer flame arrestor mesh becoming partially blocked with construction debris which was transported from the front end vent header into the vent silencer. The accumulated debris created an increased pressure drop across the flame arrestor mesh leading to a rupture of the mesh. Due to pipe size, configuration, and vent point elevation during pre-commissioning the vent header was not effectively cleaned or visually inspected for cleanliness.

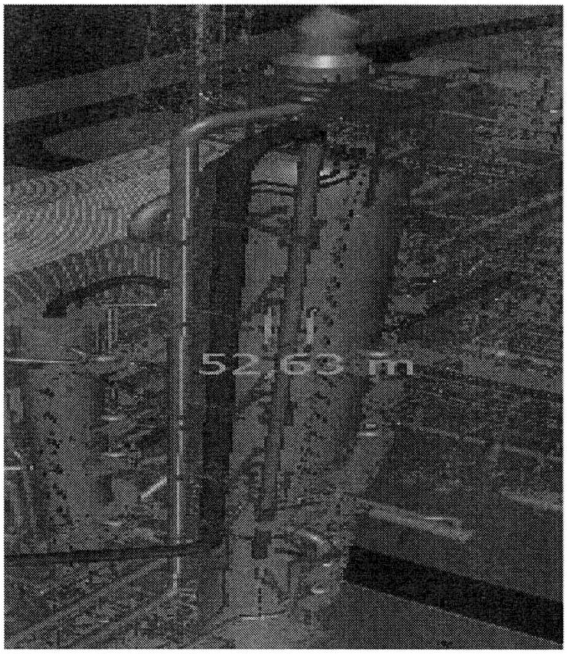

Figure 1. Hot Vent Pipe

2. Primary Reformer Catalyst Tube "Hot" spots/Zones

On the 18th December 2014 while the primary reformer was operating at a gas load of 30%, a steam to carbon ratio of 4.5, and a catalyst tube outlet temperature of 700°C (1290 F), it was noted that 77 of the primary reformer catalyst tubes exhibited "hot" spots/zones. (Figure 2) Tube wall temperature measurements of the affected tubes revealed that the tube wall temperatures were at or above the design tube wall temperature value. On the 17th February 2015 the plant was shut down for 17 days to identify the cause and to replace the catalyst in the "hot" tubes. A review of the tube pressure drop measurements made during initial catalyst loading and a tube pressure drop survey made at the time of the plant shutdown revealed numerous tubes had a pressure drop greater than +/- 10% of the tube average pressure drop. The principle cause of the hot catalyst tubes was identified as poor catalyst loading practices. A secondary cause was identified as the smaller (lower heat liberation) wall burners were installed either side of the riser tubes instead of along the side walls of the radiant box.

Figure 2. "Hot" Catalyst Tubes

3. Synthesis Gas Dryer Desiccant Milling

Each synthesis gas dryer bed is comprised of 2 layers of type 4A molecular sieve and a top layer of silica gel. Before Purifier operation is established the molecular sieve dryer beds are regenerated using dry synthesis gas sourced from the outlet of the online dryer. After Purifier operation has been established the molecular sieve dryer beds are regenerated with Purifier waste gas. The Purifier waste gas possess a molecular weight approximately twice that of the synthesis gas. After 2 ½ months of plant operation it was noted that the synthesis dryer bed pressure drops were trending upwards and the frequency of waste gas filter cleaning was also increasing. At a plant shutdown both synthesis gas dryers were opened and it was observed that the silica gel which had been installed above the desiccant hold down grid had become mounded into the center of the dryer and milled to a fine dust. (Figure 3) Some of the 4A molecular sieve was similarly affected. The silica gel was removed and operation resumed. Silica gel milling occurred primarily due to being installed in the incorrect position above the desiccant top hold down grid and secondly due to the first few synthesis gas dryer regenerations being performed at too high of a regeneration gas flow rate. The installed regeneration gas flow meter is calibrated for the higher molecular weight Purifier waste gas. During the first few synthesis dryer regenerations the indicated flow of regeneration gas was not corrected for the molecular weight difference.

Figure 3. Milled Desiccant

4. Expander Seal Pack Replacement

While performing Purifier expander pre-start-up checks on the 16th January 2015, it was noted that the expander casing had filled with water. The expander supplier recommended that the expander be removed to the main workshop for remediation and seal pack replacement. Investigation revealed that the expander casing vent line had been incorrectly piped to the front end vent header instead of the back end flare header and additionally no isolation valve has been provided in the vent line. While steaming the primary reformer at a plant start-up or shut down the front end vent header contains steam a portion of which was able to flow in a reverse direction into the expander casing where it cooled and condensed. To avoid such an event in the future the vent was rerouted to atmosphere at a safe location and an isolation valve was added.

5. Cross Connect Of Purifier Product and Waste Gas Pipe Work

While performing the Purifier process side dry out it was noted that while attempting to dry the product gas flow path that the dry out gas would flow into the waste gas pipe work and vice versa. Investigation revealed that the waste gas pipe work had been mistakenly connected to the Purifier product gas nozzle and vice versa. This error was traced to the development of the piping isometric drawings during detailed engineering. The error was corrected by reworking the installed pipe work so it connected to the correct Purifier nozzles.

6. Process Air Coil Protection Steam Flow Control Valve Damage

During the plant initial start-up after a flow of protection steam had been established to the process air pre-heat coils difficulties were encountered in controlling the front end back pressure. It was noted that the indicated flow of protection steam was approximately 60 mt/h and could not be regulated. Investigation revealed that the process air coil protection steam control valve was installed in a pocketed pipe which was not provided with any steam condensate drainage facilities. Due to water hammer, the control valve trim had become detached from the control valve stem and dislodged causing an uncontrolled flow of medium pressure steam to flow through the process air preheat coil into the plant front end. Steam traps and manual drains were provided both up-steam and downstream of the control valve and the operating procedure was modified to require the field operator to check for the presence of liquid and drain as required before lining up the control valve station.

7. High Pressure Steam Drum Relief Valve Tail Pipe

During a steam distribution system upset a high pressure steam drum pressure relief valve opened. The reactive forces on the relief valve tail pipe created by the steam flow caused the tail pipe to become dislodged from its supports and it fell against the high pressure steam drum snapping off a 1" diameter nozzle on the steam drum. The plant was shut down immediately for repairs. The high pressure steam drum relief valve tail pipe supports were modified to prevent the tail pipe from becoming dislodged in the future.

8. OASE White Carryover From LP Flash Drum

On January 23rd, 2015 severe foaming of the OASE White solution occurred in the LP flash drum resulting in a carryover of OASE White solution into the carbon dioxide product header. Approximately 25mt of OASE White concentrate was lost into the plant drainage system while an additional 25mt of OASE White solution was recovered from the carbon dioxide product header.

9. Purifier Rectifying Column Level Transmitter Calibration

During the Purifier initial start-up and operation a higher than expected rectifying column differential pressure and unstable synthesis gas compressor suction pressure were observed (Figure 4). Checking revealed that the rectifying column liquid sump level transmitters had been incorrectly calibrated. This calibration error resulted in the liquid level submerging both the process gas inlet nozzle to the column and also the bottom few rectifying column trays. The liquid level transmitters were re-calibrated to the correct value.

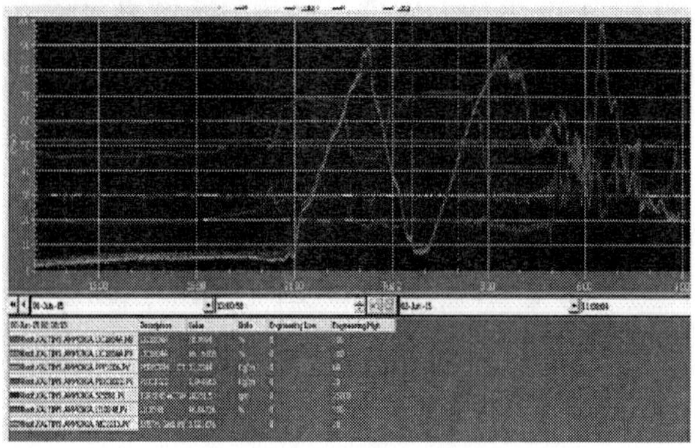

Figure 4. Purifier Parameter Fluctuation

10. Under Performance Of HTS Effluent Steam Generator (103C1/C2)

The HTS effluent steam generator (103C1/C2) are designed to cool the HTS effluent stream to the target inlet temperature necessary for the LTS (Figure 5) while vaporizing 25% of the boiler feed water flow passing through the tube side of the exchangers.

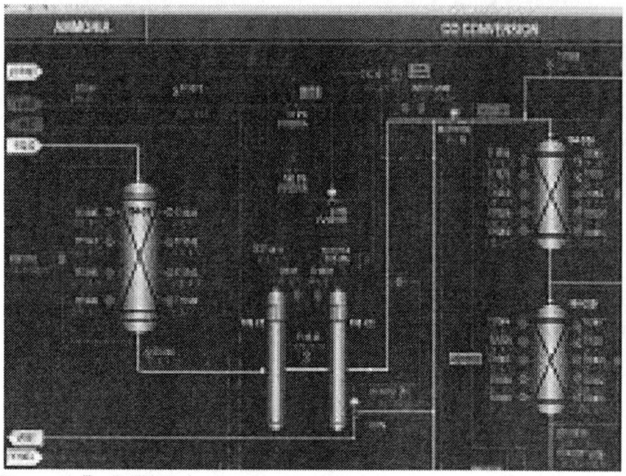

Figure 5. BFW Arrangement

However the 103C1/C2 exchangers underperformed resulting in several effects:

- Inability to cool the process gas feed to the LTS to the desired start of catalyst life temperature
- Imbalance in boiler feed water flows between the 103C1/C2 exchangers and the ammonia effluent steam generator (123C1/C2)
- A short fall in the overall medium pressure steam production and export
- A redistribution of heat load between the secondary reformer waste heat boiler and the high pressure steam super heater – this heat load redistribution being necessary in order to partially compensate for the short fall in steam generation by 103C1/C2

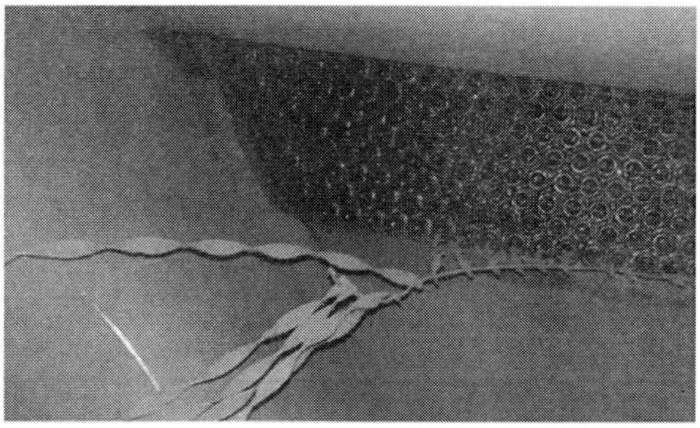

Figure 6. Twisted Tapes In Pass-6 103-C1

Numerous actions were undertaken in an effort to troubleshoot this issue:

- On the 2nd July 2015 the tube side (boiler feed side) channel covers were removed. Inspection revealed a gap existed between the partition plates for each tube side pass which allowed some bypassing of boiler feed water. The gaps were sealed by welding. Rectification of this issue failed to improve the 103C1/C2 performance.

- On the 6th September 2015 the tube side (boiler feed side) channel covers were again removed. Inspection revealed that the twisted tape inserts from 5th & 6th tube side passes of 103C1 had become dislodged. (Figure 6) These tapes were removed. Removal of the tapes had no effect upon the exchanger performance

- At other plant shutdowns further minor remedial actions were completed. All such actions were unsuccessful in improving the 103C1/C2 performance.

Additionally, extensive reviews of all engineering design documentation for this plant item revealed some alterations had been made to the original design provided by the licensor.

At the time of writing this paper the issue remains unresolved. Further actions are planned for the next plant shutdown.

11. Boiler Feed Water Leakage From HTS Effluent Steam Generators (103C1/C2) Boiler Feed Water Flanges

While lining up boiler feed water through the tube side of the HTS effluent steam generators, significant water leakage was detected from each boiler feed water connection. These flanges are ring type class 1500psi flanges ranging in size from 8" – 12". Torqueing of all the bolts in the affected flanges was unsuccessful in arresting the leakage. Inspection of the sealing surfaces of the affected flanges revealed severe corrosion and pitting which required machining in place. The corrosion and pitting occurred as a result of inadequate sealing surface preservation during the construction phase of the project.

12. Ammonia Synthesis Convertor Removable Head Flange Damage

While performing the final closure of the ammonia synthesis convertor removable head flange the sealing surfaces of the flange and associated metallic conical gasket was damaged. Investigation revealed that the metallic conical gasket had corroded during storage, and that the flange closure procedure provided by the original equipment supplier was not closely followed. The damage necessitated an insitu repair being made which involved spray metallization of the damaged area of the flange.

Pre-commissioning Activities

The main activities completed during pre-commissioning of the plant are described below:

- Cooling and sea water system cleaning, which combined manual cleaning of the large diameter sections of pipe work and water flushing using strategically placed temporary supply to return header jumper pipe work.

- Process air compressor inter-stage pipe work was cleaned by rapid decompression air blowing utilizing the CO_2 absorber as an air volume tank. Temporary pipe work was provided to route air from the CO_2 absorber to the process air compressor inter-stage pipe work. For this activity 4 portable air compressors were used to pressurize the CO_2 absorber to a pressure of 8 kg/cm^2. The activity was completed between 13th May and 11th June 2014.

- Process gas pipe work was cleaned by continuous air blowing utilizing the process air compressor to provide up to 145000 Nm^3/hr of air at a pressure of 12 kg/cm^2. Some temporary pipe work was required to route air to all locations to be blown. The activity was completed between 8th September and 9th November 2014.

- The plant steam network was blown utilizing medium pressure steam supplied by a coal fired boiler located outside the plant battery limit. A maximum steam flow of 80 mt/h was made available for blowing. Steam blows were made at a total of 80 points excluding utility stations. The cleanliness of the steam used to blow the pipe work to each steam turbine inlet was determined using copper target plates. The activity was completed within 2 months between May and June 2014.

- The shell side of the secondary reformer waste heat boiler, the high pressure steam drum, the inter-connecting riser and downcomer pipe work between the boiler and the steam drum, and the high pressure steam header (bypassing the high pressure steam superheater tube side) were chemically cleaned. The chemical cleaning procedure comprised water flushing, alkaline degreasing, acid cleaning, rinsing, passivation, and layup. The activity was completed between 13th August and 2nd September 2014.

- All the OASE White containing sections of the CO_2 removal system were cleaned by once through water flushing, followed by water recirculation using the installed pumps and capturing particles on the pump suction strainers. Oil and grease was removed from the column random dumped packed beds by using the installed pumps to recirculate 2%$^w/_w$ sodium hydroxide solution followed by water rinsing. The activity was completed between 14th September and 17th October 2014.

- Catalyst and desiccant were loaded into a total of 12 vessels including 2 synthesis gas dryers. The activity was completed between 2nd May and 21st July 2014.

Problems Faced During Pre-commissioning

1. Process Air Compressor Suction Duct Deformation

During plant pre-commissioning while using the process air compressor to supply air for the air blowing activities a lower than normal pressure developed in the 88" diameter air intake duct to the compressor 1st stage causing a partial collapse of the ducting. (Figure 7) The lower than normal pressure in the air intake duct was created by a greater than design pressure drop occurring across the air intake filter housing/elements. To lower the pressure drop across the air intake filter housing/elements to an acceptable value to allow the plant pre-commissioning and plant initial start-up/operation activities to continue the pulse cleaning nozzles were removed from the filter elements.

Figure 7. Suction Duct 101-J

2. CO_2 Removal Low Pressure Flash Drum Distributor Tray Damage

At the end of the CO_2 removal system cleaning activities, all vessels in the system were opened and inspected for cleanliness. During this inspection it was noted that 19 gas riser chimneys had become dislodged from a liquid distributor tray within the low pressure flash column. The gas riser chimneys were reinstalled.

3. OASE White Pump Suction Strainer Damage

Cleaning of the CO_2 removal system involves frequent cleaning of the OASE White solvent recirculation pump suction strainers. During this activity it was noted that the pump suction strainers required reinforcing with an additional reinforcing ring. Additional support reinforcement was given not only to the pump suction strainers but also to major compressor suction strainers.

Summary

The Kaltim 5 experience highlights the importance of a robust equipment preservation program, thorough plant pre-start-up checkout, and effective pre-commissioning practices.

Despite the challenges faced, the Kaltim-5 plant successfully met all its energy consumption, production, and product quality guarantees.

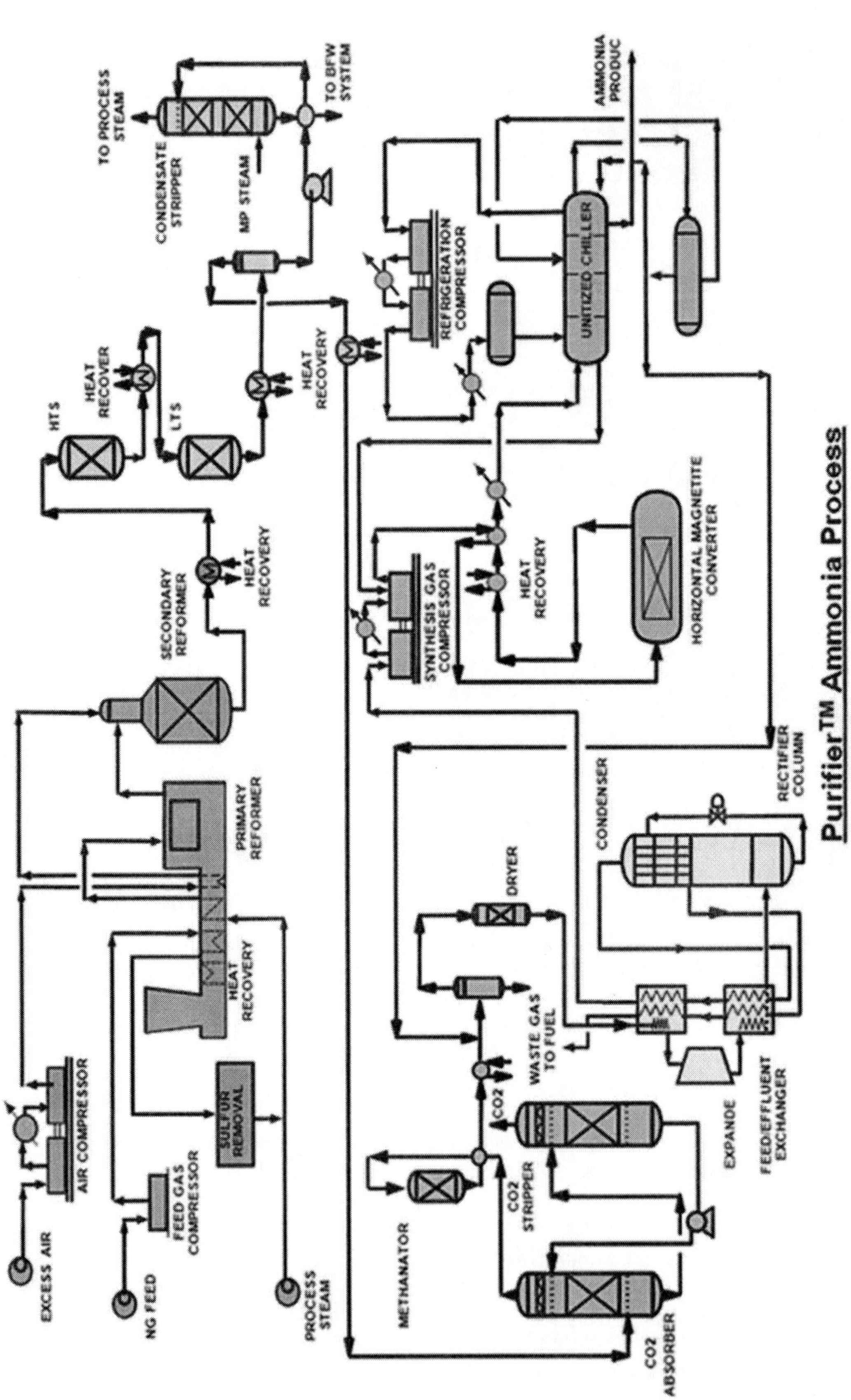

Figure A-1. KBR's Purifier Ammonia Process Flow Sheet

Toxic Gas Detection in Ammonia Plants for Shelter-In-Place

Multiple gas detection mechanisms are available in the process industry, with applications ranging from modeling gas releases, to detecting an accidental release in a process facility, which may trigger a shutdown or mitigation function. Many of those technologies are used for gas detection in ammonia plants, in order to protect personnel and equipment. Toxic Shelter-In-Place (SIP) or Toxic Safe Havens are vital for deployment of gas detection as part of a strategy to manage risks from Toxic releases, as they can help shelter or safely evacuate personnel based on the ammonia gas concentration during a release scenario. Toxic SIPs depend on reliable, timely toxic gas detection in order to allow the ventilation system configuration to be aligned to the specific sheltering mode and to alert personnel that toxic emergency procedures must be implemented. This paper discusses the application of various gas detection technologies to strengthen the SIP by providing much-needed data in the form of gas concentrations at various locations within the plant and also within the SIP to prevent personnel from evacuating when it is safer inside the SIP, or to ensure that occupants evacuate when necessary. This paper also compares different technologies in terms of response times, power requirements and measurement ranges.

Murtaza I. Gandhi, Anthony G. Sarrack
Baker Engineering & Risk Consultants, Inc.

Introduction

Many facilities in the chemical and petrochemical industry handle acutely toxic materials. One common method of mitigating risk posed by these toxic hazards is to designate occupied buildings as toxic SIPs with or without a backup plan. The purpose of this paper is to discuss toxic gas detection and its application within SIPs. This paper does not address toxic risk for outdoor populations.

Key aspects of a highly effective toxic risk mitigation strategy for SIP are as follows:

1. Timely, reliable detection of toxic gases in order to take actions such as:
    - Outdoors – isolate heating, ventilation, and air condition (HVAC) prior to impact by the toxic plume and issue guidelines for evacuation once the plume clears
    - At the SIP HVAC inlet - ensure reliable, timely isolation when the toxic plume impacts the building
    - Within the SIP – provide informed decisions for evacuation from the SIP by providing information about the toxic concentrations outside
2. Timely, reliable isolation of ventilation systems to avoid ingression of toxic gases.
3. Establish a leak-tight SIP volume to minimize infiltration.
    - Establish a secondary barrier, to minimize impact of toxic air infiltration. Use a room or set of rooms within the SIP building as the protected space(s) to address impacts due to doors being opened after the toxic plume impacts the building.
    - Address occupancy/volume/tightness to ensure that any CO_2 buildup and/or O_2 depletion issue is understood.

4. Train personnel and provide instructions to ensure that sheltering activities are performed properly.
5. Provide an effective backup/evacuation plan.
    - Provide indoor toxic gas monitoring for the SIP
    - Develop procedures that identify when to evacuate the SIP
    - Train personnel so they take appropriate actions
    - Provide clear communication channels between the SIP location and emergency response personnel
    - Maintain reliable, appropriate escape packs to allow safe evacuation

Additional enhancements can be implemented to further reduce toxic risk or to improve comfort levels within the shelter location. These include the following:

- Implement a filtration / scrubbing system to reduce toxic concentration.
- Provide reliable, efficient purging of SIP and reduce effective infiltration rate.
- Provide an air-supplied header with full face respirators to allow long-term occupancy by essential personnel (typically limited to control room locations, this is intended only for trained personnel).
- Enable heating and cooling systems to remain functional without causing any ingression or potential path of infiltration.
- Provide potable water and lavatory.
- Include sealed windows and/or cameras to better reveal outdoor conditions to the SIP occupants.
- Establish outdoor / additional indoor toxic gas monitoring.

Aspects of an effective risk mitigation strategy for SIPs are addressed in more detail below.

Toxic Gas Detection

Facilities with severe toxic hazards should be equipped with toxic gas monitoring to ensure that personnel are made aware of possible toxic gas infiltration in their work area. Effective monitoring will allow personnel to take appropriate actions in response to the hazard.

Figure 1. Toxic Gas Monitor

To be effective, toxic gas monitoring should be provided in these areas:

- Ventilation intake of the SIP
- Inside the SIP
- Outside of the SIP

The most important location for toxic gas monitoring is at the ventilation system intake of each occupied building intended to be used as an SIP. Without toxic gas monitors at the ventilation intake, a highly reliable ventilation system isolation strategy cannot be implemented. Functional requirements are summarized below:

- Short response time
- Highly reliable
- Practical concentration threshold
- Appropriate design for the toxins of interest

Because it is possible for the SIP environment to become dangerous, it is important for the SIP to also provide toxic gas monitoring for the indoor environment. It is not critical for indoor toxic gas monitors to have as short a response time as the ventilation intake monitors because indoor concentration will not rise as quickly as the concentration at the ventilation intake (outdoors). It is important that the indoor toxic gas monitors are maintained to be reliable and are appropriate for the toxins of interest (toxins with potential to cause dangerous concentrations within the SIP). It is not as important to establish interlocks or alarms on indoor toxic gas monitors, as occupants who are sheltering in place would almost certainly pay close attention to the monitors, and are likely to have necessary training and knowledge to act appropriately. In addition, actions would likely be taken as conditions degrade, so timeliness and responsiveness are not nearly as critical as with ventilation isolation.

To take full advantage of the safety afforded by an SIP, it is important for occupants to remain within the SIP while it continues to provide a safe environment. Providing proper PPE will allow occupants to evacuate the SIP before it becomes dangerous to remain within the SIP. Although indoor toxic gas monitors within the SIP do not need to have alarms, it is important for outdoor concentration thresholds to be established and understood by occupants of the SIP. Without this understanding, occupants may evacuate the SIP even though the environment within the SIP is safe, and the outdoor toxic gas concentration is highly lethal. The lack of pre-established concentration thresholds and response actions may also result in occupants remaining within the SIP as the conditions become too dangerous to stay there.

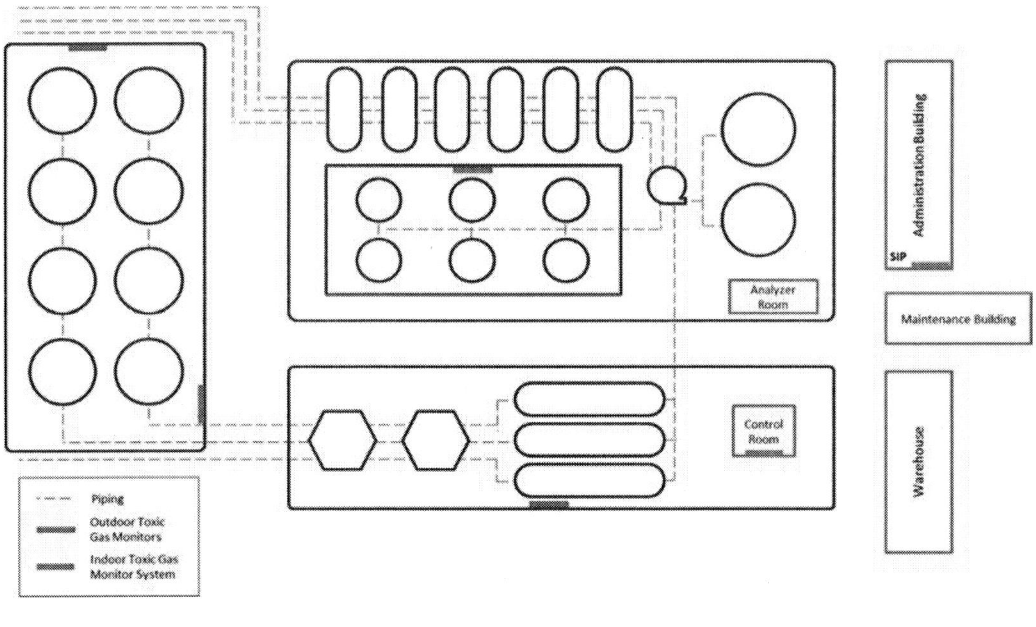

Figure 2. Plant Toxic Gas Detection System

In addition to toxic gas monitors at ventilation intake and within the SIP (shown in Figure 3), it is essential to also provide reasonably thorough toxic gas monitoring within outdoor areas of the plant as shown in Figure 2.

Providing toxic gas monitoring in areas where releases are most likely to occur (pumps, compressors, load/unloading areas, etc.) improves the likelihood that the hazard will be detected before the SIP is impacted. This allows ventilation to be isolated (reduce ingression), and, equally as important, assists in determining when it is safe to evacuate the SIP after the release has been isolated.

Outdoor monitoring should provide real time concentration readings to emergency response personnel so they can provide helpful advice regarding evacuation routes and mustering points. The information may also be used to identify the release location so it can be rapidly isolated.

It is critical to select appropriate instrumentation for toxic gas detection and concentration monitoring. For example, the detectors and associated instrumentation must be designed for the material as well as the concentration ranges that may occur. Detectors also need to be located appropriately to ensure that vapors are reliably detected in a timely manner. Point gas detectors vs. open path should be considered. Point gas detectors are contact type detectors and need the gas to come in contact with the sensor. If the sensor is not laid out correctly, the most expensive and responsive sensor may fail to detect a release. Open path detectors (typically infra-red or laser based) provide the product of toxic gas concentration and distance alone lines of interest between sensors, which helps determine if the area under the detector coverage is impacted by a gas release.

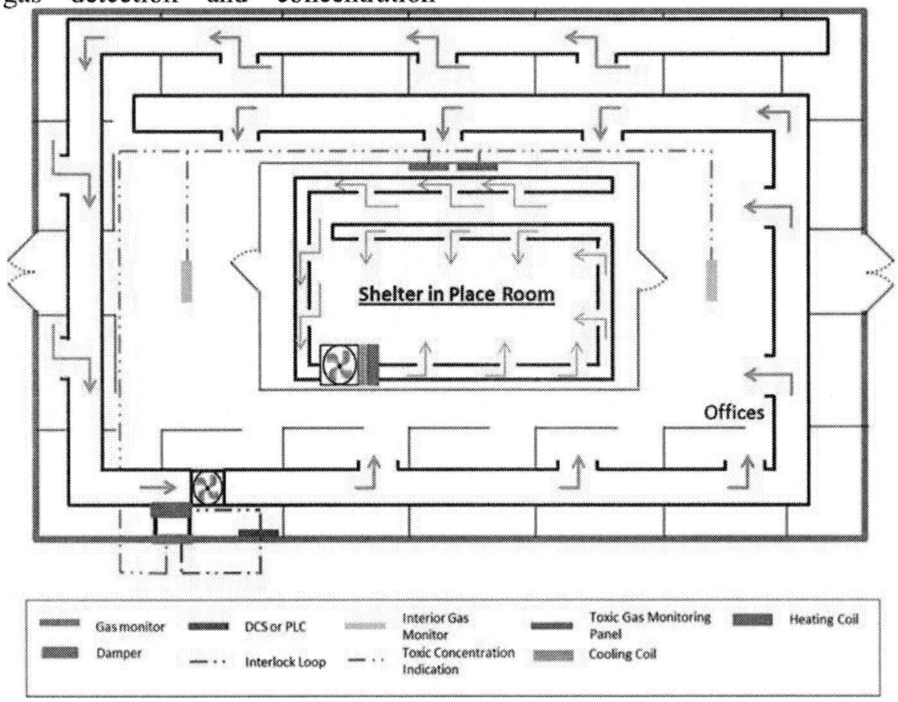

Figure 3. Toxic Gas Detection System for SIP

Open path detectors are generally more expensive than point gas detectors, but since they do not need to contact the gas, fewer detectors may be needed to adequately cover a given area and may be more practical in some cases, if an existing toxic gas system is being upgraded, and the new detection system can interface with the existing PLC/DCS in the area. Costs may be very similar for point detectors vs. open path system in terms of supporting hardware and software, as well as training of personnel.

Other aspects for selection of gas detectors are the signal generation/response time and the alarm setpoint. Response time is the time taken by a sensor to generate a signal after coming in contact with a gas. It can be thought of as the amount of time the system lags the actual gas concentration. It is an important parameter in all cases where rapid response is desired. For example, it is important for detectors on the HVAC inlet to respond quickly to ensure that the amount of toxic gases drawn in by the ventilation system is minimized. However, response timing is not important for detectors within the SIP because conditions change relatively slowly, and actions that occur in response to the alarm are not immediate.

Figure 4 shows a range of different gas concentration vs. time profiles for a hypothetical HVAC inlet toxic gas detection system. The importance of response time and alarm setpoint is discussed for each example concentration profile to clarify this point.

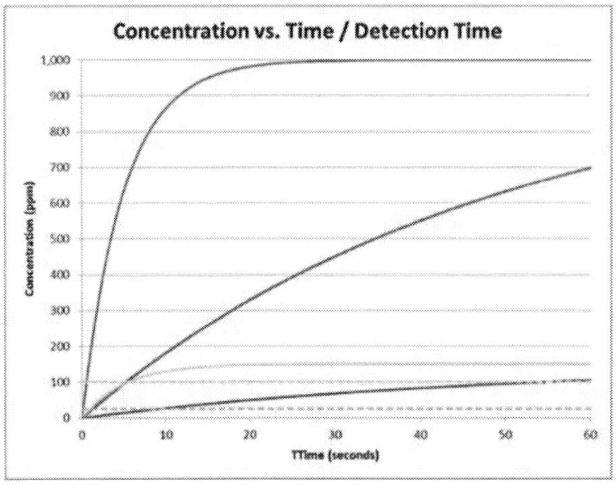

Figure 4. Example Concentration Profiles and Alarm Setpoints

For the purple line, maximum concentration is only marginally higher than the high setpoint (100 ppm), and is about five times as high as the lower setpoint (25 ppm).

If the detector has a 30-second response time, and the alarm setpoint is 25 ppm, it would alarm approximately 40 seconds after the toxic gas initially reaches the detector (10 seconds to reach 25 ppm plus 30 seconds response time). However, if the alarm setpoint were 100 ppm, it would take 85 seconds (55 seconds to reach 100 ppm plus 30 second response time).

Note the extra 45 seconds to reach the alarm setpoint in this example may be critical in a real-world situation.

The peach line shows a more rapid concentration rise, so alarm setpoint is less important, since the amount of time it takes for gas concentration to rise from the 25 ppm alarm setpoint to 100 ppm is only about 5 seconds. In that example, the alarms would be tripped at approximately 31 seconds or 36 seconds after toxic gases initially reached the target for the low and high setpoints, respectively.

The red line indicates that the rate at which concentration rises to the maximum concentration is slower, but since the maximum concentration is dramatically higher than the alarm setpoints, the time it takes to rise from the lower to higher setpoints is short (very similar to the peach case, about 5 seconds).

The blue line is an example showing that the effect of an alarm setpoint is negligible for the timing of an alarm because the maximum concentration is dramatically higher than the alarm setpoints being considered, and the concentration rises quickly to that level.

Although a lower alarm setpoint means a shorter alarm delay, it is also more likely to result in nuisance alarms. Alarm setpoints should be optimized with consideration of the types of toxic gas concentration profiles that might occur, the importance of alarm timing, and the likelihood of nuisance alarms.

General Characteristics of Toxic Gas Detectors

Toxic gas detectors are generally electro-chemical, point-type sensors. Some vendors have toxic line-of-sight gas detectors, which are not covered in this paper, typically infra-red or laser based sensors. Other technologies for toxic detection are rarely found.

Electrochemical sensors operate by reacting with the gas of interest and producing an electrical signal proportional to the gas concentration. Gas that comes in contact with the sensor first passes through a small capillary-type opening and then diffuses through a hydrophobic barrier as shown in Figure 5, and eventually reaches the electrode surface. This approach is adopted to allow the proper amount of gas to react at the sensing electrode. The gas that passes through the barrier then reacts with the sensor and produces an equivalent electrical signal.

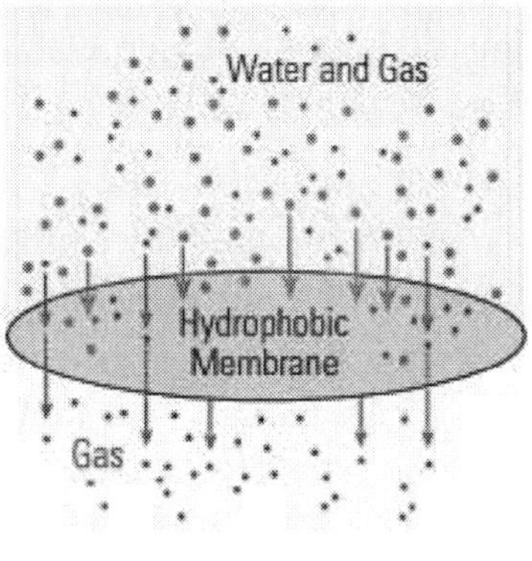

Figure 5. Hydrophobic Membrane for Gas Sensing

Some of the key characteristics to look for when selecting toxic gas detectors are described below:

Detection Range

The range in PPM, depends on the specific gas. The sensor should be chosen based on the gas of interest. For NH_3, the range is normally 0-100ppm or 0-500 ppm. Consideration should be made also for the location: e.g., the sensors inside the building can have a lower range, while those outside could be a higher range.

Response Time

T_{90} approximately 1 minute where T_{90} is the time for the concentration of the gas in contact with the sensor to reach 90% of its full scale value. Faster response time is generally advantageous, especially for external sensors. For sensors inside the room, a higher response time may be acceptable as the building prevents rapid buildup of gas; hence the approximately 1 minute response time has no impact on the overall SIP response.

Accuracy

10% of reading or 5% of full scale value are acceptable. Toxic detectors average about 10% accuracy throughout the different manufacturers.

Nuisance Readings

Nuisance readings or false alarms can reduce the overall reliability of the system. Alarm setpoints should be low enough to alert personnel of buildup of toxic gas concentration, but high enough to reduce nuisance alarms/trips—this is more important for outside sensors, as inside sensors should usually indicate a zero value.

Humidity

Sensors are designed to operate at certain relative humidity (typically 15-90% RH). Provision should be made to operate the sensor in the permitted environment. In high humidity conditions, prolonged exposure can cause excessive water buildup and create leakage. In low humidity conditions, the sensor can dry out.

Installation Location

Location of sensors should be determined using consequence based tools (e.g., the best location is where the sensor comes in contact with toxic gas for maximum possible credible release scenarios.) For SIPs, it is critical that detection be provided at the HVAC inlet to ensure timely HVAC isolation when needed. Within the SIP, the location of detectors is not important, since concentration would typically rise slowly and readily disperse throughout the protected volume. Location should also consider ease of maintenance and calibration.

Calibration

Some manufacturers require separate calibration gas and more frequent calibration. Sensors like these should be provided with calibration kits to reduce maintenance times. Other manufacturers have auto calibration where a sample gas is always present in the sensor body thereby not requiring an external calibration gas assembly. Calibration and testing of detectors should be performed at least at the manufacturer's recommended frequency. If the governing standard in the area or calculations require more frequent testing, that should be followed.

Test gases and sample gases

To achieve proper calibration of the sensor, it is important to have the correct concentration of test and sample gases. If the gas concentrations of the sample gas and the actual process gas do not match within the tolerance range, response during a release may be delayed.

Diagnostics

Sensors provide diagnostics such as end of life, out of calibration, signal issues, etc. Diagnostics should be closely monitored to improve accuracy and reduce nuisance readings.

Hazardous Area Classification

Sensor assembly should be classified according to the area of installation

Regional and National Certifications

Local certification requirements should be met.

IEC61508 Certification

Safety Integrity Level (SIL) certification is available for certain devices. If the toxic gas detection is part of an automatic action associated with a specific reliability requirement (e.g., HVAC shutdown), that certification should be considered.

Case Study

An example of a processing plant with ammonia is shown in Figure 6. There is only one building in the plant (marked with brown dashes), it is set up as Toxic SIP. There is an ammonia release source approximately 130 ft (40m) from the building. A ½-inch release from the source is expected to reach over 1,500 feet (457m) at 90% lethality levels for an assumed exposure duration of 10 minutes.

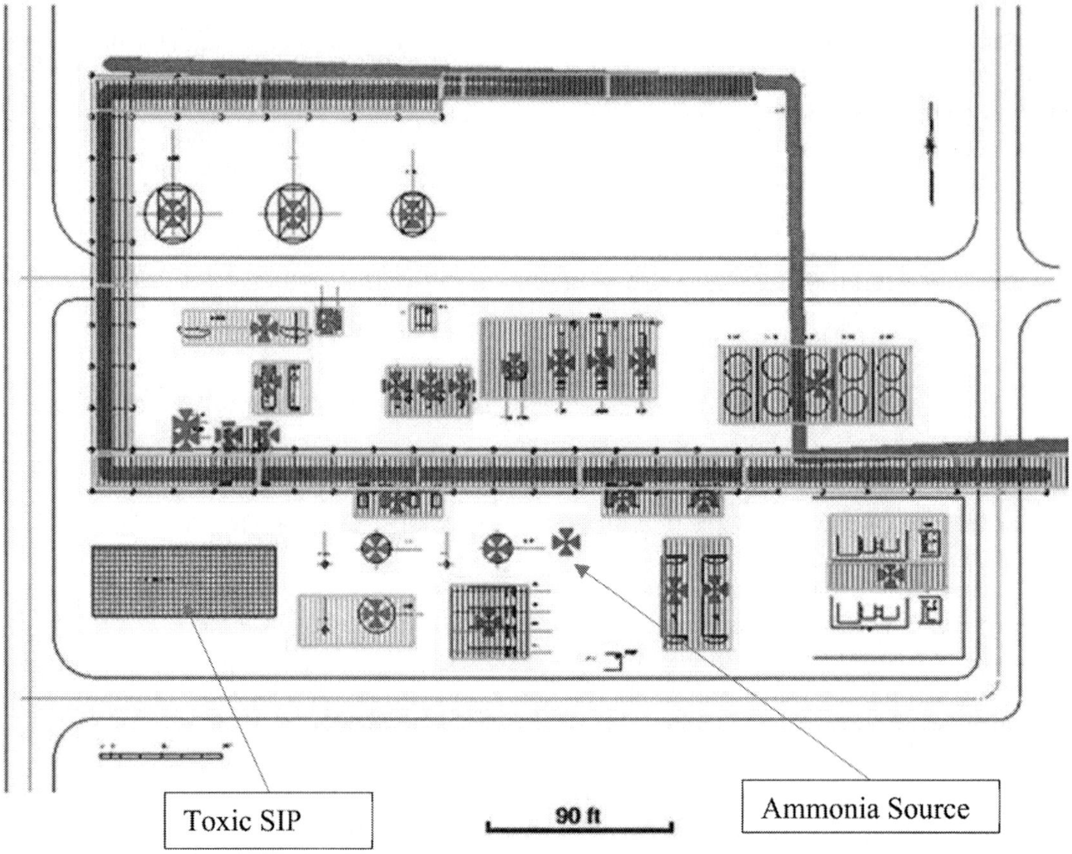

Figure 6. Example of a Plant with Toxic Source and SIP

Note: Red crosses in the figure indicate different release sources

Source Parameters:
Ammonia at Pressure of 150 psig, 70 °F, pressurized liquid

Unlimited inventory, 0.5-inch release at 3.3-ft (1m) elevation

Building Parameters:
Wood Trailer, with HVAC air changes of 0.6 ACH (air changes per hour)

Isolated building (Toxic SIP) 0.5 ACH at 5 mph wind speed

SIP Room inside building 0.15ACH
Egress delay after source isolation: 5 minutes

Case 1 (without detection):
Time to HVAC isolation: 3 minutes

Case 2 Building Parameters (with detection):
Time to HVAC isolation: 1 minute

Evacuation from SIP (Building) at NH_3 concentration: 500 ppm
PPE availability: 95%

Case #	Building: Section	Wind Direction	OV for 15 min - no HVAC isolation	OV for 30 min - no HVAC isolation	OV for 15 min - with HVAC isolation	OV for 30 min - with HVAC isolation
1	Admin - no detection	90	41.1%	76.8%	1.9%	4.7%
1	Admin - no detection	112.5	41.0%	76.8%	1.9%	4.7%
1	Admin - no detection	67.5	39.7%	75.8%	1.7%	4.4%
2	Admin - with detection	90	0.6%	0.6%	0.6%	0.6%
2	Admin - with detection	112.5	0.6%	0.6%	0.6%	0.6%
2	Admin - with detection	67.5	0.5%	0.5%	0.6%	0.6%

Table 1. Building Occupant Vulnerability – Cases 1 and 2

As shown in Table 1 for Case 1, Occupant Vulnerability (OV) resulting in a fatality for 15-minute duration shelter inside the building is approximately 40% for all the dominant wind directions. For 30- minute duration, OV is approximately 75%. In both these scenarios, all personnel are sheltering in place and not evacuating. Also, the HVAC isolation has failed to occur, since this is a manual shutdown and not based on gas concentrations in the HVAC duct.

For Case 1 with successful HVAC isolation, the OV values reduce to <2% and <5%, respectively, for 15- and 30-minute duration sheltering. Since the HVAC was successfully isolated within 3 minutes, the concentration inside the SIP room within the building was estimated to be approximately 1500 ppm. Evacuation is estimated 5 minutes after the source is isolated; total time for personnel exposure is 26 minutes.

As shown in Table 1, for Case 2, OV for 15-minute and 30-minute exposures with and without HVAC isolation is approximately 0.6%. Due to toxic gas detection inside the SIP, personnel evacuate when the indoor concentration in the SIP room reaches 500 ppm. With 95% success probability for PPE, OV is calculated to be significantly lower than Case 1.

Conclusions

Developing toxic SIPs is one strategy that can be implemented to protect building occupants in case of an unintentional release of ammonia or other toxic gas. Toxic gas detectors play a very important role in making a toxic SIP highly effective. It is important to select appropriate detectors for the use, place them in proper locations, maintain them to ensure reliability, and implement automatic HVAC isolation to support SIP effectiveness and backup plan reliability.

References

Jack Chou, I. S. (1999). *Hazardous Gas Monitors, A Practical Guide to Selection, Operation and Applications.* McGraw-Hill and SciTech Publishing.

Predicting Process Risks for Improved Safety and Operational Excellence: A Breakthrough Technology and Case Studies

Most of the incidents and unexpected process failures can be avoided if the operating team gets timely information about developing risks and takes preventative action. While advances have been made in process risk management in the past few years, in practice, there still remain significant information gaps that prevent facilities from practicing proactive process risk management.

Asean Bintulu Fertilizer (ABF), a subsidiary of PETRONAS, produces 700kMt (772,000 short tons) *of Urea annually. Due to the nature of materials involved and interdependency between its operating units, minimizing process safety and reliability risks is the highest priority for ABF's operating team. Although ABF is equipped with the latest tools in process monitoring, these tools have their limitations, particularly, lacking predictive risk capabilities.*

The use of a new software, Dynamic Risk Analyzer™ (DRA™), based on proprietary predictive risk analysis, is outlined in this paper. DRA allows the operating team to assess the risk level of its plant operations and identify early indicators of developing process risks daily. The result – processes are operated with reduced risk profiles and improved capital effectiveness.

ABF adopted the DRA system at their facility in December 2014. Since then, ABF has improved its organizational culture and workflow that involve day-to-day proactive assessment of process risks at their early stages. Real-life operations issues that were detected and addressed early on are discussed.

Aslam Khan B Mohater Khan, Shahrul Anuar B Saat, A Aziz Asri B Sulaiman
PETRONAS Asean Bintulu Fertilizer, Sarawak, Malaysia

Ankur Pariyani, Ulku Oktem, Deborah Grubbe
Near-Miss Management LLC, Philadelphia, PA, USA

INTRODUCTION

Most of the incidents and unexpected process failures can be avoided if the operating team gets timely information about developing risks and takes preventative action. Post-incident investigations show that there are several near-misses that occur before these events evolve (gradually or often, rapidly) to become abnormal situations [1-4]. While advances have been made in process risk management in the past few years, there still remain significant information gaps that prevent facilities from proactively identifying process risks. The latest process monitoring techniques provide clear insights on current operating conditions, however, they are not predictive in nature.

With new advances in information technology, a typical industrial plant operation monitors hundreds (even thousands) of parameters continuously, generating on the order of 10 million to 50 million data points every day. Over the past few years, although facilities have become "data rich", they are "information poor" – this is typically referred to as the 'big data challenge'. There is a great need of tools that can analyze this "big data" and inform plant personnel of changes in the risk profile of the plant over time with indication of potentially evolving disturbances – to help avert unplanned shutdowns and accidents.

Big data is indeed big – typically, over 5 billion data points are recorded every 6 months with just 320 tags (recording sensor measurements every second). It is often characterized by four v's: volume, variety, velocity, and variability, which change with time. Lost in the flood of this *big data* are indicators that can help plants understand the dynamically changing risks and avoid some of the losses the chemical and petrochemical industry experience every year (due to unexpected failures). Research confirms that taking a different-in-kind approach to harnessing *big data* – based on processing the information directly, from a *risk* perspective, with advanced data mining techniques – creates a wealth of insights that were previously unavailable [5]. This has significant potential to transform the way facilities operate, and to reduce unexpected disruptions.

Equipped with insights from this full range of data pinpointing changing risk levels on a dynamic basis, a plant operations team can identify when problems actually start. Going further, plant management and engineers can utilize leading indicators ex-ante to take *proactive* measures to prevent and avoid operational problems. Ultimately, this type of knowledge can help engineers and safety personnel monitor effectiveness of existing risk reduction measures and point out issues early on, so management can allocate resources to the most needed areas.

This paper provides an overview of the current process risk analysis methods and the gaps in the risk assessment landscape. It introduces an innovative approach for proactive risk assessment to help facilities operate safely and reliably. To demonstrate this vision, the use and implementation of Dynamic Risk Analyzer (DRA) software at Asean Bintulu Fertilizer (ABF) operations is discussed, along with some real-life case studies.

CURRENT RISK ANALYSES AND THEIR LIMITATIONS

Improved process risk management is the primary outcome of the widely used Process Safety Management (PSM) standard, which is promulgated by the U.S. Occupational Safety and Health Administration to maintain and improve safety, operability, and productivity of plant operations. While advances have been made in the process risk assessment area in the last decade, there still remain significant gaps causing facilities to continually look for new solutions. The current risk analysis techniques and associated limitations are as follows [5]:

a. Quantitative Risk Assessment (QRA). Typically, QRAs are conducted once every 3-5 years by most facilities. These use various data sources available to the industry, such as incident data, material safety data, and equipment and human reliability data, to identify incident scenarios and evaluate their risks by defining the probability of failure and their potential consequences. They help users identify areas for risk reduction.

Limitations: Because QRA mostly involves incident and failure data – excluding day-to-day process and alarm data that contain information on precursor events – it has limited predictive power. Interestingly, a summary report [6] by the Joint Research Centre and Denmark Risk National Laboratory of the European Commission indicates that risk estimates based on generic reliability/failure databases are prone to biases and could result in large deviations depending on data sources. Their project employed seven partners that conducted risk analyses for the same ammonia storage facility, finding "large differences in frequency assessments of the same hazardous scenarios." For these reasons, the importance of utilizing process-specific databases for *objective* risk analyses has been gaining recognition.

b. Safety Audits. Many facilities conduct safety, health and environmental audits using both internal teams and consulting companies. The frequency and effectiveness of safety audits depend highly on the resource availability for the facility. In most cases, safety professionals with support from engineers, operators and sometimes even managers periodically review operating procedures and safety records, and conduct interviews about safety practices.

Limitations: Formal, in-depth, safety audits are conducted periodically, frequency ranging from once a year (in rare cases more than once a year) to once in several years. An integral part of these audits is to review incident history and observable near-misses that are reported by employees. The latter of which depends upon the safety culture at the facility and may not always provide an accurate picture of risks.

Furthermore, these approaches do not have the capability to monitor the change in the process risk levels in real, or even near, time.

c. Operations Management and Manufacturing Intelligence Tools. Operations management and manufacturing intelligence software provide KPIs (key performance indicators) for performance monitoring of operations, and assessment of availability/effectiveness of equipment. They focus on trending, reporting, and visual analytics of a select data slice, which help users monitor the variability of different parameters in a time period (shift, day, week, etc.).

Limitations: These systems fall short when it comes to *big data* analytics, particularly when users need insights on when parts of operation are becoming riskier and how anomalies are creeping in. This requires comparison of operating conditions with their normal behavior to identify new changes, which is not the focus of these systems. With aging equipment and expected departure of many seasoned operators from the workforce, this handicap becomes even more considerable.

d. Condition Based Monitoring Tools. These tools are mainly designed for equipment condition monitoring. They identify abnormal situations in real- or near-time by comparing equipment's performance with its expected behavior and alerting the user when there is a mismatch. Both model-driven (based on quantitative process models) as well as data-driven tools (based on clustering and dimensionality reduction approaches) are available in the market that help operators take immediate corrective actions as real-time alerts are dispatched.

Limitations: Many of them require continuous efforts in maintaining the baseline. In addition, they often involve remote monitoring and diagnosis of the data at an offsite facility, which is a resource and capital intensive project.

Thus, the need is clear for an approach that helps facilities close the risk assessment gaps that continue to be problematic. Next, a new approach is outlined that identifies risk levels and drivers dynamically and can help busy plant personnel harness the insights in the big data and take appropriate actions rapidly.

TAKING A DIFFERENT APPROACH –DYNAMIC RISK ANALYSIS

Accidents are rare events, which occur when a series of failures of risk management barriers occur in succession – implying a 'chance' factor involved in their occurrences. Post-accident investigations show that there are typically several near-misses or precursors that occur over time before their occurrence [1-4]. This concept is captured in the well-known "safety pyramid". Figure 1 introduces an extended version of the safety pyramid, indicating two categories of near-misses that are precursors to accidents. First are *observable near-misses*, which refer to events that are typically notable by the operations team, such as, equipment failures, leaks, etc. Second are *hidden near-misses*™, which refer to events that can only be detected through rigorous data analysis and are typically not observable to human eye. The novel approach of identifying *hidden near-misses* in the process and alarm databases permits detection of operational problems at their developing stages. This approach forms the backbone of Dynamic Risk Analyzer (DRA) software, which analyzes process data to determine risk indicators.

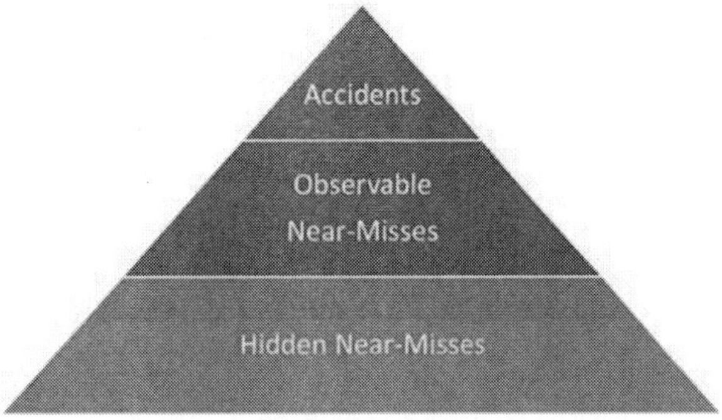

Figure 1. Extended Safety Pyramid

With new advances in information technology, a typical industrial plant operation monitors hundreds (sometimes thousands) of parameters continuously, generating extensive data sets, exceeding the billion mark within a few weeks of operation. The approach of identifying *hidden near-misses* requires advanced data-mining techniques to sift through this *big data* to find new anomalies that often cannot be detected using manual/visual data analysis or engineering models. This information can then used to estimate risk that indicate likelihood of normal conditions escalating to abnormal levels, providing insights on potential performance issues in the operations.

Figure 2 shows the typical evolution of a major event. By identifying the *hidden near-misses*, before the alarms are triggered, the techniques behind DRA software enables the operations team to address the events at their formation stage, before they become visible or threatening. The dynamic risk levels guide plant personnel to the sources of the risk – to deploy the right resources in a timely manner, to plan just-in-time maintenance, and to head off potential problems – several days or even weeks in advance. The results can be made accessible to all users (plant managers, supervisors, engineers, reliability and maintenance crew, as well as operators) to promote transparency among the operating team and to complement their existing predictive maintenance, PSM, hazard identification, and quantitative risk analysis activities.

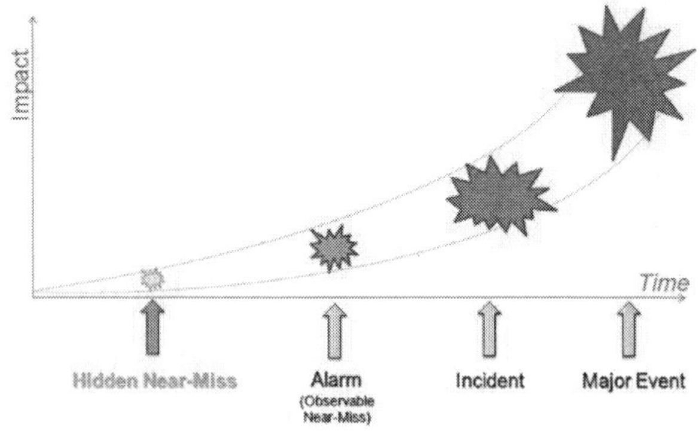

Figure 2. Hidden Near-Miss and Evolution of a Major Event

USE OF DRA AT ABF OPERATIONS

Asean Bintulu Fertilizer (ABF), a subsidiary of PETRONAS, produces 700kMt (772,000 short tons) of urea annually. Due to the nature of materials involved and interdependency between its operating units, minimizing risks related to safety and reliability is the highest priority for ABF's operating team. Unsafe plant conditions may lead to process safety incidents, which in the worst case, can cost lives, risk the business, and jeopardize the reputation of the organization. Similarly, downtime, due to unreliability of equipments, can lead to opportunity loss. ABF is equipped with the latest tools in process monitoring, such as Alarm Management System (AMS) and Plant Management Information System (PIMS), and historians, however, these tools are not predictive in nature. In real life situations, most of the time critical alarms are triggered too late for any proactive corrective actions to be taken.

In 2014, ABF decided to change the status quo and adopt a proactive organizational culture and workflow. The vision was to have a leading indicators system that can inform the operations team about plant risks early on, giving them sufficient time to take strategic corrective actions and confidently maintain safe and reliable operation.

The first step towards accomplishing such resilience was selection of new software, Dynamic Risk Analyzer (DRA), as the process risk management platform at their facility. Instilling awareness and responsibilty was given a high priority during the DRA implementation. The user base ranged from top management to the frontliners in the plant i.e shift superintendent, operation technicians, etc. Focus groups were formed to formulate detailed training program for each group, ranging from basic principles to overall vision. A series of training sessions were carried out during normal working days and shift hours to cover all identified key personnel. A simple step by step guideline on the tool was also developed so that the knowledge can be transferred and stored within the organization. Easy web-based access ensured the risk results to be accessed from anywhere within the organization's network.

During the installation and training, the focus remained on cultivating a *proactive* response culture, that is, once a process issue is identified by DRA and confirmed by the team, the next step involved coming-up with a solution to reduce or eliminate the risk.

ABF believed that shifting into proactive approach in dealing with risks and threats could not be achieved just by attending the training and learning about the system alone. It had to be embedded into ABF work flow to ensure continuity, sustenance and improvement. Over the past 1.5 years, ABF management has worked hard to achieve this vision. The leading indicators are now reported and discussed during the morning Area Based Team (ABT) meetings as a permanent agenda item. Based on the outcomes from those discussions, a mitigation action plan is devised and status updates are reported in subsequent meetings until the change is normalized or fixed. A mitigation that requires additional resources (particularly, monetary) is escalated to the ABF Management.

With this arrangement, ABF now confidently maintains the safety and reliability of its plant facilities, avoiding incidents and downtime, and importantly, remains a highly profitable organization. Figure 3 shows the risk schematics in the DRA software. Through analysis of large volumes of past and current process data, the latest plant risk indices are delivered to the operating team, along with risk status for individual tags. This allows them to drill down and conduct detailed resolution of the risk drivers on a daily basis.

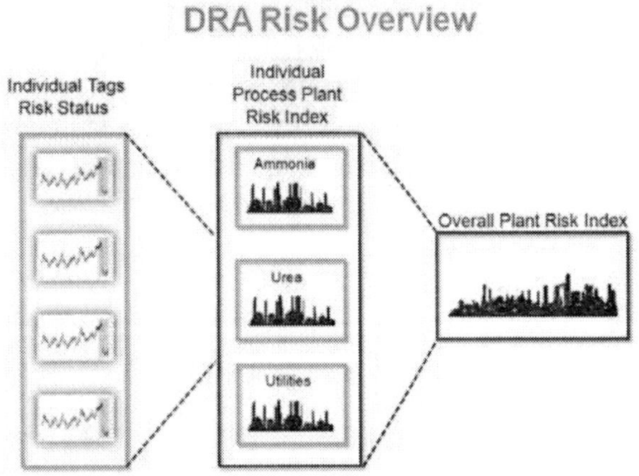

Figure 3. DRA Risk Overview

CASE STUDIES

Since embarking on utilizing DRA in daily risk assessment, ABF has benefited significantly by early identification of plant problems leading to avoidance of potential process safety and plant reliability issues. Herein three case studies are considered as examples:

(a) Early Identification of Level Control Valve Failure.

Figure 4 shows recent plot of a level control valve opening from mid-January to mid-March (60 day). This control valve is located in the tail gas recovery section operating at 115 barg.

As evident, it started experiencing a decreasing trend in early February although there was no change in the associated level indicator. This behavior was highlighted through the risk indicators generated by DRA and prompted the maintenance team for investigation.

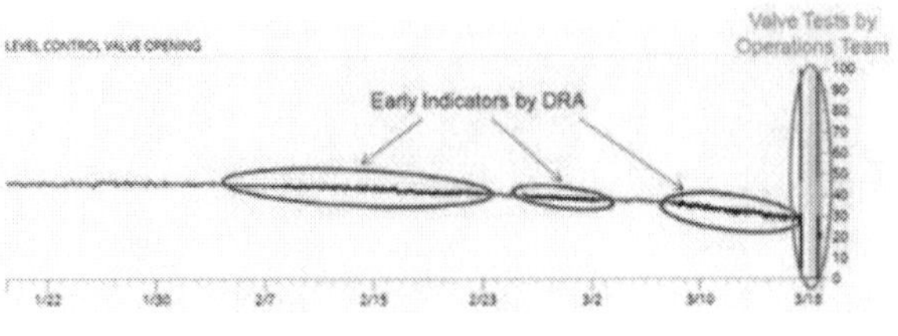

Figure 4. 60 day plot of level control valve opening – along with early indicators pointed by DRA

By pointing out this potential problem early on, this risk information provided the maintenance team with ample time to make the necessary preparation to service the valve and hence, avoid any major problem with regards to production and safety as the high pressure gas could breakthrough to next equipment which is operating at much lower pressure.

Figure 5 shows the valve plug with some erosion (picture taken during servicing).

Figure 5. Valve plug

(b) *Early Identification of Faulty Controller for Ammonia Tank Base Heater Temperature.*

Maintaining base heater temperature above 0°C (32°F) is essential in ensuring the integrity of the ammonia storage tank. Low tank base temperature may result in cracks, which can cause potential release of ammonia into the environment (major *loss of primary containment*). In April 2015, one of the ammonia storage tank base heater temperature indicator started showing risk indicators – see the 60 day plot in Figure 6. This triggered the

Area Based Team (ABT) to review the temperature indicator. The investigation led to finding of faulty temperature controller and subsequently, its rectification. Their findings also triggered checks on other temperature indicators and controllers – ensuring their reliability as well.

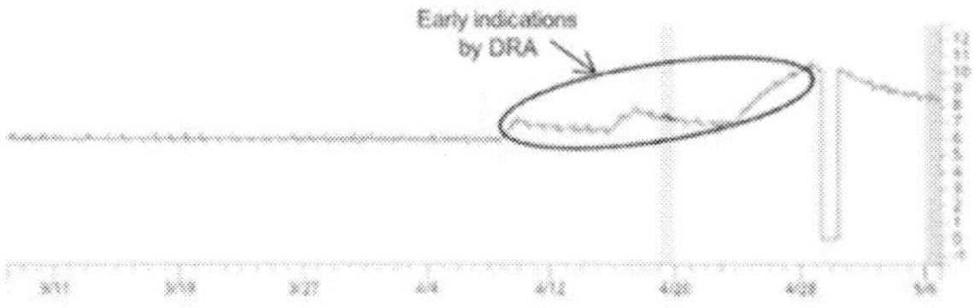

Figure 6. 60 day plot of ammonia tank base temperature

Ammonia tank base temperature readings, although available in PIMS and DCS, were not normally monitored as critical parameters. Prior to this, changes in the ammonia tank base temperature (before reaching the alarm levels) would not be easily detected as that depended on the keen eyes or vigilance of panel operators.

(c) ***Early Identification of Cooling Tower Fan Failure.***

Although rotating machine vibration is one of the critical parameters monitored in the process plant, most of the increasing vibration trends are identified once the readings reach the high alarm. When the high alarm is triggered, it is only a matter of time before the process experiences a sudden failure either due to actual rotating machine failure or triggering of high-high fail-safe protection system. As a result, this allows very limited time for any intervention plan to be worked out properly and implemented.

Figure 7 shows the 120 day plot of cooling tower fan vibration. The recurring risk indications by DRA signifying gradual increasing trend in the vibration readings prompted the operations engineers to investigate these early warnings indicating potential failure of one of the cooling tower fans. This led them to plan a local shutdown for the fan for checks and rectification. At the end, this early identification of the increasing vibration allowed for better work planning and execution, subsequently reducing the fan downtime in the long run.

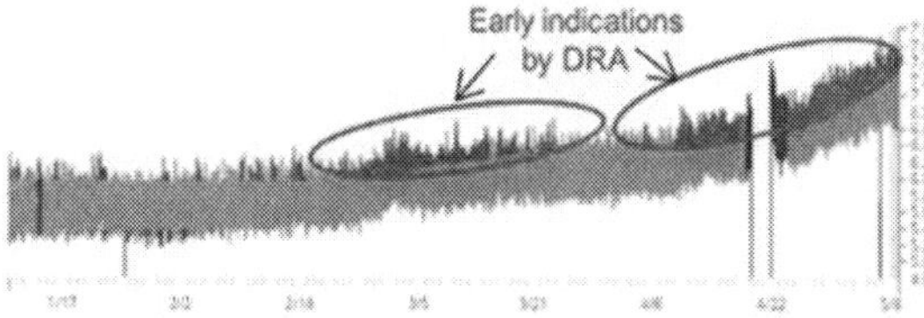

Figure 7. 120 day plot of cooling tower fan vibration

In general, there are many such process conditions, which are difficult to detect through the use of routine trending and visualization techniques – that is, if they are tracked to begin with. The use of DRA technology can provide tremendous value in highlighting issues that are lost to other approaches and not observable to the human eye.

CONCLUSION

This paper outlined a case study of Asean Bintulu Fertilizer (ABF) – its transformation process and steps in achieving a proactive risk management culture and workflow through the use of *dynamic risk analysis* technology. To demonstrate the impact of this change, three real life use cases are presented where the engineers detected hidden operational issues early on.

After running DRA for more than a year, the following observations have been made by the ABF team:

a. Closer co-operation exists between Operations and Maintenance now, as both parties have observed and worked on some important process issues in the past year.

b. People are becoming more proactive as they believe that the problems are more focused now and they can identify them much earlier.

c. Shifting into a proactive mindset in dealing with risks could not be achieved alone via knowledge and know-how of the subject. It had to be embedded into ABF work process to ensure continuity, sustenance and continuous improvement. DRA has played an important part in ensuring ABF's journey to achieve this goal.

REFERENCES

[1] Kleindorfer P. R., *et al.*, "Accident Epidemiology and the U.S. Chemical Industry: Accident History and Worst-Case Data from RMP*Info," *Risk Analysis*, 23 (5), pp. 865–881 (2003).

[2] Kleindorfer P., Oktem U.G., Pariyani A., and Seider W.D., "Assessment of catastrophe risk and potential losses in industry," *Computers and Chemical Engineering*, 47, 85-96 (2012).

[3] Pariyani A., Seider W. D., Oktem U. G., and Soroush M., "Incidents Investigation and Dynamic Analysis of Large Databases in Chemical Plants: An FCCU Case Study," *Ind. Eng. Chem. Res.*, 49, 8062-8079 (2010).

[4] Phimister J. R., Oktem U. G., Kleindorfer P. R., and Kunreuther H., "Near-miss incident management in the chemical process industry," *Risk Analysis*, 23(3), 445–459 (2003).

[5] Pariyani A., Oktem U. G., and Grubbe D.L., "Process Risk Assessment Uses Big Data," *Control Engineering*, June Issue (2013).

[6] Lauridsen K., Kozine I., Markert F., Amendola A., Christou M., and Fiori M., "Assessment of uncertainties in risk analysis of chemical establishments," Summary Report on ASSURANCE project, Risk National Laboratory, Roskilde, Denmark (2002).

Validating Multiple Credits from Your DCS... or Not

The recently released Center for Chemical Process Safety (CCPS) "Guidelines for Initiating Events and Independent Protection Layers for Layers of Protection Analysis" has generated a new round of discussion regarding the amount of credit that can be taken from a DCS. Both new and old CCPS guidelines provide A) a conservative approach where a DCS loop, alarm, or shutdown can only show up once in a scenario, either as an initiating event OR as a protection layer, but also provide for B) a less conservative alternate approach that allows for a DCS loop, alarm, or shutdown to show up twice in a scenario, in a combination of initiating events and/or protection layers.

The newest CCPS release includes additional guidance for requirements that should be confirmed before the less conservative alternate approach can be claimed, such as administrative controls, and assessment of common cause failures. While the current CCPS book suggests possible considerations, readers are left to determine for themselves the most reasonable approach to use. This paper identifies some specific considerations and methods for validation that end-users can consider.

Peter G. Hereña
Baker Engineering and Risk Consultants, Inc. (BakerRisk)®

Introduction

Layers of Protection Analysis (LOPA) is a semi-quantitative risk analysis method that has been practiced under various names since the late 1980s and was adopted in the late 1990s as a widely recognized method for assessing performance requirements for Safety Instrumented Systems (SIS). The 2001 seminal book on the topic is CCPS' Layers of Protection Analysis (ref. 1). The core principle of the LOPA method defines an initiating event as one that results in an unwanted consequence (a process safety hazard). Once the initiating event is defined, one or more Independent Protection Layers (IPLs) that can prevent or mitigate that defined hazard can be identified. An assessment is then performed to determine if sufficient protection layers exist, such that a certain acceptable risk threshold is passed.

Industry has found LOPA to be an effective risk identification method to be used in parallel with other types of purely qualitative process hazard analyses, such as the Hazard and Operability (HazOp) method, which is particularly effective for major process safety hazards that can occur at extremely low frequencies. LOPA is also linked with the familiar concept of the "defense-in-depth" or "onionskin diagram" (Figure 1) which asserts that numerous and diverse types of safeguards will lead to more effective overall risk reduction.

One frequently discussed issue among LOPA practitioners is how to measure the risk reduction effectiveness provided by industrial control systems, also known as Basic Process Control Systems (BPCS). The BPCS can have varying designs, but process plants typically utilize some type of "DCS" system to provide regulatory control for normal operations. The broader BPCS term will be used in this paper.

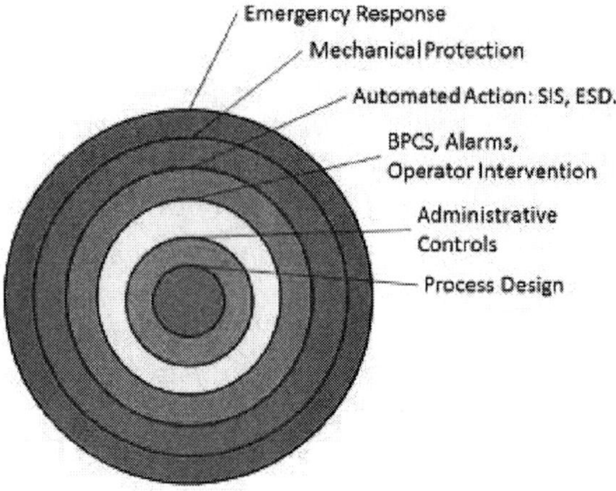

Figure 1. The LOPA Onionskin Diagram

Layers of Protection Analysis (LOPA)

Traditional methods such as HAZOP might require a team to qualitatively weigh all factors at once when determining the risk posed by a certain hazard. While most teams are fairly capable of identifying the frequencies of common events, estimation of rare events like explosions or catastrophic releases pose a much tougher challenge. Teams tend to overstate the frequency of such events to extremes of likelihood to occur/not occur. Teams also can struggle with the concept of an event that could happen at a statistical rate of one in a hundred thousand or one in a million years. LOPA can be used as a more objective method of estimating the frequency of a rare, severe event. By quantifying the initiating event frequency and the IPL effectiveness more realistically, a more consistent and defensible risk judgment can be made.

This is not to say that LOPA is free from potential pitfalls. Problems of consistency can occur, especially when the assembled teams do not have sufficient guidance in determining IPL effectiveness or initiating event frequency. A team might place too much faith in a particular protection layer that, in reality, is quite weak, or they might apply excessive additional protection layers without understanding that some types of protection layers are inherently very reliable. For this reason, CCPS recently published a book - "Guidelines for Initiating Events and IPLs in LOPA" (ref. 3) - which provides conservative reliability and frequency numbers that can be applied in LOPA studies.

Companies have varying methods by which to apply IPLs to quantify risk. The most common determines that for each applicable IPL "credit", the risk is reduced by one order of magnitude, and the Probability of Failure on Demand (PFD) is also reduced. Table 1 describes the relationship between LOPA credits, PFD, and risk reduction factors. Table 2 provides examples of typical number of credits applied for some common protection layers.

LOPA Number of Credits	PFD	Risk Reduction Factor (RRF)
1	0.1	10
2	0.01	100
3	0.001	1,000

Table 1. IPL Equivalencies in LOPA

Independent Protection Layer (IPL)	Typical Credit Provided
Safety Instrumented System (SIS)	1 to 3, depending on SIL rating
Basic Process Control System (BPCS) Control loop	1
BPCS Alarm	1
Pressure Safety Valve (PSV)	2

Table 2. Some Example IPLs and Typical Credits

IPL Independence

One of the more contested topics among LOPA practitioners is the definition and characteristics of **Independence**. In truth, there is no IPL that is absolutely independent from another. For example, all IPLs at a site can be subject to human error in design, management failure to provide maintenance resources, or site environmental conditions.

There are clear cases where credit cannot be taken because of insufficient independence, such as when an alarm is taken from the same transmitter that fails as part of a BPCS control loop (Figure 2). If the transmitter LT-2 has failed such that the signal to the BPCS remains constant, the actual process condition could go critical but the alarm LAH-2 would not activate. Thus, no likelihood reduction credit should be allowed for the alarm in that example.

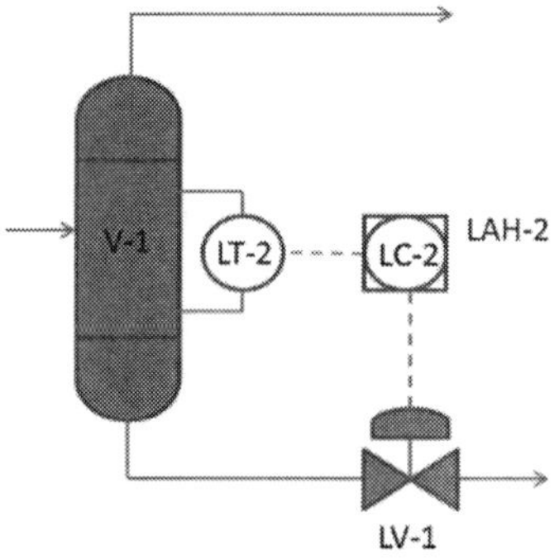

Figure 2. Example Feed Separator

In contrast, consider the separator (Figure 3) where the level is again controlled by a BPCS in loop LC-2, and the hazard of concern is loss of level in the tank. In this case though, there is an independent shutdown system I-3 with separate level instrument and separate valve to prevent loss of level. If the primary level control loop were to fail, the I3 shutdown system should still be available to prevent the low level hazard. Although in reality there are human, environmental and management factors that can simultaneously defeat both systems, most practitioners would consider the shutdown system sufficiently independent to take account of the system in reducing scenario frequency.

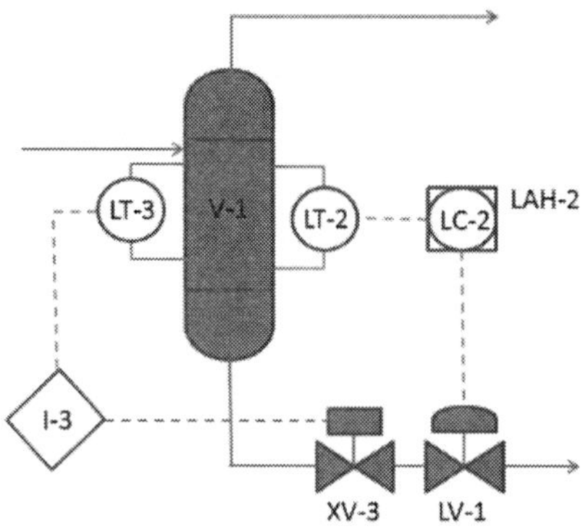

Figure 3. Example Independent IPLs

However, there is a grey area that is not so clear. In a third example, let us consider the previous separator, but the shutdown control function is also performed through the BPCS instead of a completely different logic solver (Figure 4). Does the shutdown BPCS loop qualify as sufficiently independent?

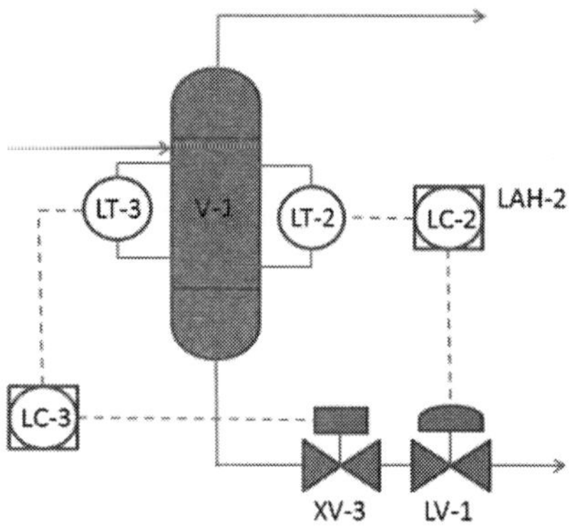

Figure 4. Control and Shutdown in Same BPCS

Or consider another example such as in Figure 5 where an alarm comes from a separate sensor as the BPCS loop that failed, but the control could share some common processor components with the failed loop. Many companies, teams, and experts would have different technical arguments, with effective points to be made for taking full credit for the added alarm, and equally effective points on the side of denying any additional credit for it.

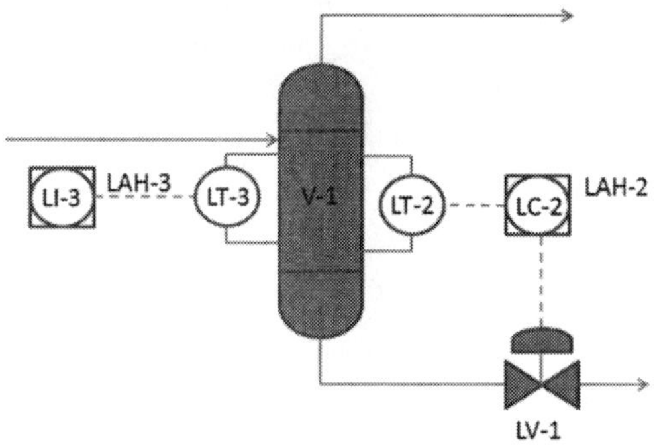

Figure 5. Control and Alarm in Same BPCS

Evaluating BPCS Credits in LOPA: "Approach A" vs "Approach B"

A philosophical difference in applying BPCS credits is inherent in both the original CCPS LOPA book, and the more recently released CCPS Guidelines for IPLs in LOPA, which includes additional guidance on the subject. The books present two different approaches—identified therein as "Approach A" and "Approach B"—for operating companies to consider. The more conservative Approach A allows the BPCS to be credited a single time in the scenario, while the less conservative Approach B allows the BPCS to be credited twice in a scenario. The two approaches are described as follows:

Approach A:

- If the initiating event is in the BPCS, no BPCS IPL credits may be taken unless all components (transmitter, I/O, logic solver, output) are independent of the initiating event.
- If the initiating event is not in the BPCS, a maximum of a single BPCS IPL credit may be taken for the scenario. Additional BPCS IPL credits are prohibited unless all components (transmitter, I/O, logic solver, output) are independent of the initiating event.

Approach B:

- If the initiating event is in the BPCS, one BPCS IPL credit for the scenario may be taken, provided the loops have been designed to minimize common cause failures, and administrative controls are strictly followed
- If the initiating event is not in the BPCS, a maximum of two BPCS IPL credits may be taken for the scenario, provided the loops have been designed to minimize common cause failures, and administrative controls are strictly followed.

To provide an example of the difference between Approach A and Approach B, consider again the system depicted in Figure 5. If Approach A is followed, and if the cause of the hazard is control loop failure, then no credit can be taken from the alarm because it shares some components with the control loop (the I/O or logic solver). If Approach B is followed, then credit could be taken from the alarm if an assessment was performed to indicate proper design, common cause faults were minimized, and the system is resistant to unauthorized modifications. Approach B allows for more risk reduction credit from the BPCS, which can lower the corresponding requirements of other protection layers such as SIS, but it does so at a cost of additional performance assumptions to the BPCS.

Note that neither reference allows a third credit for BPCS under any circumstances when applying LOPA as a risk analysis method. Doing so would presume the BPCS can have the same level of performance as a top end SIL-3 rated safety PLC! Taking additional BPCS credits should only be attempted if using a very advanced model, which would include a combination of fault tree, human reliability analysis, and a detailed evaluation of failure modes and effects. LOPA is simply not up to the task of those quantitative methods.

Considerations in Claiming Approach B

Listed below are some of the many potential concerns with claiming CCPS LOPA Approach B with respect to BPCS credits. Discussion follows for each concern listed below.

1. Common modes of failure between the loops may be overlooked.
2. The BPCS may not be designed with sufficient reliability or diagnostics to claim a high performance factor of two IPL credits.
3. Maintenance practices, management of change, and security access deficiencies may simultaneously defeat multiple BPCS protection layers.
4. The Process Safety Information (PSI) of many BPCS loops and alarms is not well documented.
5. The IEC standard IEC-61511 limits the credit that may be taken by a BPCS to a single order of magnitude risk reduction (a single credit), or an initiating event frequency of no less than approximately 0.1 /yr.

Consideration 1: Evaluate Common Failure Modes

Upon inspection, often the protective devices of a system are not as independent as initially believed. As an example, a fired heater (like a reformer) has many types of alarms and shutdowns to prevent hazards in the firebox, but all of those protection layers depend on the logic solver and the fuel gas valves closing, in response to shutdown conditions.

To assess the impact of common failure modes, performing a quantitative calculation using a Fault Tree Analysis (FTA) or Failure Modes Effects Analysis (FMEA) are two excellent options. A FTA is a logical model describing how system subcomponents can individually fail in a way that results in an unwanted event. It is capable of modeling complex interactions where some devices perform dual service (example, an I/O card that communicates an alarm and a control loop signal). A FMEA defines system failure modes and their effects on the overall operability of a system. It is sometimes easier to apply because it traces incipient failure conditions more directly to their impacts. FTA and FMEA are nearly always quantitative reviews, providing either a probability of failure of system, or a frequency of an unwanted event. Quantitative FTA and FMEA are reproducible, tunable to account for good control system design decisions, and capable of defining system and testing requirements.

The details of the FTA or FMEA will depend on the specifics of the BPCS design. Some types of manufacturers and control system models have higher inherent independence between components. If it is not practical to perform an analysis for every combination of BPCS protection layers, at the minimum, some representative models can be established to define necessary BPCS characteristics. For example, a FTA or FMEA can help the user determine if the two BPCS protection layers must be on different I/O cards in order to take credit for both.

Consideration 2: Design the BPCS with Effective Diagnostics

A BPCS with strong diagnostics will allow potential dangerous failure modes to be discovered and corrected before a demand is placed on the IPL. Of particular concern are the failure modes that would prevent the IPL from performing its primary safety requirement. The user should assemble a diverse team of process, operations, instrumentation, and software subject matter experts that can perform a FMEA or qualitative assessment, listing common BPCS failure modes and asking:

- Would this fault result in a dangerous operating mode?

- Are there any BPCS diagnostics that could detect the fault?
- Is that diagnostic mode utilized (enabled)?
- What is the operations/maintenance response to the fault?
- If the fault is detected can it be repaired within a reasonable timeframe?

Failure modes should consider the signal to and from field devices, Input/Output (I/O) component failures, motherboard, backplane and rack failures, card communication failures, and the Programmable Logic Controller (PLC) itself. Any major gaps should be addressed, and the essential diagnostic requirements should be documented. Below is a minimum list of faults the review team should consider:

- Out of range (high and low) signal from the field
- Field diagnostic failure signal
- I/O card failure
- I/O point failure
- I/O card communication failure
- I/O card power supply out of acceptable range
- PLC self-check failure
- Frozen/hung PLC
- PLC memory failure
- Logic failure
- PLC temperature out of range

Consideration 3: Confirm BPCS Administrative Controls Will Be Effective

BPCS administrative policies include the effective practices for system design, maintenance, bypass, security, and management of change. A good reference text that discusses best practices for control system design is CCPS Guidelines for Safe and Reliable Instrumented Protective Systems (ref. 4). Although these requirements are geared toward safety systems, the concepts apply to a well-designed BPCS.

The administrative policy assessment will almost necessarily be qualitative; however, there will be discussion of some potential avenues to measure program effectiveness. The team of subject matter experts (SME) should review how personnel will interact with the BPCS throughout its lifecycle. Table 3 below includes some example administrative programs and possible concerns:

Administrative Program	Question
Design	What checks are in place to ensure proper field equipment specification is done?
	Is there an alarm management program in place, and if so, is it effective?
	What is the baseline alarm burden to the operator?
	What are peak alarm burdens to the operator during a significant process disturbance?
Maintenance / Testing	What is the frequency of I/O, communication card, or logic solver failures?
	How are test results documented?
	When a component is found to have failed during a routine test, what countermeasures are put into place until the repair is complete?
	Is maintenance results information periodically reviewed to assess equipment failure rates?
	Are maintenance/testing procedures in place, and if so, are they followed?
Bypass / Manual Mode	Can control loops or alarms be bypassed without approval?
	Have bypass control measure historically been effective?
	At any time, approximately what percentage of control loops or alarms are left in bypass or manual mode?
	What is the typical duration for a control loop or alarm to be left bypassed or in manual mode?
Security	Can alarm or trip setpoints by changed without approval?
	Can BPCS software, logic or configuration be changed while operating? What controls exist?
	Has a cybersecurity assessment been performed to prevent unauthorized access from outside parties?
	Have security measures been historically effective?
Management of Change	What types of change to the BPCS would trigger an MOC?
	Have previous PSM audits identified MOC concerns that apply to BPCS effectiveness?
	How are PLC and I/O firmware updates controlled?

Table 3. Assessing Human Factors in BPCS Effectiveness

As a team discusses the answers to the question, they can use this information in several ways.

- A ranking or weighted grade scale can be applied to each question, which is tallied to yield a numerical "score" of the site BPCS control policy. Multiple credits may be taken if a score exceeds a threshold value
- If the team identifies obvious program gaps, they can recommend that the gaps be closed before Approach B for IPL credit is taken.
- The team might identify specific systems or units where Approach B assumptions may not apply. Multiple credits from the BPCS may not be taken for those systems or units until upgrades are performed.
- The team might identify a minimum checklist of questions that the LOPA facilitator should confirm prior to invoking Approach B.

Consideration 4: Ensure Process Safety Information is Complete

A LOPA, or any other type of process hazards analysis, should confirm that a minimum set of Process Safety Information (PSI) is in place. The LOPA team in this case should confirm that BPCS-specific PSI is available. Important control PSI includes, at minimum:

- Field instrument specification sheets
- BPCS loop drawings
- A control narrative or logic description for each major unit or piece of equipment
- List of alarm and trip setpoints
- Cause and effect diagram or equivalent explanation of automated BPCS actions
- If the system is complex, a Safety Requirements System in accordance with IEC 61511 Section 10 may be necessary

Consideration 5: Limitations per the IEC 61511 Standard

IEC 61511 (ref. 5) is an international standard related to the design of Safety Instrumented Systems (SIS) for the process industries. IEC 61511 outlines SIS design requirements and some limitations when considering the effectiveness of automated systems in preventing major process releases. IEC 61511 clauses 8.2.2 and 9.4.2 limit the initiating event frequency and number of BPCS credits that can be taken for a single scenario. It requires that, when multiple credits are taken from a BPCS as part of a risk assessment like LOPA, the BPCS is designed in accordance with IEC 61511. Although this may seem like an impossibly high bar to cross, by diligently following the considerations outlined in this paper, the operator will be a long way toward justifying multiple BPCS credits in accordance with CCPS Approach B. Many of the IEC 61511 fundamentals are conceptually outlined in the previously discussed considerations.

How BPCS Planning Affects Site Policy

An outcome of the above considerations can be used to drive the needs to increase BPCS effectiveness. Some outcomes, as described earlier, can lead to increased focus on BPCS in better alarm rationalization, or minimum hardware requirements, managing bypasses more effectively, collecting more complete BPCS failure rate data, or strengthening other programs that are related to BPCS.

Some control system guidance documents, such as the ISA Technical Report 3, "Mechanical Integrity of Safety Instrumented Systems," (ref. 6) recommend periodic spot auditing of effectiveness of SIS programs. This concept also applies to BPCS, especially when a high safety performance is tied to the proper functionality of the BPCS. The auditing can be carried out standalone, or as part of other audits. For example, if the site already has a bypass management audit system, the results of that program can drive additional BPCS enforcement or training needs.

Other industry guidance, like EEMUA 191 "Alarm Systems – A Guide to Design, Management and Procurement", specify predictive and reactive metrics (benchmarks) that can be applied to monitor the health of BPCS alarm systems over time. Although the guidance is written in consideration of alarms, the principles apply more broadly to any type of BPCS loop, alarm or shutdown action. Some of these metrics include:

- Average alarm rate
- Peak alarm rate
- Average number of standing alarms
- Average number and duration of shelved/silenced alarms
- Nuisance alarm rate
- Fraction or number of alarms bypassed at any time

Conclusions

The application of BPCS credits in LOPA is a topic of long-standing discussion. Some operating companies strictly limit the number of BPCS credits in risk reduction that can be taken, preferring to implement high-performance prevention systems like SIS. Other operating companies may not find it feasible to ignore the real-world effectiveness of BPCS functions in preventing process incidents. Instead they prefer to take credits that are justifiable, and reduce the requirements of other IPLs.

This paper outlines some of the considerations that should be discussed prior to the LOPA meetings, which will help frame the limits of risk reduction afforded by the BPCS. It also provides some example discussion topics to be shared with a technically diverse team of subject matter experts who can assess the site's effectiveness in BPCS design, operation, and change management. The reader is encouraged to use these discussion topics as a starting point, to be expanded as needed, based on site history and process complexity. Proper application of these consideration topics will underscore where a LOPA team may take multiple credits in a scenario for their BPCS, and where they should limit themselves to only a single credit without applying additional BPCS modifications.

References:

1. Layers of Protection Analysis – Simplified Process Risk Assessment (CCPS, 2001)
2. Straight Choices: The Psychology of Decision Making (Newell et. al, 2015)
3. Guidelines for Initiating Events and Independent Protection Layers in Layers of Protection Analysis (CCPS, 2015)
4. Guidelines for Safe and Reliable Instrumented Protective Systems (CCPS, 2007)
5. IEC 61511 – Functional safety - Safety instrumented systems for the process industry sector (International Electro-technical Commission)
6. Technical report ISA-TR84.00.04-2012 Mechanical Integrity of Safety Instrumented Systems (ISA, 2012)
7. EEMUA 191 "Alarm Systems – A Guide to Design, Management and Procurement", (Engineering Equipment Materials Users Association)

Ammonia distribution and handling improvements

Following an incident in 2011, an independent engineer made recommendations which had to be implemented to Orica's Kooragang Island Site pressurized ammonia storage systems to gain government approval to restart our plants.

Subsequently further changes were made to the ammonia distribution and handling systems to improve their inherent safety, reduce the ammonia exposure risk, odor impact on surrounding communities and ensure the site's social license to operate. The changes are detailed in the paper below.

Peter McGrath, Scott Sandhoff and Ajay Joshi
Orica Australia Pty Ltd

Introduction

Orica Australia Pty Ltd operates a site on Kooragang Island in Newcastle, Australia which includes an Ammonia, three Nitric Acid and two Ammonium Nitrate plants. Operations commenced in 1969. The site is surrounded by the Port and City of Newcastle. Industrial neighbors are immediately adjacent to the site boundaries and the port's general cargo and coal loading facilities are nearby. The nearest residence is 800 m away.

Background

Prior to 1998 the site's ammonia distribution and handling system was a combination of the original design, many cycles of plant debottlenecking and various minor improvements to address particular issues.

In 1998 an incident occurred when the control valve on the Ammonia plant refrigeration purge failed to open during start-up causing a refrigeration relief valve to open for a short period of time. The odor was detected by a sensitive neighbor approximately 2.5 km (1.5 miles) from the site, resulting in concern and two people being transported to hospital for assessment. The environmental regulator prosecuted the sites previous owner and a penalty was imposed by the court.

As a result of this incident a site program was started to identify all the potential points where releases of ammonia could occur. A detailed study of each emission point quantified the frequency, the amount released and the likely downwind concentration in populated areas. A risk grid was devised to help decision making for the various emission points. For those emissions which did not meet the required frequency-consequence standards, improvements were devised to reduce either the frequency or the consequences, or both. Across the site this resulted in several significant modifications, which mainly involved directing vents to high points. A paper was published on this project at the 2001 AIChE Safety in Ammonia Plants and Related Facilities Symposium [Ref 1].

The 2011 event

Summary

In 2011, while transferring ammonia from the atmospheric storage tank to a pressurized ammonia bullet, a relief valve on the bullet operated briefly a number of times. This resulted in ammonia being discharged from a 48 meter (157 foot) high vent.

Ammonia odor was detected approximately 1.6 km (1 mile) from site in an industrial area of a Newcastle suburb. It caused significant community outrage and the environmental regulator prevented the use of pressurized ammonia storage vessels at the site until the incident had been investigated and actions implemented to prevent a recurrence. As a result of this the production of ammonium nitrate ceased for approximately 5 weeks until the actions had been completed.

The Outcome

Following the incident, Orica committed to an upgrade of the site's ammonia storage and handling systems to ensure the community will not be impacted by site events or operations involving the potential release of ammonia to the environment. This project was given the name Ammonia Management Improvement (AMI) Program.

This paper describes the process to gain regulatory approval to restart production, options considered to reduce the site's ammonia release risk profile and the final implemented solution. The main features of this solution are:

- elimination / simplification of ammonia handling systems
- reduction of pressurized ammonia inventories
- capture of routine ammonia vent flows
- destruction of emergency vent flows
- improved ammonia detection and isolation capability
- multilayered control and protection systems

Details

In November 2011 the Ammonia plant was off line and in accordance with plant procedures three pumps were being used to supply ammonia to the nitrates plants from the atmospheric storage tank. One pump was being used in a manual batch process to periodically fill No.1 ammonia bullet, as the pump's minimum flow was higher than the demand from the bullet.

The pump discharged into the bottom of the ammonia bullet and the filling rate exceeded the rate at which vapor could condense in the tank. This resulted in an increase in pressure, which lifted the relief valve briefly several times.

The relief valve discharged into a high point stack and ammonia odor was detected in an industrial area of a Newcastle suburb, 1.6 km (1 mile) from the factory. A number of workers reported symptoms consistent with exposure to elevated concentrations of ammonia.

A contributing factor to this event was that the nitrates plant distributed control system (DCS), allowed operators to disable alarms. At the time of the event, the three high pressure alarms on the ammonia bullet and flow indicator on the ammonia vent system had been disabled at various times during the previous two days. Operations personnel were subsequently unaware that elevated pressure had occurred in the ammonia bullet and that the relief valve had operated.

Orica was made aware of the incidents when contacted by the local fire authority. Orica immediately ceased the filling operations and brought the nitric acid and ammonium nitrate plants offline. The environmental regulator issued Orica with a notice that prevented it from using all pressurized ammonia feed systems onsite until it was satisfied that the event wouldn't reoccur.

Gaining approval to restart

The incident resulted in public outrage. The government through its regulatory agencies set up the Orica Start-up Committee, to oversee the actions required prior to Orica using the pressurized ammonia bullets again. The committee consisted of representatives from Safe Work NSW (the government department responsible for overseeing industrial safety and major hazard facilities), Environmental Protection Authority (EPA), NSW Department of Planning, Newcastle City Council, Port Stephens Council, NSW Ministry of Health, NSW Police and Fire and Rescue NSW.

The committee required Orica to engage an independent engineer to review the pressurized ammonia feed systems on site and recommend improvements to prevent a re-occurrence. Once Orica had completed the improvements, the independent engineer audited their implementation against his recommendations, before recommending to the committee approval for startup. When the start-up committee was satisfied the prevention notice by the EPA was revoked and the restart of the nitrates plants was able to commence.

The independent engineer's recommendations required to be completed prior to restart were to:

- install a high integrity trip valve operated by the trip system on the ammonia bullets
- upgrade the existing trips on the ammonia bullets
- reconfigure alarms for improved trip monitoring
- introduce more stringent procedures to ensure that alarms could not be disabled without proper review
- change the batch filling procedure to better manage and match pump outputs and production capacity
- train all operators in the modified equipment and updated operating procedures.

This work was completed to the satisfaction of the independent engineer and the start-up committee, with production resuming in December 2011.

Medium and long term actions were also recommended by the independent engineer. Progress on these actions had to be communicated to the regulators on a regular basis. These actions included:

- committing to the upgrading of the nitrates distributed control system (DCS) to include effective alarm management tools.
- updating procedures for emptying and refilling the ammonia bullets to prevent overpressure during start-up and shut-down
- reviewing and making significant changes to the emergency procedures to ensure they dealt with ammonia releases
- conducting a layer of protection analysis on the ammonia vessels to determine the required integrity and developing an action plan to address any additional improvements
- installing a high rate of change level alarm on the feed tanks that provides early warning of potential overpressure on the DCS.

Reducing the ammonia risk profile

After the plant was restarted Orica continued working with the EPA and Safe Work NSW to further improve the management of storage and handling of ammonia at the site.

A multi-layered approach was implemented to reduce the site's ammonia venting risk profile, based on the inherently safe design principles of:

- elimination
- capture or destruction
- simplification
- mitigation

Elimination

Pressurized pipelines and storage
Originally the site had pumped pressurized ammonia liquid from the ammonia plant and ammonia storage to the nitrates plants which had a pressurized bullet for feed continuity. When extra plants were added, this network was expanded and the number of pressurized bullets was increased.

Initially investigations examined the option of removing all pressurized storage and handling of liquid ammonia on site to reduce the potential for a release. The option of transferring vapor from the ammonia plant and from ammonia storage to the aitrates plants was examined.

However when this option was examined in detail it was found to present significant challengers, which were:

- two of the nitric acid plants are of Weatherly design and as a result need high pressure ammonia gas for feed. Hence, the vapour system would have to be at high pressure. A leak of high pressure gas is only slightly better than a leak of low pressure liquid, so the site's risk profile would not be changed significantly. Also due to the high operating pressures, condensation can occur in these pipes as a result of ambient temperatures which then requires heat tracing or liquid removal equipment, adding complexity and cost.
- to ensure the Nitrates plants didn't trip if the ammonia plant came off line, the Ammonia Storage vaporiser had to be kept operating at all times. Yet there was no available waste heat at storage so steam would have to be continuously used to keep it operating, increasing the sites energy consumption. This was in marked contrast to the nitrates vaporizers which used waste energy for vaporization.
- the nitrates plants are located a significant distance from the ammonia storage tank and ammonia plant, so the pipes would have to be large to minimise pressure drop. This which would have required extensions to the existing pipe racks and would have been very costly.

The option of converting to ammonia vapor supply was therefore not progressed.

The current ammonia liquid supply system was then re-examined to identify ways of reducing the risk associated with its use. The optimum solution determined was to:

- operate the liquid distribution system at lower pressure and install pumps at nitrates to raise the pressure for each plant
- the amount of pressurised liquid ammonia stored on site was reduced. This was accomplished by removing the three carbon steel pressurised ammonia bullets and replacing them with one small stainless steel bullet operating at low pressure. The inventory in the new bullet was minimised to cover the feed requirement during pump changeover from the ammonia plant to ammonia storage only. This change in design philosophy reduced the amount of pressurised liquid ammonia storage on site from a maximum of 111 to 15 tonnes, significantly reducing the site's risk contours.
- this new ammonia bullet was made of stainless steel to eliminate failure modes

- and reduce the inspection frequency to once every 10 years. The long gap between inspections is needed because taking the bullet out of service would cause significant business disruption as it supplies all nitrates plants.
- Excess flow valves were installed on the outside of the ammonia bullet (on each plant supply) where they could be isolated and checked regularly, without affecting operation of other nitrates plants.

Figure 1
The new pressurised ammonia bullet at nitrates with the new pumps in the foreground.

Pumps

The site's ammonia pumps were all single mechanical seal type which had caused many significant ammonia releases in history. They were changed to seal-less "canned" motor pumps, except for the atmospheric ammonia storage tank pumps which were changed to double sealed pumps.

The double seals eliminate as far as possible, the potential for a leak. We couldn't source seal-less pumps for the atmospheric ammonia storage tank which could pump with low enough suction head to operate with the tank near to empty. This maximizes the tanks useable inventory. The pumps to be installed have suction inducers, so theoretically they can pump the tank empty.

Control systems

The nitrates DCS which was very old and had no options to manage alarms, was completely replaced. As part of this project the alarms in the nitrates plants were all reviewed, prioritized and suppressed where appropriate. This has resulted in much improved alarm rates. Alarm disabling by operators is no longer possible. Lower priority alarms can be shelved for short periods of time only.

Level of protection analyses (LOPA) were performed on all the ammonia control systems to minimize over-pressure events which could lead to relief valve operation. This analysis identified opportunities for improvements to many of the control systems and resulted in the installation of additional trips where required.

New ship loading arm

A ship load of ammonia is exported from the site every fortnight. The original arm from 1969 was still being used for this activity. It only had single seals on the joints and this was the opportunity to transition to the safety features of a modern arm.

A new arm with double seals, position monitoring, quick release couplings and a self-driving mechanism was purchased. These changes improved the integrity of this operation.

Figure 2
The new ship loading arm

Capture or destruction

Prior to 1998 ammonia venting was undertaken at many locations across the site and the vents were generally not very high. After the 1998 event, all the relief valve discharges and routine venting were collected together and discharged to high point vents, with one vent per plant area. However, the incident in 2011 demonstrated that the use of high point venting did not meet modern standards and a new approach was required. A review of available technologies identified that flares would be the best option to prevent the community being affected by ammonia odors as a result of a release.

The regulators do not allow the use of flares for the treatment of routine venting of ammonia, which can result from activities such as blowing down equipment.

Orica Kooragang Island has a strong market for aqueous ammonia which is manufactured on site. To utilize this opportunity the ammonia from routine venting is directed to ammonia scrubbers, so that the resulting aqueous ammonia can be sold.

The discharges from relief valves are directed to flares. Due to the separation distances between the plants on the site, it was found that a centralized flare was not feasible. Rather three separate flares were installed, in nitrates, ammonia and ammonia storage plant areas.

When the flares were being designed, Orica also reviewed the sizing of all the ammonia relief valves on site, as historically many of them had been significantly over sized. The review enabled the rationalization of the size of several relief valves, which in turn reduced the maximum ammonia flow that the flares are required to treat. This enabled a reduction in the flares' sizing.

Figure 3
The new Nitrates flare

Simplification

The review of the sites ammonia systems identified many opportunities to reduce the risk of an ammonia release.

New ammonia bullet

The new ammonia bullet was designed for a pressure higher than the maximum discharge head of the feed pumps, so the pressure can't exceed the design pressure of the vessel. It was also designed with the liquid inlet into the top of the bullet, so that backflow of liquid ammonia to the Ammonia Storage tank from the bullet can't occur in the event of non-return and trip valve failure.

Ammonia Pipelines

The ammonia distribution system over the years had grown very complex, with multiple redundant connections and issues which had to be managed. These were designed out.

The last of the liquid ammonia distribution piping, which was still the original carbon steel was replaced. This eliminated external corrosion as a potential failure mode.

The ammonia pipelines on site were predominantly designed with a flange rating of 300# but some sections were 150#. The site engineering standard was set at 300# and some upgrades to pipework of the low rated sections were undertaken.

Thermal relief valves had propagated on ammonia piping over the years. When the systems were analyzed, they were found on all the vapor pipes, duplicated on the same section of pipe and in some other locations where they were not needed. The regulatory burden of inspecting them all was quite significant. The number was rationalized and the type was changed from the simple screwed fitting to a more robust and reliable type of valve (the smallest conventional relief valve on the market).

Figure 4
The new thermal relief valve

Ammonia pumping

Liquid ammonia pumping to nitrates is required from the atmospheric ammonia storage tank when the ammonia plant is off line. It had become very complex as the storage pumps were as originally installed, however the ammonia consumption on site had doubled. Three operating pumps and heaters were now required to keep the uprated nitrates section at full operating rates. These pumps will be replaced with one pump and heater capable of handling the total duty.

For ship loading of cold ammonia, the heaters could not be bypassed so trip defeats had to be enacted whenever a ship was loaded (every fortnight.) and three pumps were needed. This will be simplified by installing a dedicated ship loading pump connected directly to the wharf pipeline.

Isolation valves

The manual isolation valves on the ammonia system were drilled ball valves. This type of ball valve has a hole to the valve cavity to relieve pressure from the seals in one direction only.

The issue with this valve type is they only isolate in one direction and sometimes it is necessary to isolate in the other direction. In addition they cause uncertainty when preparing isolations as its unknown, which direction they seal? Where the system was being modified they were replaced with energized seat type valves which don't suffer from this issue.

Instrumentation
The ammonia storage instrumentation featured a lot of the original pneumatic system and during the review several other areas of instrumentation were identified as requiring upgrades to ensure reliability. These systems were upgraded.

Mitigation

Some mitigation was added to further minimize the impact of an event.

Ammonia truck loading
The ammonia truck loading system had been modified once from the original; however there were several issues which were addressed, including:

- upgrading emergency water sprays over the loading points to maximise coverage
- installing a spring to close valve on the disconnection blow down point to minimise the likelihood of the valve being left open
- installing more ammonia detectors in the vicinity of the loadout area
- installing a quick release coupling on the truck loading arm to provide better protection from a drive away event
- connecting the site's trip system to the trucks system, to allow our trip system to shut the truck isolation valves
- installing "dead man" logic on the loading control panel to ensure the driver stays in the loading bay area at all times

Fixed fog monitors
Fixed fog monitors supplied with water from the fire system were installed around the new ammonia bullet to enable quick knock down of any ammonia release.

Ammonia detectors
Extra ammonia detectors were installed in the plants, around pipeline valve stations, around the new ammonia bullet and at the wharf to detect any ammonia releases.

Conclusion

Significant changes have been made to the ammonia distribution and handling systems at Orica Kooragang Island to further improve their inherent safety, reduce the ammonia exposure risk and potential odor impact on the neighboring communities and ensure the site's social license to operate.

The project has been carried out over several years. One of the major challengers was coordinating the changes to the plant while minimizing the impact of plant production.

References

1 "Incitec Ammonia Venting Study to reduce Environmental Risks" 46th Annual Safety in Ammonia Plant and Related Facilities Symposium. Peter McGrath & Rod McConnell

Ammonia tank inspection findings and follow-up tank modifications at two 1967 ammonia tanks from OCI Nitrogen

In 2015 OCI Nitrogen inspected two ammonia tanks located in Rotterdam, The Netherlands. Both tanks consist of a single inner steel tank with an outer concrete tank around it. An internal inspection was performed since at the same time the tanks also had to undergo various modifications to bring them to more modern standards. During the inspection, cracks were found in the bottom plate welds of the tanks. The cracks originated from original manufacturing defects. This paper will go into detail on the repair of the cracks, and all the other tank modifications that were done to these old tanks to bring these up to more modern standards.

Chiel Deij
OCI Nitrogen Ammonia Terminal Rotterdam, The Netherlands

Harrie Duisters
OCI Fertilizers, Geleen, The Netherlands

Introduction

Located at Geleen in The Netherlands, OCI Nitrogen (former DSM Agro) operates two ammonia plants, each with a name-plate capacity of 1360 MTPD (1500 STPD) of ammonia. One plant (AFA-2) was designed and constructed by Bechtel; the plant was commissioned in 1971. The other plant (AFA-3) was designed and constructed by M.W. Kellogg based on Kellogg's reduced energy ammonia technology; the plant was commissioned in July 1984. Current capacity of both plants is about 1550 MTPD.

In 2010 OCI Nitrogen acquired the ammonia terminal in Rotterdam that has two ammonia tanks each with a capacity of 15,500 metric tons. The terminal is used for import/export of ammonia in to the West European market and as a buffer for the Geleen plants operation. When there is a major turnaround, ammonia can be supplied to the Geleen site from the Rotterdam terminal via barges. Vice versa, when the Geleen site has an oversupply of ammonia and the site tanks are full, ammonia can be sent from Geleen to Rotterdam. The Geleen site is connected to a harbor via a 4 km underground pipe-line system so barges can be loaded and unload-ed. In this way the Geleen and the Rotterdam tanks are connected and form an integral part of the OCI Nitrogen ammonia business operations.

This paper describes the findings from an internal inspection of the Rotterdam tanks. Also practical information is shared on how tank operation and tank integrity has been improved at the OCI Rotterdam ammonia tanks by installing various tank modifications.

OCI Rotterdam atmospheric ammonia storage tanks

In Rotterdam, the Netherlands, OCI Nitrogen owns and operates two 15,500 metric tons ammonia tanks. The tanks are comparable in construction and size. The tanks were built by Graver in 1967. Each tank consists of an inner steel tank with a concrete wall around the tank (second containment).

The volume of each tank is 22,500 m^3 (5.94 million US gallon), the diameter is 30.4 meter (101 feet) and the cylindrical height is 32.0 meter (107 feet). The single wall steel tank is surrounded by a second containment concrete tank with rain cover. The steel tank is insulated with foam glass on the outside. The reinforced concrete tank has a 350 mm (14 inch) thick wall and has a diameter of 34.4 meter (115 feet) . The annular space between the outside of the insulation of the steel tank and the inside of the concrete tank is 1.55 meter (5.17 feet) wide.

The steel bottom plates of the tank are 6.35 mm (0.25 inch) thick and there is no annular bottom plate ring. The wall has 13 courses of steel plates ranging in thickness from 29.2 mm (1.15 inch) (bottom part) to 8.0 mm (0.3 inch) (top part). Each plate is 9.58 meter (32 feet) long and 2.44 (8 feet) meter high. The first course is 2.74 meter (9 feet) high.

The ammonia is stored at boiling point (-33 °C, -28 °F) at atmospheric pressure. The steel tank roof is insulated with a polyurethane insulation and the wall with foam glass insulation. Also under the tank, bottom insulation material is present. To prevent frost heave, the tank foundation is 1.5 meter (5 feet) above ground based on concrete piles.

Figures 1 and 2 show photos of the tank.

Figure 1. The OCI ammonia storage tanks in Rotterdam: Steel inner tank with concrete second containment.

Figure 2. Top view of the ammonia storage tank

Inspection

In 2015 we inspected the first tank, TK1002. The type of inspection was "Out of Service Inspection" (OSI). We choose OSI because of all the modifications in the tank. The inspection plan was defined by OCI and Lloyds Register (authority). The inspection of the tank bottom was part of the inspection plan and includes vacuum testing, floor scan and weld inspection (magnetic testing). During vacuum testing a leak was found in the welds of the bottom plates, see Figure 3. For the investigation we cut out a 30 cm (12 inch) diameter disc from the tank bottom. At the backside of the disc we saw the backing strip, see Figure 4. The crack was caused by bad welding of the backing strip and as a result high tension in the bottom plate weld.

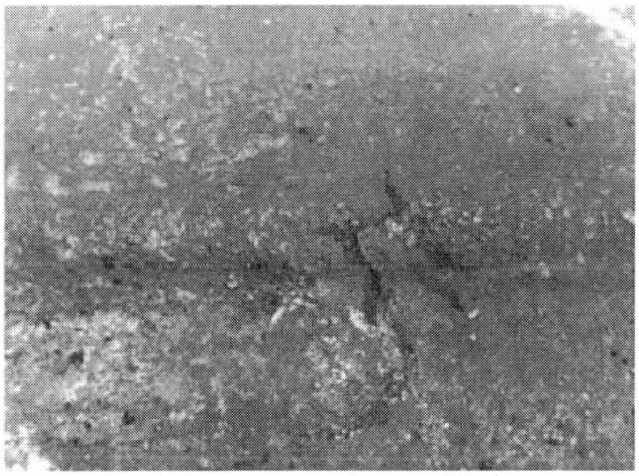

Figure 3. Leak in bottom plate weld

Figure 4. Backing strip with weld

To make sure the rest of the tank bottom has no other cracks we tested the bottom plate welds with Time of Flight Diffraction (ToFD) technique. The test indicated many cracks in the length of the welds. We cut three more discs out of the tank bottom and did a macroscopic investigation, see Figures 5 and 7.

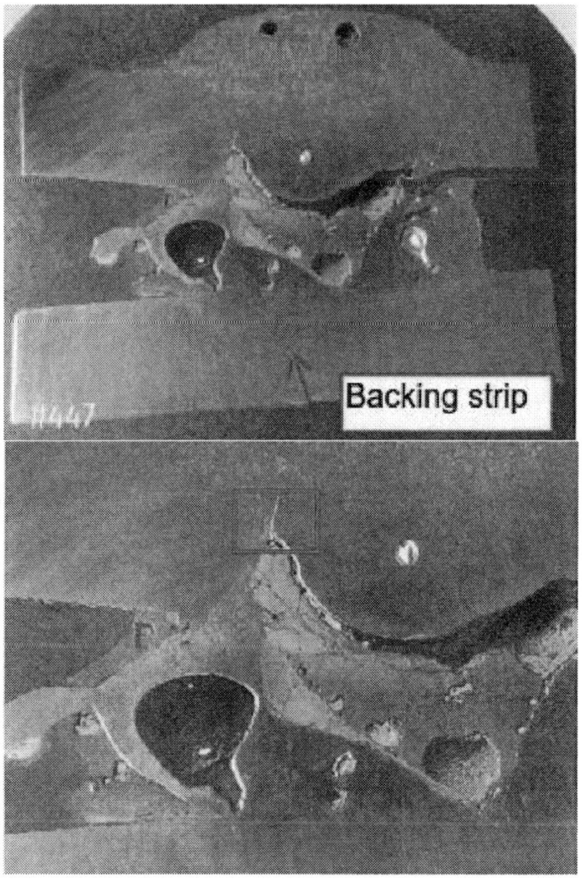

Figure 5. Macroscopic photo of a crack in the weld in the bottom sample

All the samples showed cracks in the welds. To ensure that the cracks were not caused by Stress Corrosion Cracking (SCC) we did a microscopic investigation on the samples to find the tip of the crack, see Figures 6 and 8. The microscopic results show that all of the cracks were due to fabrication defects. All of the cracks were caused by pores, slag inclusions, lack of fusion and insufficient weld penetration. Figures 5 and 6 show a crack caused by lack of fusion. The failure mode and the time based effect in growth was defined based on these microscopic laboratory tests. Based on these tests we concluded that the size of the cracks had not increased since 1967 and thus were fully contributable to original manufacturing defects.

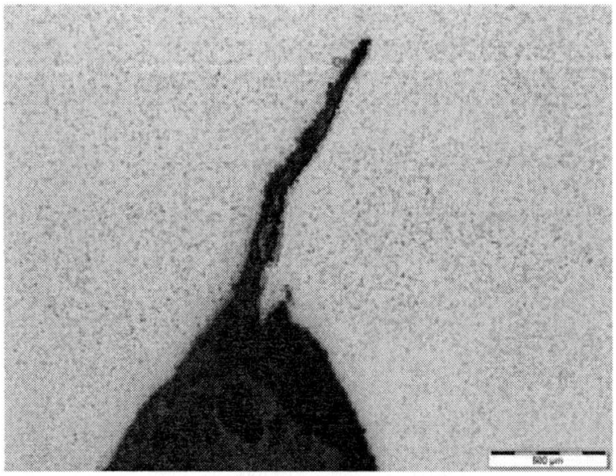

Figure 6. Tip of the crack in microscopic sample caused by lack of weld fusion

Figure 7. Macroscopic photo of a crack and inclusions

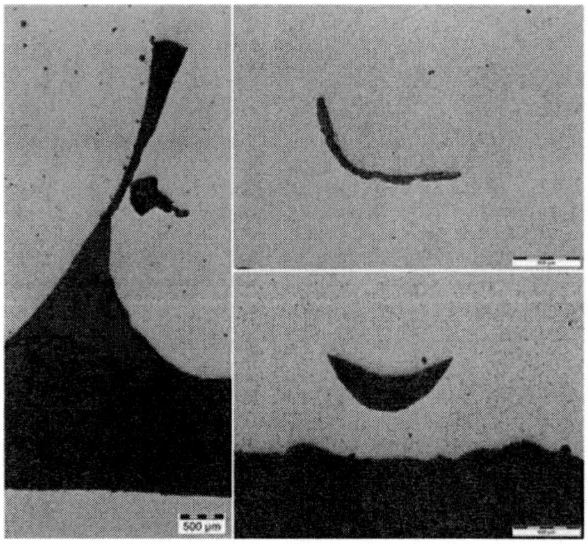

Figure 8. Microscopic photo showing lack of fusion and pores (sample 35), lack of fusion (sample 36 and 37)

The best method to continue with the tank operation is to leave the cracks to rest. The cracks will not propagate or increase and thus will have no influence on the integrity of the tank bottom. Also the cracks are present in neutral locations without significant stress or strain loadings.

Tank modifications

To bring the tanks up to more operational standards (Dutch regulations on the storage and transport of ammonia, published in report PGS12) OCI modified the old tanks and brought the integrity to a higher level.

The modifications that were done to the tanks are listed in below paragraph.

OCI engineered and installed for each tank two big stainless steel square manholes (Figure 11) with stairs in the bottom of the annular space. Those two big 80 cm x 200 cm (31 inch x 78 inch) manholes gave us safe access to the annular space. This access is thus from underneath, see Figures 9 and 10. All the contractors could work safe and the emergency exit was secured. These two big manholes remain in the annular space when the tank is in service. With these two manholes OCI is able to inspect the annular space frequently in a safe way. Of course special attention is given to the fact that these manholes need to be liquid tight. Both manholes are made out of stainless steel and sealed by an Ethylene Propylene Diene Monomer (EPDM) gasket.

Figure 9. Entrance to tank foundation

Figure 10. Entrance to manhole

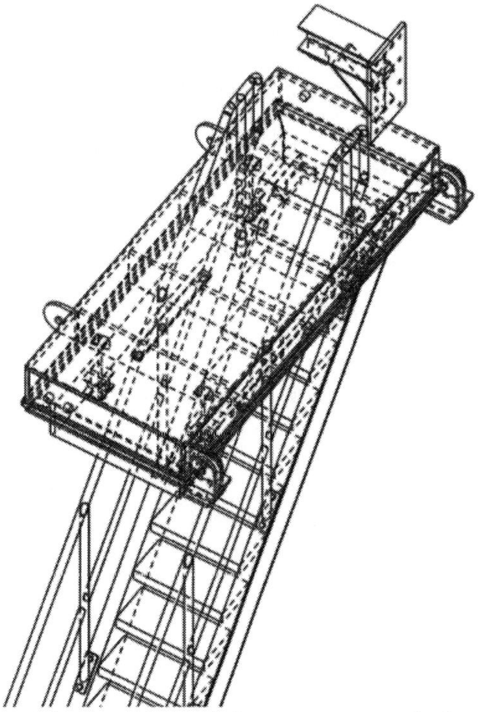

Figure 11. Drawing of the square manhole together with access stairs

OCI engineered and installed a stiffening clamp on the 14" discharge line between the inner tank and the concrete second containment wall, see Figure 12. If the tank instantly fails and vertically ruptures from bottom-up, the discharge line will stay in position and will not force valves and pipelines out of the annular space, which may have resulted in damage to the concrete secondary containment. Thus this modification will ensure that all of the ammonia stays in the annular space (second containment), a new "sliding vane pump" has also been installed to enable pumping ammonia out the second containment to the other tank, in the event of the tank failure.

Figure 12. The new clamp (the silver part on the left) on the tank discharge line

Figure 13. Servo level gauge

OCI engineered new instrumentation on the roof and removed all the existing nozzles from the tank wall. This modification decreases any risk of leakage due to bad connections on the tank wall. The new instrumentation includes a redundant level indicator, guided radar and servo level gauge, and an independent level switch. Also a temperature and pressure indicators are reinstalled on the roof. Figures 13 and 14 showing the new Honeywell ENRAF servo level gauge.

Figure 14. Servo level gauge on roof

The insulation on the inner tank was more than 40 years old and was in bad conditions. The insulation came loose and much ice was growing between the tank and the insulation. Therefore OCI removed all the existing insulation on the tank wall. Then we sandblasted and coated the 3400 m^2 (36,584 ft^2) surface on the inner tank. After coating, a new 10 cm (4 inches) foam glass insulation and a vapour barrier was installed on the tank wall, see Figure 15.

Figure 15. Foam glass insulation on the steel tank in annular space

OCI installed a lift to get safe access to the roof, see Figures 16 and 17. A fly-over from one tank to the other gives access to the other ammonia tank. The lift is specially engineered and fire safe. In case of an emergency, a chairlift is installed to leave the elevator safely.

Figure 16. Entrance to the permanent lift.

Figure 17. Lift installation on the tank wall

All of the existing valves were replaced with new ones that are fire safe, failsafe and made out of stainless steel. Due to the limited space, and the need to provide double isolation, a high performance butterfly valve was installed in the annular space combined with a gate valve.

Figure 18. New valves and pressure relief valves with ammonia detectors on the tank roof

An ammonia detection system (Honeywell) was installed around the tank and in the annular space. Every position where there is a potential probability of leakage (flanges, etc.) we installed this gas detection system. This detection system triggers the Emergency Shutdown Devices (ESD) and an automatic scrubber installation will be activated. Figure 18 shows NH_3 detection devices at the outlet of the pressure relief valves.

Conclusion

After seven months completion of all of these modifications both ammonia tanks are now equipped with significantly improved levels of safety. Integrity for the OCI Terminal in Rotterdam is the most important motivation for our environment and future activities. Our terminal is reborn!

4,700 mtpd Single-Train Ammonia Plant based on Proven Technology

At AIChE's 2005 Ammonia Safety Symposium, Uhde (now thyssenkrupp Industrial Solutions) presented a study on the technical and economic feasibility of a 4,000 mtpd single-train ammonia plant. At that time, the largest ammonia plants were operating at around 2,200 mtpd and there was only one single plant of 3,300 mtpd under construction. Today, the trend to higher plant capacities is still unbroken. Two plants with a design capacity of 3,300 mtpd are in operation, and two more are under commissioning.

This paper presents a study on cost savings for the 3,300 mtpd plant and for a 4,700 mtpd single-train Ammonia Plant using proven technology. By thyssenkrupp's dual-pressure ammonia process some critical equipment items can be built at a size which is less than one would get just by scaling up a conventional plant, thus minimizing the scale-up risk. This ingenious and proven concept is also applied in this study and is combined with the design and operating experience of the two largest ammonia plants in the world.

Klaus Noelker, Christoph Meissner
thyssenkrupp Industrial Solutions AG

The step to 3,300 mtpd ammonia plants – a 10 year success story

The trend to larger ammonia plants is a fact which is illustrated by the data from Figure 1. Motivation for this trend is the so-called economy of scale: The cost per unit of output decreases for increasing production rate. This trend holds both for the investment cost for a large production facility (e.g. an ammonia plant) and for its operating costs as parts of these costs do not vary with plant capacity, leading to lower costs per ton of product from larger plants.

In 2006, ten years before this paper was written, the first 3,300 mtpd ammonia plant went on stream at Saudi Arabian Fertilizer Company (SAFCO). This was a remarkable step increase in capacity as at that time the largest plants had a nameplate capacity of maximum 2,200 mtpd only.

Such a step increase in capacity of course offers huge cost savings by the economy of scale, but it does not occur very often and needs to be well planned in order to avoid the risk associated with it.

In this text, the term "risk" is not used to denote a safety hazard. What is meant is a risk to the success of building a plant according to the new concept. Such project risks could be:

- cost or time overrun by unforeseen difficulties in engineering, fabrication, transport or construction;
- not reaching the desired daily capacity;

- not reaching the desired economics of operation due to unreliable operation, frequent failure etc.
- There are parts in an ammonia plant where the scale up of 50% has not been very critical, because referenced equipment having the target capacity exists in other applications. Examples are the primary reformer, where larger ones had already been in operation in methanol plants, or process air compressors, where similar ones had already been used in air separation plants.

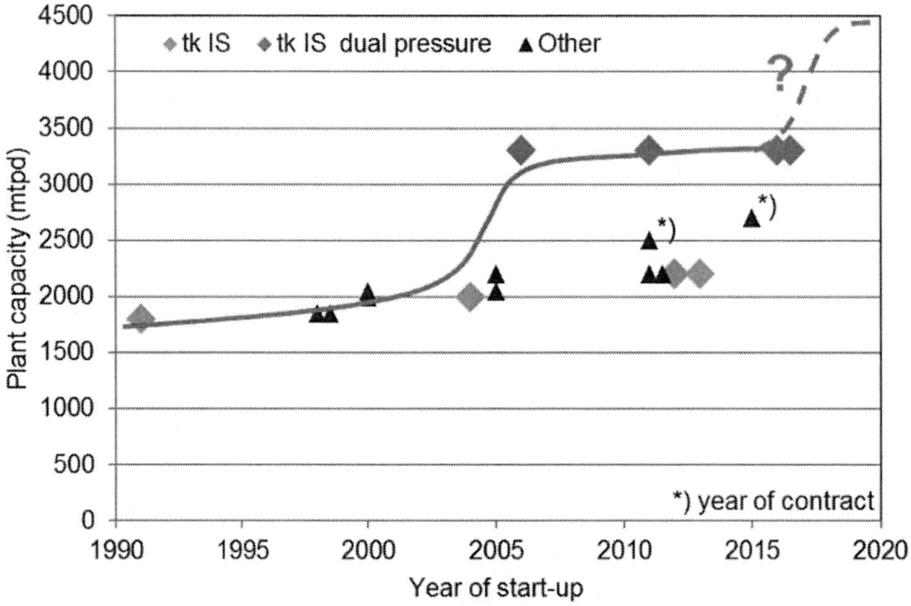

Figure 1. Selection of large-scale ammonia plants with plant capacity and year of start-up. Red dots: Uhde Dual Pressure plants by thyssenkrupp, Blue and black dots: Uhde and other licensor single pressure plants.

Therefore, the considerations during the design of the first 3,300 mtpd plant were:

- Keep as much as possible from the existing referenced process where scale up is not critical.
- Develop a way to avoid the scale up risk in those places where it is high.

Analysis revealed that in particular the synthesis gas compression and the synthesis loop are the areas with high scale up risk. The solution here was to insert a once-through ammonia synthesis upstream of the loop synthesis, operating at the outlet pressure of the LP casing, then to separate the ammonia produced there and finally to send the remaining synthesis gas into the conventional synthesis loop (refer to Figure 2). The desired effects are:

- The syngas compressor load is lowered and the same model as in a 2,200 mtpd plant can be used.
- Synthesis loop equipment and piping sizes are identical to that of a conventional 2,200 mtpd plant, thus avoiding dimensions for which no references exist.

Both contributed in lowering the scale up risk to an acceptable level.

Thus, not a completely new process but a quite conventional one was developed with the only major difference being the newly introduced once-through synthesis.

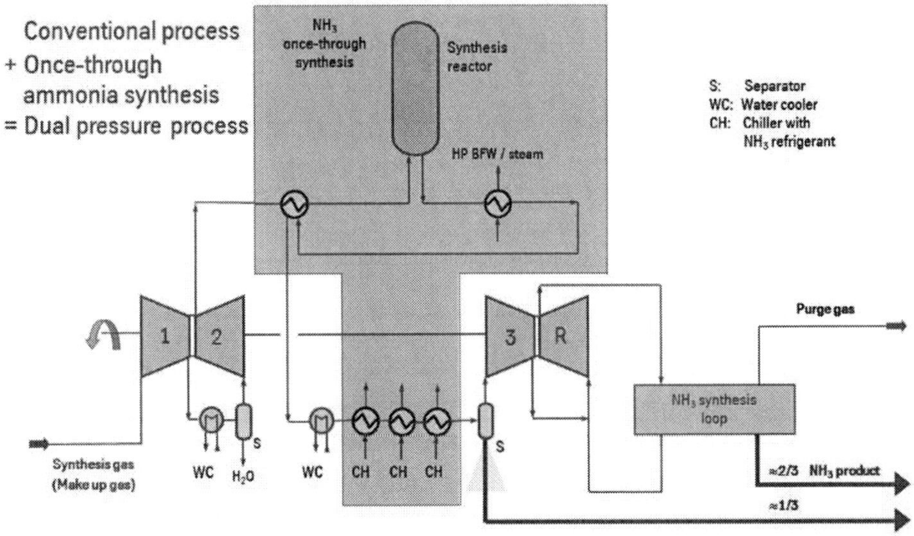

Figure 2. Simplified scheme of thyssenkrupp's dual pressure ammonia synthesis.

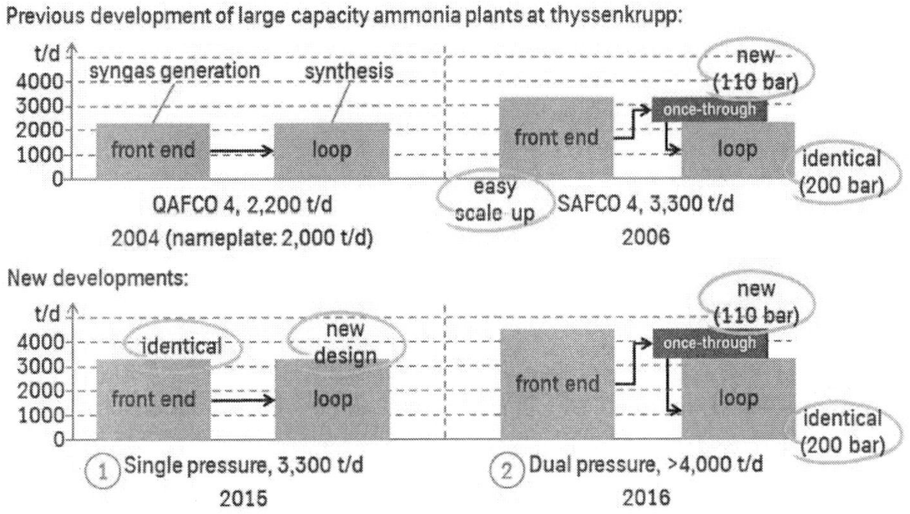

Figure 3. Strategy to minimize the scale-up risk by using the once-through synthesis: Upper row: Executed projects from the past. Bottom row: New developments discussed in this paper.

Since ammonia production takes places at two pressure levels (once through and loop), it is also referred to as dual pressure process. The considerations leading to its development are discussed in more detail in [1]. The principle of this scale-up is shown in Figure 3, upper part.

After SAFCO 4 in 2006, in 2011, a second plant with the same nameplate capacity went on stream for Ma'aden Phosphate Company in Ras al Khair, Saudi Arabia. In recent time both plants have exceeded production figures of 3,400 mtpd. Two more plants of the same size are under commissioning as this paper is written

(CF Donaldsonville, USA and Ma'aden 2, Saudi Arabia) (red dots in Figure 1).

New developments

This success shows that the decision was correct to develop the large plant concept. The reference plants by thyssenkrupp show that an ammonia plant at 3,300 mtpd can be built and operated. The fact that the trend to higher capacities is unbroken is also shown by other ammonia projects with capacities above 2,200 mtpd. Examples are Kaltim-5 by PT Pupuk Kalimantan Timur, Indonesia, with 2,500 mtpd and EuroChem at Kingisepp, Russia, with 2,700 mtpd. These projects were awarded to companies offering the KBR ammonia process, which means they will not make use of the dual pressure process (black dots with asterisk in Figure 1).

Despite its success, the dual pressure process also has a weak point, meaning that there is room for improvement: For the sake of risk-free scale up, it adds equipment to the standard process, mostly another ammonia converter and the cooling train associated to it. Hence, it does not fully exploit the potential of the economy of scale. Further savings are possible by deleting the additional equipment and integrating its function into that of the standard process.

Therefore, thyssenkrupp has launched two research and development (R&D) projects on this subject:

1. Explore the cost saving potential by building a 3,300 mtpd ammonia plant using the conventional process, however keeping the scale-up risk at acceptable level.
2. When a concept for a conventional 3,300 mtpd plant is available, it is the logical consequence to investigate the possibilities to use its synthesis gas compression and synthesis loop and check to which capacities it can be expanded by adding a once-through synthesis.

These two developments are schematically shown in Figure 3, bottom row, and are discussed in the following two sections.

Cost optimization of the 3,300 mtpd plants

Concept

The cost optimization of the 3,300 mtpd plant mostly consists in deleting the once-through synthesis and increasing capacity of the loop synthesis and the syngas compressor. Attention is focused on addressing measures discussed in the following sections to minimize shaft power of the syngas compressor, keeping it small in size to make use of existing reference installations:

- synthesis loop with high conversion and low circulation rate
- suction chilling to 4 °C at all stages
- synthesis gas drying unit
- cryogenic hydrogen recovery unit

Ammonia synthesis

The biggest changes are located in the ammonia synthesis loop which has to be increased by approximately 50% compared to the reference plants.

The ammonia synthesis loop of a 2,200 mtpd plant as well as that of a dual pressure 3,300 mtpd plant has a pipe size of 20". For the conventional concept, this is being increased to 24", the next available standard size. Keeping the velocities and other parameters unchanged would allow for 3,170 mtpd with this pipe size. More production is achieved by higher conversion per pass and by lower circulation rate.

Several configurations of ammonia converters and waste heat boilers have been investigated. The one which is found to be the most suitable one is to distribute the catalyst volume into four beds in two converters. Both ammonia converters have two catalyst beds and an internal heat exchanger for temperature control of the bed inlet temperatures. The synthesis loop of the dual pressure plant is already equipped with two ammonia converters (three beds in total) – that means no equipment is added, it is only designed in a different way. High catalyst volume and highly active catalyst $KATALCO_{JM}$ 74-1 from Johnson Matthey is provided. This additionally promoted high activity catalyst has originally been developed for low-pressure applications, but it is well-referenced and provides a performance enhancement at usual synthesis loop pressures of around 200 bar. The second converter houses more catalyst than the first one. The diameters of both pressure vessels are intentionally kept within a self-imposed limit of 3.2 m, which is the size of the largest converters designed by thyssenkrupp so far (see Table 7). Fabrication of larger diameter vessels is of course possible, but this procedure ensures there will be no fabrication and transport issues.

Same as in the reference plants, downstream of each ammonia converter an HP steam generator is installed. This ensures best recovery of the reaction enthalpy. The two waste heat boilers are similar in size and capacity.

The gas / gas heat exchanger downstream of the second waste heat boiler, preheating the converter inlet gas, can be built in two parallel units if highest energy recovery is desired. The remaining synthesis loop equipment is less critical and no major risk is expected with its size increase.

Syngas compression

For economic reasons, the general configuration of the syngas compression train, consisting of one turbine, two compressor casings, and no gearbox shall be maintained. This means that concepts with separate turbines for LP and HP compressor casing or with three compressor casings are not taken into consideration. It is essential that there is a suitable operating reference for the syngas compressor – anything else would not be welcomed by the market.

Changing a dual pressure plant (Figure 2) to a single pressure (conventional) synthesis loop plant has the following implications on the synthesis gas compressor:

- Shaft power increases because the full amount of synthesis gas has to be compressed up to the outlet of the third stage and the recycle gas flow rate increases.
- The higher flow rate present in the third compressor stage leads to higher distance between bearings, making the machine more susceptible to vibrations.
- The higher shaft power demand might require a larger turbine with lower speed. If the speed gets too low this has to be outbalanced by the compressor by either having larger impeller diameters or more impellers. In both cases this would lead to a different compressor model, meaning that there might be no operating experience with it in an ammonia plant. This again would increase the risk involved, which is against the intention of the project and should be avoided.

The good news is that the same model of the syngas compressor of the existing 3,300 mtpd plants can be used for the simplified plant. Data are compared in Table 1.

As seen from Table 2, changes in the LP casing against the reference plant are absolutely marginal and not critical. Impellers 5 to 8 even get slightly smaller (by 30 mm in diameter). The distance between bearings (bearing span) remains unchanged. The LP casing power in the new 3,300 mtpd single-pressure plant is a little less than in the reference because the pressure profile is slightly different.

In the HP casing, the impeller sizes of the newly proposed plant are slightly larger (by 25 to 30 mm in diameter), except for the recycle wheel which remains unchanged. The main difference however, is its bearing span which due to the higher flow rate passing through the HP casing increases by 152 mm (plus 10%), but remains much smaller than in the LP casing. The approximately 50% increase in flow rate manifests itself in higher shaft power.

Other than in most of the conventional 2,200 mtpd plants, a synthesis gas drying unit has been added to the flowsheet. Such a unit is already part of the 3,300 mtpd dual-pressure reference plants. (The plant can operate at lower rate in case the drying unit is not available.) The drying unit allows feeding the syngas to the synthesis loop directly upstream of the ammonia converter, leading to lower converter inlet ammonia concentration.

Plant / capacity (mtpd)	Shaft power in normal operation (kW)	Normal speed (rpm)
2,200 conventional, as built	25,864	9575
3,300 dual pressure, as built	28,600	9674
3,300 conventional	33,200	9701
4,700 dual pressure	38,078	8859

Table 1. Characteristic data of syngas compressors for different plant sizes.

Plant / capacity (mtpd)	LP casing			HP casing			
	impeller 1-8 dia.	bearing span	shaft power	impeller 1-7 dia.	recycle imp. dia.	bearing span	shaft power
	mm	mm	kW	mm	mm	mm	kW
3,300 dual pressure, as built	560	1823	17895	475 and 450	425	1485	10701
3,300 conventional	560 and 530	1823	16869	500	425	1637	15015

Table 2. Syngas compressor details for actually built 3,300 and newly proposed 3,300 mtpd single pressure plant.

Other units

In most conventional plants, the hydrogen recovery is a membrane unit. While typically this is a cost-effective and reliable unit, its disadvantage is that it delivers both outlet streams at unfavorable pressures: The permeate stream, produced at low pressure, is the hydrogen-rich stream, to be returned to the synthesis loop. The retentate stream, produced at high pressure, is the methane-rich stream, to be sent to the low-pressure fuel system. In the opposite, a cryogenic hydrogen recovery unit (as in service in the existing 3,300 mtpd plants) returns both streams at suitable pressure levels. It avoids loading the first stages of the syngas compressor with the hydrogen-rich stream from it and therefore is selected for the new concept.

The plant front end remains mostly unchanged. That means there is no scale up versus the existing 3,300 mtpd plants. Changes are only implementation from lessons-learnt items from the existing plants.

Overall economics

The economic comparison is carried out with both plant versions designed to the same specific energy consumption of 6.74 Gcal/t (24.3 MMBTU/st) ammonia. This accounts for combined consumption of natural gas (expressed by its LHV) and electricity, with credit for 52 bar (750 psi) steam export. The figure can be varied in both directions to reflect the economic conditions of a particular project.

The overall cost saving versus the dual pressure plant for engineering, supply and construction is estimated to be around 3.0 percent for a typical U.S. Gulf Coast location.

Application of dual pressure process

Concept

The next logical step on the way to higher ammonia capacities is to combine the 3,300 mtpd synthesis loop with a once-through synthesis upstream of it. This can increase the capacity to beyond 4,000 mtpd.

A similar exercise has been presented in [2] at a time when operating experience of any of the 3,300 mtpd plants was not yet available. Even then, not many limitations were found at capacities in excess of 4,000 mtpd.

In this paper, the target capacity is determined by the maximum amount of gas which can be handled by a synthesis gas compressor whose design data is close enough to referenced machines in successful operation. The capacity determined this way is 4,700 mtpd (Figure 3, bottom row).

This is an impressive figure and one might ask whether there are really projects of that size. In the recent past, there were two projects with similar ammonia capacity: Sorfert, Algeria (2 x 2,200 mtpd, thyssenkrupp Uhde process) and QAFCO 5 and 6, Qatar (2 x 2,200 mtpd, Topsoe process). Of course, the plants have been built in two identical trains each. But the example shows that there are projects where a single train plant above 4,000 mtpd would economically make sense, because it can reasonably be assumed that there is high cost savings potential due to the economy of scale.

Reforming

One means to unload the synthesis gas compressor is to operate the front end at higher pressure. In the concept presented here, the reformer outlet pressure is increased by 5 bar compared to the 3,300 mtpd plant.

This has a couple of consequences, some favorable, some less favorable: The higher pressure favors the absorption in the CO_2 removal unit and helps to keep the absorber small. By the higher pressure, load is shifted from the syngas compressor to the process air and natural gas compressors, where the step to higher capacities is less risky due to existing references.

The primary reformer is still smaller than methanol plant reformers. Its dimensions are compared with the other presented plants in Table 3. The combustion air preheater at the end of the flue gas duct is split in two parts for reason of size.

<table>
<tr><th>Plant capacity (mtpd) and type</th><th>Reformer box volume (m³)</th><th>No. of reformer tubes</th></tr>
<tr><td>2,200 conventional, as built</td><td>3025</td><td>288 5" tubes in 6 rows</td></tr>
<tr><td>3,300 dual pressure, as built</td><td rowspan="2">4445</td><td rowspan="2">408 5" tubes in 8 rows</td></tr>
<tr><td>3,300, single pressure</td></tr>
<tr><td>4,700, dual pressure</td><td>7088</td><td>570 5.12" tubes in 10 rows</td></tr>
</table>

Table 3. Primary reformer data.

Also the scale up of the secondary reformer is not very critical. Table 4 shows characteristic data of several projects. Part of thyssenkrupp's standard design is the central riser pipe for the incoming reformed gas acting as a support for the refractory arch in the bottom. Therefore, its span is only half of the vessel diameter and half of that of competing designs.

<table>
<tr><th>Plant capacity (mtpd) and type</th><th>Diameter (m)</th><th>Cylindrical height (m)</th></tr>
<tr><td>2,200 conventional, as built</td><td>4.5</td><td>17.4</td></tr>
<tr><td>3,300 dual pressure, as built</td><td rowspan="2">5.0</td><td rowspan="2">19.4</td></tr>
<tr><td>3,300, single pressure</td></tr>
<tr><td>4,700, dual pressure</td><td>5.5</td><td>20.5</td></tr>
</table>

Table 4. Secondary reformer data.

The steam-to-carbon ratio has been increased to 3.35 (measured at feed / steam mixing point) compared to 2.95 for the reference plant. This increase is caused by requirements to maintain the partial pressure of the reducing components H_2 and CO at the inlet to the HT shift reactor below a certain limit in order to avoid an adverse effect on the catalyst. The increase in steam flow of course has some impact on the process; however, its energy efficiency does not suffer from it. The higher throughput through the front end leads to higher HP steam production.

Process air compression

The process air compressor proposed is an integrally geared compressor, driven by a steam turbine. The geared compressor is not a must, also single-shaft machines are available, but the higher investment pays out by the higher efficiency. References in this size exist from air compression units in other industries. Turbine-driven gear type compressors are also successfully employed in ammonia plants up to 2,200 mtpd by thyssenkrupp and others.

Boiler feed water preheating

The boiler feed water heater upstream of the low temperature CO shift gets quite large. It can be built in one vessel but it seems to be cost-saving and beneficial to pressure-drop saving to install two parallel heat exchangers instead.

CO_2 removal

For this study BASF's OASE white® process (formerly known as aMDEA) has been selected for CO_2 removal because of thyssenkrupp's vast experience with this technology from its other large capacity plants.

Table 5 shows characteristic data of some of the CO_2 removal equipment. The LP flash vessel is the vessel with the largest diameter in the plant and is listed because it is most critical for transportation to site. For this reason, an effort was made to lower its diameter. Cost-effective packing material for the flash vessels in the reference plants as well as the new concepts is PRM 50. The only exception is the LP flash in the 4,700 mtpd plant where IMTP 70 has been selected with licensor approval because it allows for a diameter reduction of 0.8 m.

Most of the recently built large ammonia plants are serving export-oriented fertilizer production and are located close to the sea, therefore, transportation of large vessels to site should not be critical. As seen from Table 5, the change in diameter from 3,300 to 4,700 mtpd is only 0.3 m.

Plant capacity (mtpd) and type	Absorber i. dia. lower / upper part x cyl. height (m)	LP flash i. dia. (m), packing type
2,200 conventional, as built	5.0 / 3.3 x 42.6	6.1, PRM 50
3,300 dual pressure, as built	6.1 / 4.1 x 42.6	7.6, PRM 50
3,300, single pressure		
4,700, dual pressure	7.1 / 4.6 x 42.6	7.9, IMTP 70 (8.6, PRM 50)

Table 5. Dimensions of CO_2 removal equipment.

Three motor-driven pumps with rated capacity of 60% of the total flow are foreseen for the semi-lean solution circulation, two of them supported by expansion turbines installed in the rich solution flow from the absorber. This is the same concept as in the 3,300 mtpd reference plant. All other pumps are installed as pairs with 2 x 100% capacity.

Syngas compression

For the equipment selection of the synthesis gas compression, the same considerations as for the 3,300 mtpd plant are applied. Table 6 shows the main data. Despite the high capacity, the main dimensions are not severely larger than those of the 3,300 mtpd single-pressure plant: The bearing span of the LP casing is of course larger due to the higher volumetric flow rate processed by the impellers, but only by 5.9%. In contrast to that, the HP casing bearing span is even 3.5% smaller than that of the 3,300 mtpd single-pressure plant (see Tables 2 and 6). The reason is that the mass flow to the HP loop is approximately the same for both plants, but the dual pressure plant has a lower gas inlet temperature since the gas is coming from the first ammonia separator (see Figure 2) and hence a lower volume flow. The fact that the dimensions are close to each other suggest that the rotor dynamics are not much different for both machines and can help to build trust that selecting the 4,700 mtpd process does not involve high risk for the machine.

Plant / capacity (mtpd)	LP casing		HP casing			Shaft power
	impeller 1-8 dia.	bearing span	impeller 1-7 dia.	recycle imp. dia.	bearing span	
	mm	mm	mm	mm	mm	kW
4,700 dual pressure	600	1930	520	450	1580	38100

Table 6. Syngas compressor details for the proposed 4,700 mtpd plant. Refer to Tables 1 and 2 for the 3,300 mtpd plants.

Ammonia synthesis

The ammonia synthesis loop is identical to the 3,300 mtpd single pressure plant. The upstream once-through synthesis is equipped with a converter having the same dimensions as the ones in service in the 3,300 mtpd dual pressure plants. Due to its higher inlet flow rate it produces approximately 1,400 mtpd ammonia. This concept of combining equipment already being used elsewhere ensures no new risk is created by the increase in total plant capacity. The converter dimensions are listed in Table 7.

Plant capacity (mtpd) and type	Once-through converter: i. dia. x cyl. height (m)	Loop converters: i. dia. x cyl. height (m)
2,200 conventional, as built	–	3.0 x 11.3 2.9 x 10.2
3,300 dual pressure, as built	3.2 x 18.9	3.0 x 12.5 2.9 x 12.2
3,300, single pressure	–	3.2 x 11.6 3.2 x 19.4
4,700, dual pressure	3.2 x 18.9	

Table 7. Dimensions of ammonia synthesis equipment.

Overall economics

There is a high MP steam export from the process plant of 136 t/h. Most of the large ammonia plants built in the recent years are located within fertilizer complexes, where urea plants or other consumers for the export steam exist, so the steam export should be welcome and should lead to lower auxiliary boiler firing.

The amount of CO_2 produced is sufficient for production of approximately 7,500 mtpd urea, about twice of the size of many urea plants being built today.

For the purpose of the study, the specific energy consumption is kept at the same level as for the 3,300 mtpd plant. Compared to a two-train plant, the concept presented has a clear cost advantage. Of course there are some elements in the cost calculation which do not have a cost regression with capacity, for example the number of heavy lifts during construction. But despite that, the savings by having less but larger equipment prevails as expected.

Summary and Conclusion

Ten years ago, thyssenkrupp's dual pressure ammonia process had opened the door to ammonia capacities of more than one million tons per year. Repeated projects of that size demonstrate that there is a market for such high production rates. Two projects have been mentioned in the paper where producers have demanded even higher capacities at a single site and built them in the form of two-train plants.

The result of the detailed studies for improvement is:

- By deleting the once-through synthesis and correspondingly increasing the size of the synthesis loop, the 3,300 mtpd plant can now be offered with less equipment than the operating reference plants. This saves investment cost without adding risk.
- For capacities above that, the one-through synthesis can be added to this type of plant, then achieving up to 4,700 mtpd capacity.

The discussions with equipment vendors and own engineering resources confirmed there are no obstacles to both versions of the plant, but clear cost advantages against the benchmarks.

Acknowledgement

The authors would like to thank all equipment vendors involved in this study for their support.

References

[1] D. Lippmann, J. Larsen, "Extending Ammonia Plant Capacity Limits with Proven Technology", Nitrogen Conference 2002, Doha, Qatar, 2002

[2] J. Ruether et al., "4,000 mtpd ammonia plant based on proven technology", 50th Annual Safety in Ammonia Plants and Related Facilities Symposium, Toronto, Canada, 2005

Development and Commercial Success of Online SCR Catalyst Cleaning: Remove Fouling Material without Downtime

Selective Catalytic Reduction (SCR) units are used in a variety of steam methane reformers (SMR) and fired heaters to reduce NOx emissions from flue gas. An SCR unit functions by passing the flue gas through a catalyst bed and using ammonia to convert NO_x into nitrogen and water. During operation, the catalyst beds can become fouled resulting in increased pressure drop and reduced production rates. To restore production rates, the debris on the catalyst must be removed. Typically, this removal has required shutting the SMR or heater down so that personnel could enter the flue gas duct and vacuum the catalyst face. Due to the requirement to shut down the plant, removal of fouling material has been costly and required interruptions to production.

With assistance from Air Products, Cetek designed and developed a novel online Hot Vacuuming System to remove the fouling material while keeping the plant running and maintaining production. This patent-pending system shifts the paradigm for managing SCR fouling from one where a plant needs to be shut down in order to clean the SCR catalyst to one where the fouling material can be safely removed while the furnace remains fully operational. This new method improves plant reliability and reduces plant operating costs.

After completing several small scale tests at operating SMR plants, Cetek completed a full scale project at a world-scale Air Products SMR in the summer of 2014. Over 90% catalyst face was successfully vacuumed, which reduced the pressure drop to near design levels and allowed production rates to be increased. This patent-pending process is now commercially available and offers a cost-effective means of maintaining efficient plant operation.

David Graham
Air Products and Chemicals, Inc.

Paolo Brunello
Cetek, Ltd.

Introduction

Nitrogen Oxides (NO_x) are formed during the combustion of hydrocarbon products by three different processes. Thermal NO_x forms using diatomic nitrogen from combustion air by high temperature oxidation. Fuel NO_x forms from nitrogen present in the hydrocarbon that is converted to NO_x. Finally, prompt NO_x is formed by a series of complex reactions not covered by the other two methods. When released into the atmosphere NO_x poses many environmental and health consequences. When mixed with atmospheric moisture, NO_x reacts to form components of acid rain. Additionally, the reaction with volatile organic compounds in the presence of sunlight can form tropospheric ozone and smog.

Due to these concerns, regulations exist to limit the amount of NO_x released to the atmosphere. In order to lessen the amount of NO_x released to the environment, NO_x is reduced to diatomic nitrogen and water. In typical applications, precious metals, such as platinum, are used as a reduction catalyst. Two processes can be used to reduce the NO_x: selective catalytic reduction (SCR) and selective non-catalytic reduction (SNCR). Both processes use urea or ammonia as reducing agent. The simplified reduction reactions are shown below:

$$4NO + 4NH_3 + O_2 \rightarrow 4N_2 + 6\,H_2O$$

$$2NO_2 + 4NH_3 + O_2 \rightarrow 3N_2 + 6\,H_2O$$

The reducing agent improves the overall conversion of NO_x. SNCR occurs between 1000-1400°F (538-760°C), while SCR occurs between 600-750°F (315-400°C). The SCR temperature limitation is due to the catalyst used for reduction. In SCR, catalysts are impregnated in honeycomb-like cartridges in which NO_x and other gases flow through, as shown in Figure 1.

Figure 1: SCR Catalysts[1]

In order for the reduction of NO_x to occur, the channels of the SCR catalyst must remain open so that gas can flow through. However, it is common that these channels become blocked with debris, as shown in Figure 2. In steam methane reformers and fired heaters the SCR units are downstream of the convection section. The radiant sections of these units can be lined with refractory ceramic fiber or insulating brick/castable.

Figure 2: Fouled SCR Cartridge

Refractory ceramic fiber is an amorphous substance. When subjected to high temperatures, as in a fired heater, the fibers can devitrify and become friable. The flue gas flow can disturb the devitrified fiber, carrying crystalline dust downstream to the SCR catalyst bed. This dust contributes to the overall fouling of the cartridge surfaces. Flue gas can also act as an abrasive to insulating brick/castable, and the resulting dust can be transported downstream to the face of the catalyst.

The fouling on the catalyst creates a greater resistance to the flue gases passing through the SCR resulting in an increase in the pressure drop across the catalyst bed and a reduction in the efficiency of NO_x conversion. As a result production rates have to be cut until the fouling is eliminated, as shown in Figure 3. In addition to reduced production rates, the fouling can cause extra load on the ID fan and higher power consumption.

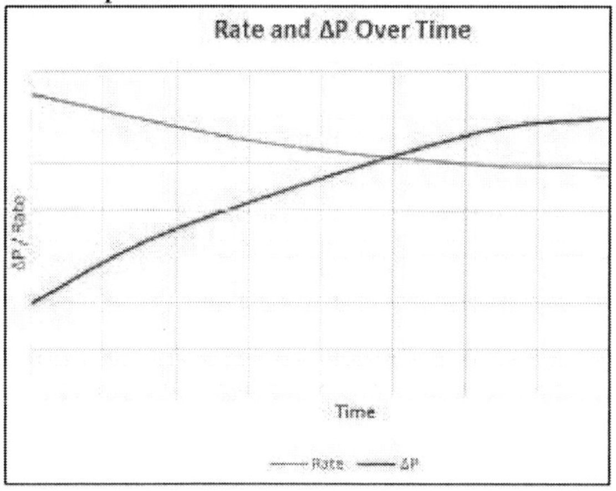

Figure 3: Production Rate and Pressure Drop

To maintain production, various cleaning methods are available and have been described in the literature[2]. For example, sonic horns have been described which attempt to dislodge fouling material using low frequency, high pressure sound waves. Likewise, soot blowers have been described as an alternative method to dislodge fouling material. These blowers can either be powered by air or steam and are pointed at the surface to blow particles off the catalyst bed. While these methods might be suitable for the short term, the upstream cleaning of SCR catalysts does not remove the fouling material from the system. The fouling material is still present and is readily recaptured by the catalyst causing the same issues. In order to remove the debris from the system in the past, the furnace needed to be taken off-line, sacrificing production. Personnel could then enter the unit and vacuum off the cartridges.

Issues and Design

Cetek and Air Products & Chemicals completed a joint development project to create an online vacuuming process to remove the fouling debris from the system without taking the furnace offline and interrupting production.

In order to have access to the fouled catalyst faces, access needed to be created while the furnace is operating. Cetek has an extensive history of creating access during operation. It was determined that access needed to be created directly above the upstream face of the catalyst bed. The minimum number of access points were positioned to maximize the number of cartridges that could be reached.

Due to the fragility of the catalyst a proprietary vacuum head needed to be developed to make contact with the cartridge while vacuuming. The head had to be durable enough to withstand the high temperatures, but soft enough not to damage the cartridge as it passed over it. Cetek developed a low friction head that does not cause damage while vacuuming. The head is mounted to a vacuum lance that is manipulated by a technician. The lance is connected to a counterbalance to offset the weight and improve control. The counterbalance is suspended above the access point on a moveable trolley. The trolley moves with the motion of the lance to reduce strain and improve the fluidity of motion for the operator.

A venturi was used to induce vacuum. It was capable of withstanding the elevated temperatures (>600°F / 315°C) of the entrained flue gas. The venturi further reduced the temperature of the fouling material and flue gas during vacuuming since they were diluted with cooler compressed air. The cooled flue gas is vented safely after passing through the high pressure, high temperature filter. The fouling material is collected for appropriate disposal.

Results

Cetek performed a full scale online vacuuming project at an Air Products & Chemicals steam methane reformer in Louisiana, USA. The unit had a horizontal convection section, downstream from which was a SCR Unit, containing about 200 catalyst cartridges. A fouled section of the SCR unit is shown in Figure 4 below.

Figure 4: SCR Unit

In order to have access to the upstream face of the SCR cartridges, Cetek created seven access ports on top of the transition between the convection section and the SCR. These access ports are created using Cetek's Hot Access Creation (HAC) services. The access ports are approximately a foot wide and are near the face of the SCR catalysts. The ports were spaced to maximize the number of reachable cartridges while minimizing the number of access points. Access port examples are shown in Figure 5 below.

Figure 5: Access Point Locations

The project was completed in a planned 5 days. Over 90% of the cartridges in the SCR were cleaned (see Figure 6). After cleaning of the SCR catalysts, the ports were closed using Cetek supplied Access Plugs. These plugs are removable for potential future cleanings.

Though more access point locations could have been created to reach the remaining 10% of cartridges, diminishing returns are realized. Creating additional access points to reach a few extra cartridges will not have a noticeable effect on the overall impact of the project.

Figure 6: Cleaning of SCR Cartridge

As shown in **Error! Reference source not found.** below, the pressure drop across the SCR catalysts was significantly reduced. The final pressure drop was close to design for new SCR catalyst cartridges. Air Products & Chemicals estimates that the payback for this project is less than 1 year, due to the recovered hydrogen production and from the electricity savings on the ID fan.

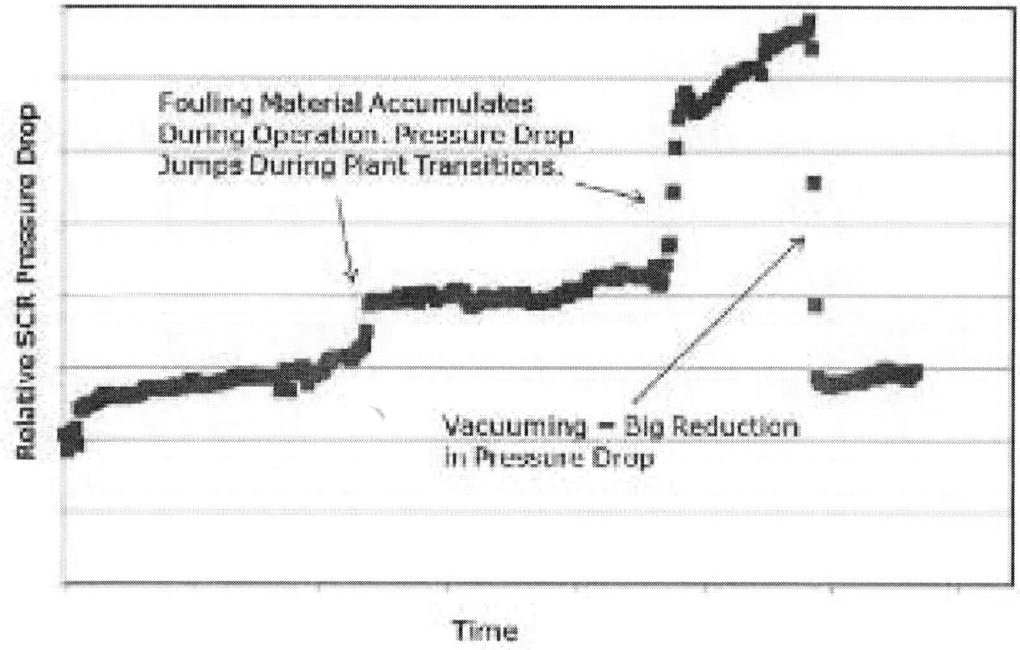

Figure 7: Pressure Drop with Vacuuming

Cetck's Hot Vacuuming System (HVS) has proven to be superior to the former cleaning methods. With this online approach, Cetek is not only able to displace the fouling material from the surface of the catalyst, but remove the debris entirely.

References:

1). DNT, De-NOx Technologies, LLC. Retrieved from http://www.de-nox.com/SCR.htm

2). Buecker, Brad. (01 November 2011). "Selective Catalytic Reduction: Operational Issues and Guidelines." Power Engineering. Retrieved from http://www.power-eng.com/articles/print/volume-115/issue-11/features/selective-catalytic-reduction-operational-issues-and-guidelines.html

Catalyst Reactive Hazard Assessment in Ammonia Flowsheets

Good engineering design involves the use of appropriate risk assessment tools to ensure the safe operation of the process units. Where a new or improved catalyst is involved then further tools such as a Reactive Hazard Assessment or Technical Risk Assessment should be used.

We will use, as an example, the conversion of the existing axial shift vessels at the Coffeyville Resources Nitrogen Fertilizers LLC Kansas ammonia facility, which were successfully converted to an axial-radial configuration using both Johnson Matthey radial flow technology and a new sour shift catalyst in July 2015.

Julian Morris
Johnson Matthey

Marc Gilberston
Coffeyville Resources Nitrogen Fertilizers, LLC

Introduction

Some recent world-wide incidents[1] reinforce the need to undertake technical risk assessments for any proposed technology or catalyst changes to a facility's flowsheet.

Meaningful process risk assessments should be considered for application in any of the following scenarios:

- where a new catalyst is to be used
- where an existing catalyst is to be used in a new application
- where an existing catalyst is to be operated outside of the proven operating window
- where there is new process equipment
- where there is a change in process conditions

The paper describes how Johnson Matthey used the principles of Reactive Hazard Assessment, in collaboration with Coffeyville Resources Nitrogen Fertilizers LLC (CVR) for a revamp of the process involving the shift conversion reactors.

Process EHS Assessment

The Environment, Health, and Safety (EHS) risk assessment techniques adopted by Johnson Matthey draw their inspiration from hazard analysis and operability studies (HAZOP), the multistage hazard study processes, and other techniques developed by Imperial Chemical Industries in the late 1960s.

A seven stage process is adopted in which each phase is linked to a particular project stage (Figure 1).

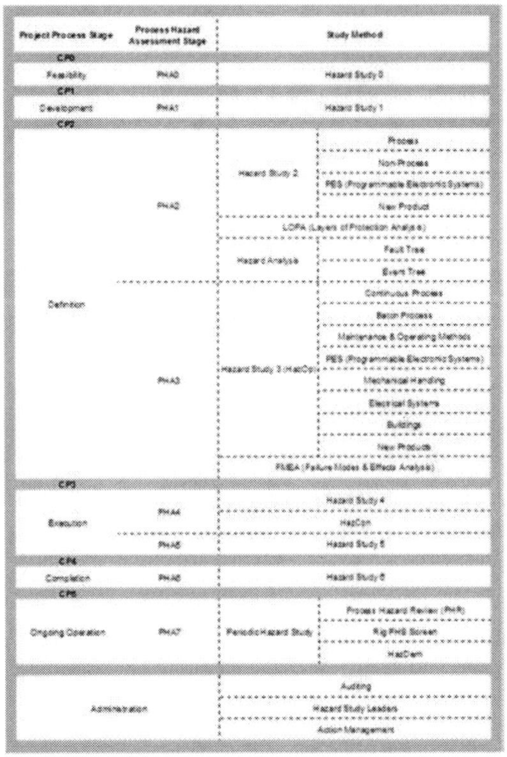

Figure 1: Project Hazard Studies

The seven specific project stages are summarized below:

- Hazard Stage 0 - Concept stage
  - Identify hazard characteristics for elimination
- Hazard Stage 1 - Development stage
  - Understand process and materials impacts on EHS
- Hazard Stage 2 - Front-end engineering design and project definition
  - Identify significant hazards and implement measures to eliminate/reduce risk.
- Hazard Stage 3 - Front-end engineering design and project definition
  - Review design for operability issues
- Hazard Stage 4 - Construction/design verification
  - Review at end of construction phase
- Hazard Stage 5 - Pre-commissioning safety review
  -EHS review
- Hazard Stage 6 - Project close-out/post start-up review
  -Post completion review
- Hazard Stage 7 – On-going operation and EHS assurance
  - Review at end of construction phase

Further information around the HAZOP process is discussed in [2].

New Catalyst and Technology into an Existing Facility

Construction began on the CVR plant in Coffeyville, Kansas, U.S.A. in January 1998. The Process Information Package for the hydrogen production facility (gasification and hydrogen purification) was developed by GE Energy (formerly Texaco Development Corporation). The engineering contractor for the gasifier and ammonia facility was Black & Veatch Pritchard. Much of the equipment for coke handling, gasification and syngas cooling was relocated from Daggitt, California where it was part of an integrated gasification combined cycle power plant demonstration project. The project was sponsored by the U.S. Department of Energy and known as Coolwater. Initial ammonia production began in late 2000, with a nominal design production rate of 1100 STPD.

The facility was built and initially operated by Farmland Industries, Inc. In March 2004 the facility was sold to an investment group.

A block flow diagram of the plant flowsheet is shown in Figure 2. More detail on the technology and operation in use at the CVR site can be found in an earlier paper [3]

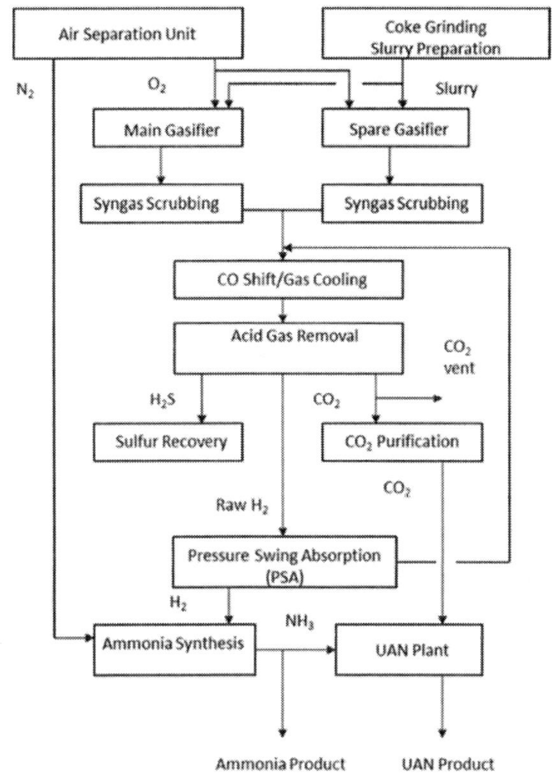

Fig 2: Gasification Block Flow Diagram

Background to the Shift System

The CVR plant uses high pressure gasification of petroleum coke in a quench gasifier that produces a saturated synthesis gas, see Figure 3. Hence it is convenient to use sour shift catalyst without first removing the sulphur. The shift system historically consisted of two axial flow adiabatic reactors with intermediate indirect gas cooling. The shifted synthesis gas then passes to a single acid gas removal system where both H_2S and CO_2 are removed. The H_2S is sent to a Claus plant for sulphur recovery and the CO_2, after some purification, is sent both to the urea plant and for export via pipeline to a regional enhanced oil recovery field. The synthesis gas is sent to a pressure swing adsorption (PSA) system for hydrogen purification and the waste gas from the PSA, which contains mostly unreacted CO, is recycled to the shift system.

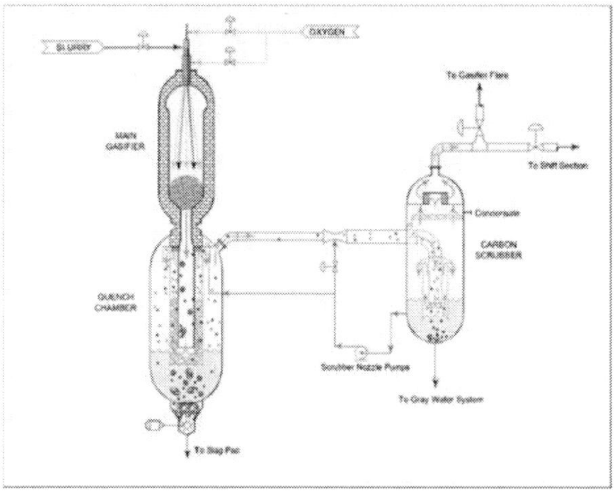

Fig 3: Upstream Gasification System

The gas from the high pressure gasifier is sent to the sour shift system at 45 bar (650 psi), see Figure 4. A lower steam to carbon ratio can be used than in a traditional sweet shift system, as over-reduction of the cobalt molybdenum catalyst is not a concern.

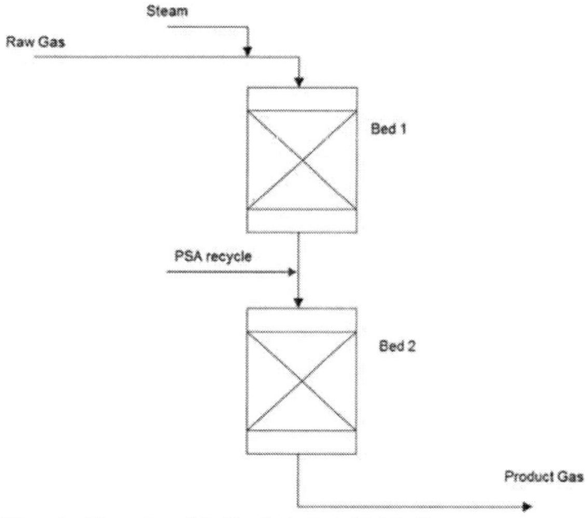

Fig 4: Sketch of 2 Bed Axial Shift System

The catalyst that has historically been used in the first reactor is **KATALCO$_{JM}$**™ K8-11. This is a cobalt/molybdenum catalyst on a specialised support to provide high physical strength to withstand potential plant upsets. The catalyst historically used in the second reactor is designated **KATALCO$_{JM}$** K8-11HA - which is a modified catalyst with higher activity and its use has allowed the reactor to be short-loaded by about 30%. The smaller quantity of catalyst has allowed some saving in pressure drop as well as giving a saving in total installed cost. Despite the smaller catalyst quantity, the reactor can be operated at a lower temperature and this gives a more favourable CO shift equilibrium and greater CO conversion.

The first stage of sour shift, with a nominal catalyst volume of 45 m^3 (1600ft^3) typically operates with an inlet temperature of around 275°C (530°F) and an exit temperature of around 440°C (820°F) and achieves a close approach to CO shift equilibrium. After cooling to around 270°C (520°F) the gas goes to the second stage, with a nominal catalyst volume of 68m^3(2400ft^3) where the CO is lowered to less than 1% while raising the temperature to about 285°C (545°F).

Background to the Revamp Project

During commissioning and the subsequent early years of operation, the gasification plant suffered from numerous system trips that resulted in liquid carryover into the downstream sour shift catalyst beds. The subsequent catalyst damage resulted in increased pressure drop during operation. The plant remedy was to operate the sour shift catalyst beds short loaded to achieve a target plant turnaround cycle time of 2 years. In short loading the catalyst beds, the resulting CO slip was slightly higher than previously due to the decreased catalyst volumes. However, this was seen as a tradeoff between larger catalyst volumes with lower CO slip but a higher pressure drop, and smaller catalyst volumes with a higher CO slip and an overall lower pressure drop in the axial bed system.

Over the first decade of operation plant reliability improved significantly, with both increased operator knowledge and process improvements to the following areas;

- Ongoing upgrades to the GE gasification system, including a new downstream separator in 2010,
- Improvements to the acid gas removal units, with new improved solution.
- Increase to the number of PSA beds from 10 to 12 in 2014, and;
- Revamp of the ammonia converter in 2010

All of the above confirmed that the plant was now capable of producing above 1300 STPD at SOR (Start of Run) conditions.

In 2014, Johnson Matthey and CVR investigated the benefits of a sour shift vessel retrofit concept that would reconfigure the present axial flow reactor.

In this new design, the inlet gas enters the vessel via an axial flow inlet distributor, and the gas is then directed to the outside of the vessel where it flows axially downward through a set of perforated inlet distributor channels installed against the side of the reactor. The gas then flows radially across the sour shift catalyst, where the exothermic reaction occurs. The gas exits the catalyst by passing through a perforated center piping system and then flows axially down the collector system to the outlet of the vessel. This new axial radial design has been implemented by Johnson Matthey in over 10 new gasification applications in China.

An indicative vessel schematic is shown in Figure 5.

Fig 5: Schematic of Axial-Radial Shift System

A technical review indicated that the existing sour shift catalyst beds could be retrofitted to an axial-radial design with the following benefits;

- Lower pressure drop in the shift reactors leading to higher pressure at the suction of the syngas compressor, with appropriate power savings, and increased production capacity;
- Opportunity to increase the catalyst volumes in the reactors to achieve greater CO conversion than with the previous reactor configuration. The greater CO conversion leads to increased hydrogen available to the synthesis loop and hence increased ammonia production.
- Lower rate of pressure drop increase (due to greater flow area to catch low levels of contaminants in the feed gas) which is likely to allow the catalyst change cycle to be extended from 3 years to 6 years, being 3 plant turnaround cycles of 2 years each.
- Use of **KATALCO**$_{JM}$ K8-11R: a cobalt–molybdenum sour shift catalyst which has been specifically designed for use in axial-radial reactor systems. It uses a smaller particle size than more traditional materials and this gives increased catalyst effectiveness with little impact of higher pressure gradients.

Given the above process benefits, approval was received from CVR to proceed with the revamp project in December 2014.

As part of the process design, a risk assessment review was undertaken by Johnson Matthey.

Catalyst and Technology Risk Assessment

When developing new processes, risk assessment techniques are required. This is accomplished by the use of a technology risk assessment (TRA) process in which the following is subject to critical review

- Experimental basis – Research data underpinning the project is reviewed to ensure that the results and conclusions can be extrapolated to a full scale application. Particular attention is given to considering the entire operating envelope (both intentional and under-fault conditions).
- Design Basis – technical and engineering assumptions and methodology are reviewed to ensure that they are appropriate.
- Operational and EHS learning relating to this or similar products/technologies – often facilitated by review of papers from events such as the AIChE Ammonia Safety Conference.

The product lifecycle is also reviewed – in the case of a new catalyst; considerations include aspects such as:

- Raw materials
- Catalyst manufacturing
- Charging
- Initial heating
- Activation
- Start up
- Normal Operation
- Shut down
- In situ storage
- Restart
- Discharge
- Disposal
- Side reactions & byproducts
- Contaminants
- Ex situ storage
- Transportation

Aspects similar to those normally covered in a HAZOP are also addressed such as the consequences of deviations in:

- Temperature
- Pressure
- Composition
- Flow velocity
- Impurities

The TRA meetings are most productive if used to review information that has already been circulated, hence the normal practice is that all participants apply their best endeavors to collate and share all pertinent information before the meeting. This means that the focus of the meeting is on interpreting and drawing conclusions from data rather than trying to quickly assimilate and understand something which has just been seen for the first time.

The composition of the review team is adjusted to reflect the needs of individual projects but would often include functional experts, product experts, research chemists, design engineers, project, technology and commercial managers.

The basic process is as follows:

STEP 1: Identify all technology "hazards",

STEP 2: Assess the likelihood and consequence of each hazard being realized, without any additional control measures and score against a word model. The results can conveniently be displayed in a standard risk matrix (Figure 6).

	Likelihood					
Consequence	Remote	Highly Unlikely	Unlikely	Possible	Quite Likely	Likely
Catastrophic						SEVERE
Massive						
Major				HIGH		
Moderate			MEDIUM			
Minor						
Slight	LOW					

Fig 6: Standard risk matrix

STEP 3: Identify the basic causes and means of reducing the risk factor, likely improvement, and the difficulty (manageability) of the technology hazard. The results may be displayed on a balloon diagram (Figure 7)

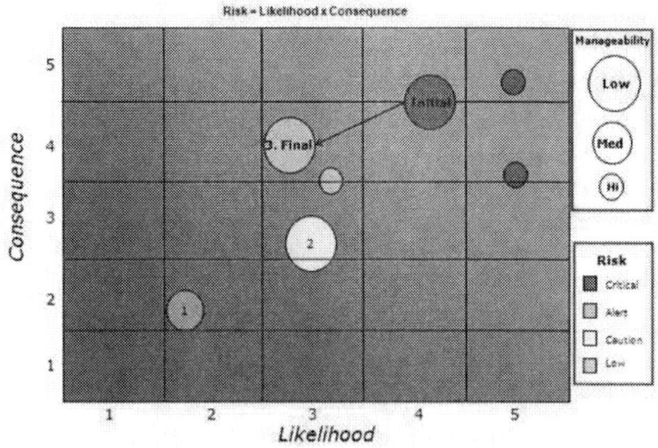

Fig 7: balloon diagram

STEP 4: Assign clear actions and accountabilities.

STEP 5: Progress and complete the actions - the technology risk is only reduced when the mitigating actions have been implemented.

This tool is appropriate for all new and revamp plant reviews, where a line by line process review is required. However the HAZOP process can be somewhat restrictive, so Johnson Matthey undertakes a supplemental broader risk assessment process, namely a Technical Risk Assessment (TRA), when evaluating risk assessment with new or revised flowsheets or new or improved catalyst in those flowsheets.

The Johnson Matthey TRA team for this project consisted of:

- Technology experts
- Project manager
- Catalyst development manager
- Process design engineer
- Commercial manager
- Johnson Matthey independent expert (experienced engineer, familiar with developing and commercializing other processes but who had not previously been involved with this project)

The TRA review focused on the conversion of the existing axial flow beds to the new axial-radial flow design, with a new catalyst **KATALCO$_{JM}$** K8-11R.

In the original design the shift beds had experienced problems related to bed fouling and water condensation with high H_2S concentrations. While these issues were noted, there were no specific actions required for the new design. The possibility of the highly exothermic methanation reaction occurring in low flow areas was discussed and a mitigation strategy was employed to minimize this possibility, with the use of inert support material in the bottom of the bed to ensure continual flow.

No major catalyst hazards were identified in the change to the new catalyst size, apart from the safety considerations associated with catalyst loading. The catalyst is pre-sulphided prior to delivery and, once sulphided, contact of the catalyst with air should be avoided as it may cause rapid overheating, generating very high temperatures that may damage the catalyst and the vessel. The reaction will also generate sulphur dioxide that could cause corrosion in downstream equipment.

Johnson Matthey recommended the installation of CatTracker® technology to accurately measure and monitor the catalyst temperatures in the new system. This consists of flexible multipoint thermocouples with up to 9 sensing points located at precisely defined locations within stainless steel tubing. The flexibility of the CatTracker enables the thermocouples to be fitted at installation and formed around center pipes and other internals. This enables temperature readings to be obtained at the desired positions that the reactor design recommends rather than where mechanical limitations permit.

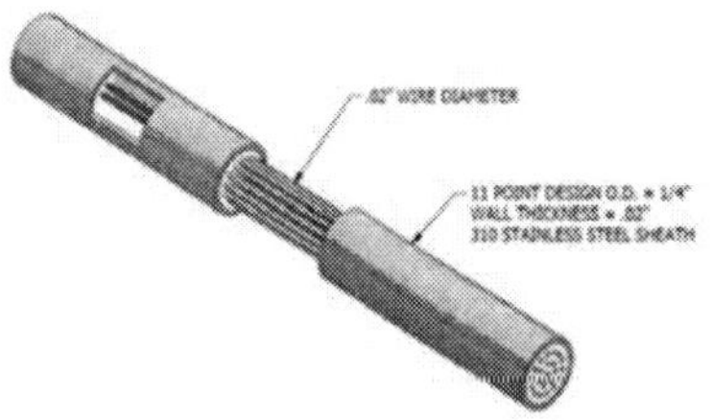

Fig 8: CatTracker Sensing Points

The revamp required that the new axial-radial flow internals, including a bottom skirt support, scallops and outlet distributor be installed into the existing vessels, which were located at the top of the tall gasification structure (See Figure. 9).

Fig 9: View of Gasification Structure

Detailed planning was required to ensure that all the new internal parts were manufactured to fit through the existing vessel manways. A series of photographs of the installation process are shown in Figures 10, 11 and 12

Fig 10: Loading Inlet Distributor Channels

One challenge that occurred during the installation period was due to an inlet nozzle protruding from inside the vessel that did not allow for the fitting of the scallop against the vessel wall. This 2" nozzle was not noted on the vessel drawings, as this vessel had been modified to sour shift catalyst service from its original duty at the Coolwater project site and the drawings had not been updated. The nozzle was subsequently ground out, however, the scallop needed to be modified to ensure a tight fit against the vessel wall.

Another issue required modification to one of the internals to account for a different outlet flange type to the one stipulated on the vessel drawing.

Fig 11: View of Bottom Support Skirt

Fig 12: View of Outlet Collector and scallops

Given the complexity of the revamp, credit goes to the design and installation crews. The learning here is that even in well documented projects there is always a possibility of surprises.

The revamp was completed on-time and on budget and the plant was re-started in August 2015.

Post Revamp Operation

The first stage of the axial-radial sour shift bed operates with an inlet temperature of around 248°C (480°F) and an exit temperature of around 430°C (804°F) with an increased volume to 91 m^3 (3221ft^3). It achieves a close approach to CO shift equilibrium of essentially zero. After cooling to around 250°C (485°F) the gas goes to the second stage, with an increased volume of 75m^3 (2652ft^3) where the CO is lowered to significantly less than 1% while raising the temperature to about 270°C (520°F). As can be seen in the temperature profiles, the beds quickly reach equilibrium with the high activity **KATALCO$_{JM}$** K8-11R catalyst, as indicated by the sharp temperature profiles. The start of run bed temperature profiles for Bed 1 and Bed 2 are shown in Figure 13.

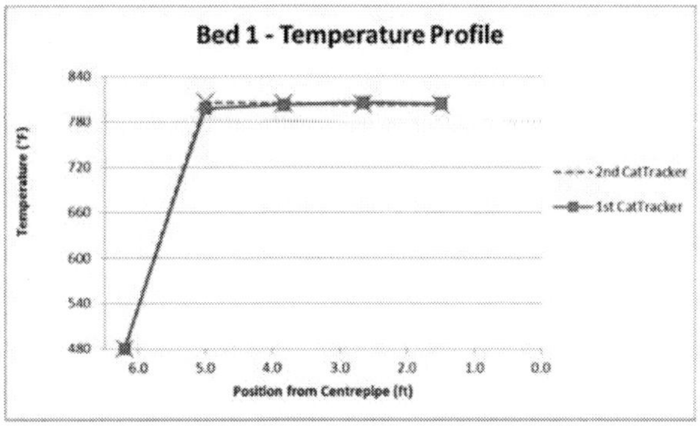

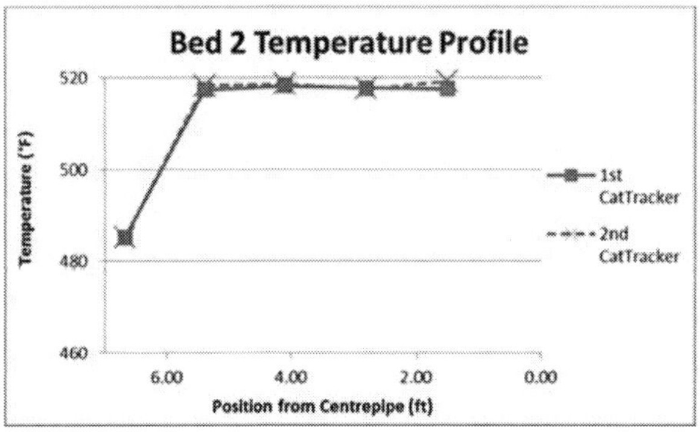

Fig 13: Bed 1 and Bed 2 Radial Temperature Profiles

The plant has achieved savings in overall pressure drop and additional production as a result of the Johnson Matthey axial-radial revamp project.

A plot of Bed 1 normalized pressure drop is shown in Figure 14 for the previous catalyst charge from October 2012-July 2015. The benefits of the axial-radial technology revamp are demonstrated with a normalized pressure drop of < 0.7 bar (10 psi) since start up in August 2015 with a nominal volume of 200% of the previously charged volume.

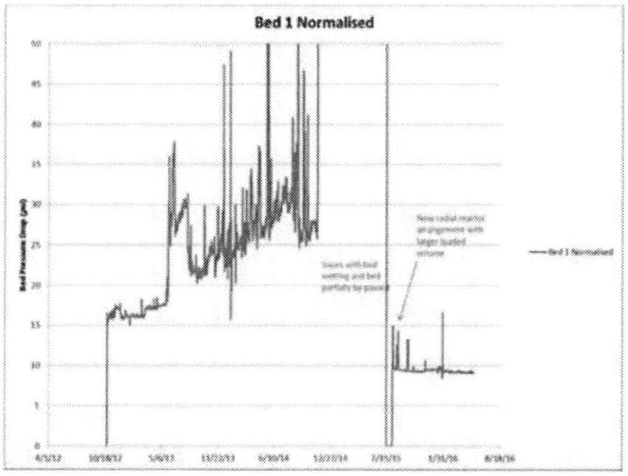

Fig 13: Plot of Normalized Pressure Drop across Bed 1 with two charges

After start up the plant ran very stably with new record ammonia production. Over the ensuing months the pressure drop in Bed 2 slowly increased without any clear indication to highlight the cause of the increase. There was no associated increase in pressure drop in the Bed 1. The plant decided to take a short turnaround in early 2016, with an inspection of both sour shift beds indicating that the bed inlet distributors had become detached and were lying on the top of the catalyst beds.

A visual inspection showed no catalyst bed movement was apparent in Bed 1 so a new modified inlet distributor was installed and the vessel buttoned up.

In Bed 2, the visual inspection indicated that the top support media had been disturbed due to high gas velocities and the support material was partially blocking the inlet distribution channels. This was the likely cause of the high pressure drop. The support medium was replaced in Bed 2, and a new modified inlet distributor was also installed and the plant bought back on-line.

A root cause analysis has indicated that faulty welding of the inlet distributors is the likely cause of the inlet distributor issue.

The plant continues to demonstrate higher production capacity as a result of the Johnson Matthey radial flow revamp. This is illustrated in the total flow to the sour shift beds being at record levels in Figure 14 below.

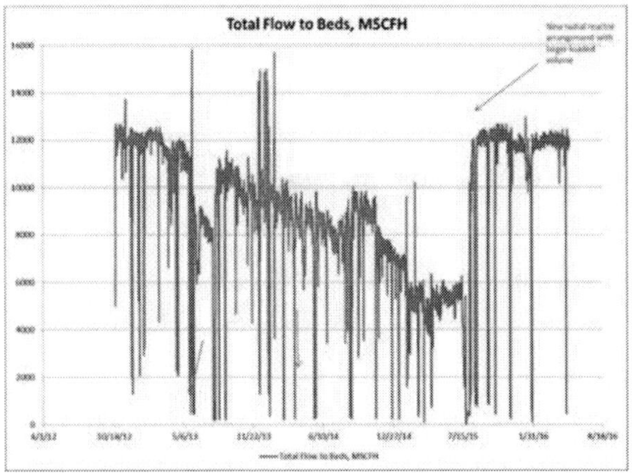

Fig 14: Total Flow to both sour shift beds

Conclusion

Johnson Matthey employs a variety of risk management processes to assess and mitigate risk both in our own catalyst manufacturing plants and in the products and technologies we supply to our customers.

Johnson Matthey's innovative sour shift axial-radial flow technology has been subjected to detailed risk assessments which concluded that the inherent risks are both well understood and well managed, and this is demonstrated with the large number of units in successful operation.

References

[1] Dutch Safety Board, Explosions MSPO2 Shell Moerdijk, July 2015

[2] EHS & Technology Risk Assessments for the Nitrogen Industry, J Pach, AIChE Vôl x, 2016

[3] Petroleum Coke Gasification Based Ammonia Plant, N E Barkley, AIChE Vol x, 2006

KATALCO is a trademark of the Johnson Matthey group of companies.

CatTracker is a registered trademark of Daily Thermetrics

Safe Pre-commissioning and Commissioning of a Multi-Product Syngas and Ammonia Facility

This paper will discuss the various technical and safety challenges associated with pre-commissioning and commissioning of a new multi-product syngas facility. It also focuses on the subtle nature of developing an effective site safety culture to enable safely working in excess of 10,000,000 man-hours with a multinational workforce in the Middle East.

Omer Hashmi
Selas Linde North America

Nrup Patel
Linde Engineering India

Fernando Rivera
Linde Jubail Industrial Gases Factory LLC

Introduction

The Linde Group's Engineering Division has engineered, delivered and constructed a new state-of-the-art multi-product syngas and ammonia facility at the new SADARA Chemical Company complex in Al Jubail, Kingdom of Saudi Arabia (KSA). Upon successful commissioning and start-up, Linde Group's Gases Division will own and operate the plant to supply hydrogen, carbon monoxide and ammonia to SADARA.

The syngas ammonia facility is comprised of a dual-train Syngas Generation section feeding a single train of ammonia production. Additionally, a refrigerated atmospheric ammonia storage tank of 20,000 ton capacity is included to ensure a continuous supply of product to the client during any scheduled or unscheduled downtime. Other process units enable recovery and purification of additional gas products, such as high-purity carbon monoxide (CO).

A multinational team of commissioning engineers from various Linde Engineering (LE) offices combined on this project to execute safe and efficient pre-commissioning, commissioning and start-up of this facility.

Process Description

A dual-train concept was chosen for syngas generation in this facility. The two trains are physically identical; however, one train generates clean export steam according to customer's specification, whereas the other train generates process steam for internal use only. The steam system of the latter recycles the process condensates of both trains to reduce effluent flows for environmental reasons as well as economic efficiency.

Natural gas from battery limit is compressed and distributed to the reforming section of the two syngas trains. Downstream of the reformers, the hot syngas is cooled down by a series of heat exchangers before being fed to the carbon dioxide (CO_2) removal amine wash process. The separated CO_2 is compressed and recycled back to the process as feed to the reformer on an as-need basis.

A Temperature Swing Adsorption (TSA) unit then removes all moisture and any residual CO_2 from the syngas, in preparation for the cryogenic separation and purification of CO within the coldbox. The purified CO is compressed and sent to battery limit as one of the products, and the remaining process gas is directed to a Pressure Swing Adsorption (PSA) unit to produce pure hydrogen (H_2).

The product H_2 is then mixed and compressed with high pressure nitrogen (N_2) from battery limit and fed to a Casale ammonia converter. The converted gas is passed through heat exchangers for energy efficiency before being condensed and refrigerated for storage as liquid ammonia product. The product ammonia is then vaporized before it is finally distributed to battery limit. Figure 1 summarizes the described process in a block flow diagram.

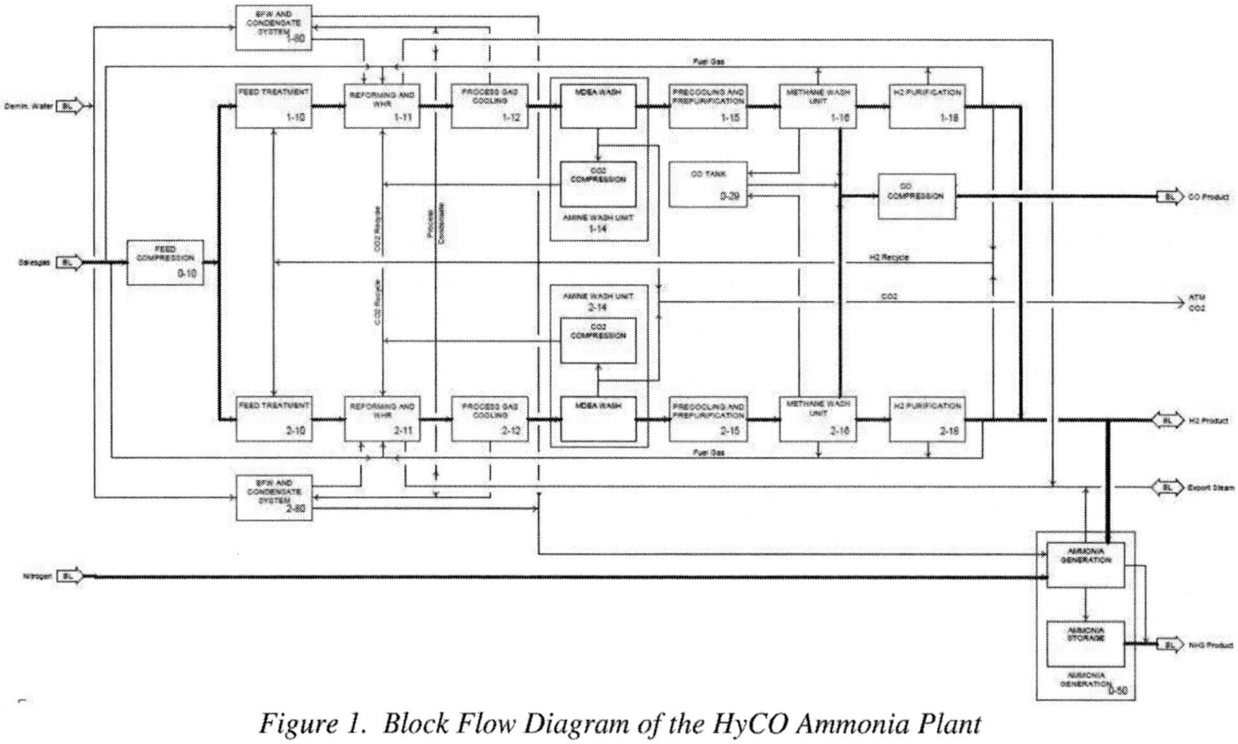

Figure 1. Block Flow Diagram of the HyCO Ammonia Plant

Planning and Workflow

During the course of an engineering, procurement, construction (EPC) project, Linde Engineering divides a plant into separate "hand-over-systems" (HOS) based on Linde's vast experience in pre-commissioning and commissioning of chemical and petrochemical plants. Each HOS is handed over from construction to pre-commissioning as soon as a system is "ready for pre-commissioning". Similarly, upon completion of the planned pre-commissioning activities, a system is declared "ready for commissioning".

For the Al Jubail syngas ammonia facility, the plant was split into nearly seventy (70) different HOSs. The size and complexity of these individual systems varied greatly ranging from a small system within the syngas cooling chain to the large intricate coldbox.

Each hand-over-system had its own procedure for executing the required pre-commissioning activities. However, some activities such as checking the construction against the P&ID, inertisation of piping and equipment, and carrying out tightness tests were common for all HOSs.

Right from the onset of construction on this project, no activities were executed without a proper permit to work. One of the key factors in the safety success of Linde here is the thoroughly implemented "permit to work" system. The different pre-commissioning and commissioning activities required different permits, which included, but were not limited to, a General Work Permit, a Hot Work Permit, a Confined Space Entry Permit and a Working at Height Permit.

Each one of these individual permits requires its own series of steps to be followed in order to plan, supervise and execute the task safely. As an example, Figure 2 shows the general workflow of how a task requiring a confined space entry permit is managed.

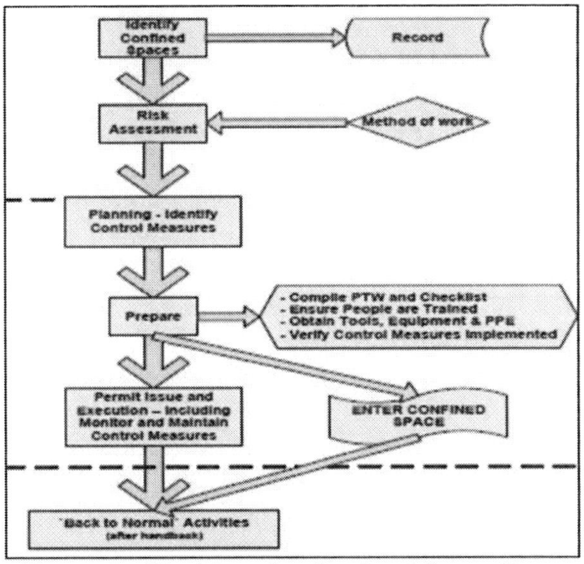

Figure 2. Workflow for Confined Space Entry Work Permit

The confined space is first recognized as an enclosed area and the hazards are assessed. A risk assessment is then performed to identify hazards, analyze and evaluate the risks, identify control measures and assess the method of work. Some critical considerations during this assessment include, but are not limited to, isolating energy sources, sufficient Lockout/Tagout (LOTO), purging of systems and equipment, lighting requirements, atmospheric testing for oxygen content, ventilating with air blowers, providing a means for access and/or any additional personal protection equipment (PPE).

A clear definition of "how" the work will be performed is also a part of this permit issuing process. Items such as number of workers required inside the confined space, personal gas monitoring requirements, tools and equipment to be used, emergency control and arrangements etc. are all valuable parts of preparing such a permit. Prior to the activity, the roles and responsibilities of each concerned individual were clearly explained to allow the task to proceed as smooth as possible. Upon completion of the permitted task, the system is returned back to its "normal" operating state.

A noteworthy item in the permit to work system for this project was the emphasis on adequate training and timely refresher courses on safe execution of critical tasks such as confined space entry. The Linde team, along with the construction workforce, was able to establish a number or interconnected codependent teams training, intelligently challenging, discussing and working together to achieve excellence in safety by taking ownership and responsibility for themselves as well as for others.

Based on the size, complexity and criticality of the HOS, and the nature of activities, the commissioning management team assigned manpower and developed a time schedule. Additionally, great importance was placed on vendor packaged units such as compressors and pumps to ensure similar activities were also executed for those units with an equal emphasis on safety.

Establishing a Safety Culture

One of the key success factors in this story was the establishment at site of a positive safety culture. Rather than seeking to "catch and punish" personnel who might be performing an unsafe act, Linde sought to switch the mindset of every person on site from the traditional reactive / punitive / independent mode, to a pro-active, supportive, and inter-dependent mode. This is accomplished when team members not only look after their own safety performance, but the safe actions of others as well.

This kind of a safety culture has been best described by "The DuPont Bradley Curve", shown in Figure 3, which was designed in 1995 to benchmark the path to a world-class safety performance.

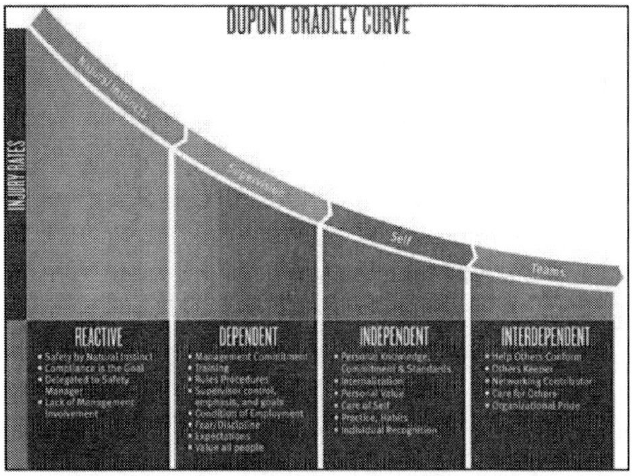

Figure 3. The DuPont Bradley Curve

In this project, the stakeholders, the project management team, the execution team, the construction and the site team, all the contractors as well as the client, all worked together towards achieving the unified goal of a world-class safety performance. The project as a whole has been able to achieve in excess of 10,000,000 manhours without a Lost Time Incident (LTI), and counting.

An example that depicts the camaraderie of the teams in this project to achieve such a milestone was the transfer of the site safety responsibility from the Engineering Division to the Gases Division after Mechanical Completion (MC) of the project. Despite the transfer of responsibility being achieved only after MC, both Linde Gas and Linde Engineering have worked together from the beginning of this project as one team not accepting low standards of safety or risk taking. The two entities truly believed and strived for long term sustainable improvement to achieve the target of zero injuries.

Pre-commissioning

Given the large size and the diverse nature of the various sections of this plant, pre-commissioning posed not only technical challenges, but also scheduling concerns.

Additionally, due to delays in the supply of some major utilities, the safe execution of scheduled tasks required added planning and watchful efforts. One such effort was the organization of compressed air supply by means of a temporary air compressor and a dryer station. Such temporary sources, while very helpful, also meant increased operational and safety concerns during the various pre-commissioning activities of the plant. Each pre-commissioning activity brought about its own unique safety challenge thereby requiring the entire team to constantly improvise and work together to maintain the site safety culture.

Initial Check of HOS against the P&ID

Once a system was deemed "ready for pre-commissioning", a commissioning engineer checked the system for consistency of construction against the P&ID. While this activity sounds fairly straightforward, one of the biggest safety challenges was dealing with the scaffolding that covered most of the plant at the time, especially the pipe rack, as seen in Figure 4.

Figure 4. Scaffolding around main pipe rack

Site rules required 100% tie-off using a safety harness when on a temporary structure. In fact, for even permanent structures such as platforms and vessels, until the safety department had thoroughly inspected the integrity of the structure itself, a safety harness was mandatory for all personnel. This was followed strictly both by Linde personnel as well as the construction contractor's workforce, as seen in Figure 5.

Figure 5. 100% tie off on elevated structures

Cleaning of Equipment and Piping

Cleaning of equipment and piping is one of the most critical activities in pre-commissioning of any plant. In the case of this facility, the dusty weather and surrounding desert meant that extra attention had to be paid to cleaning of lines.

After a preliminary check confirmed that tie-ins were as per the P&ID, and that correct valves and piping components had been installed, the lines and equipment were cleaned by air blowing, Figure 6, water flushing or sand blasting.

Figure 6. Air blowing of lines

Safe execution of cleaning requires a lot of planning and preparation. The blow-out points were temporarily barricaded and a horn was sounded to ensure that those in the vicinity were not caught off-guard or without any form of ear protection. In general, the team managing and supervising such line blowing activates utilized ear plugs and/or ear muffs rated between 95 and 105 decibels, as in Figure 7a. In the case of steam blowing, to minimize the noise generated, the steam was blown into the atmosphere via a silencer, as shown in Figure 7b.

Figure 7a. Use of ear muffs for protection

Figure 7b. Steam Blowing via Steam Silencer

Inspection of Equipment

Most large sized equipment needs to also be cleaned and inspected to minimize complications during commissioning and start-up. Reactors, separator vessels, drums and furnaces need to all be checked thoroughly. The challenge with this kind of equipment is that these are all confined spaces and they must be dealt with extreme caution.

As part of the pro-active interdependent approach for maintaining a world class safety culture in the facility, as an absolute minimum, each individual entering any confined space was required to carry a personal gas detector with alarms to ensure continuous gas monitoring, a ventilation source was made available to ensure sufficient oxygen content, a watchman was available outside the confined space maintaining an entry/exit log sheet, and an adequate lighting source was organized as well.

One of the key success elements of this project was the training and ownership of designated watchmen for confined space work. Their duties, which they were duly trained in, included an awareness of activities inside the confined space, verbal communication with authorized entrants to ensure their well-being, maintaining an entry-exit log, remaining vigilantly placed outside the confined space during entry operations, including when entrants leave for breaks unless all openings are properly secured to prevent unauthorized access. These watchmen were a great example of the interdependent safety culture that the Linde Group achieved in this project. Most of their responsibilities focused on the safety of other personnel involved in the task but their responsible ownership of safety was a highlight of this project.

In certain special cases of confined space activities, additional safety efforts were required. In the case of the reformer refractory inspection for example, the danger is compounded because not only are the people working in a confined space, but also working at height. Even though it may not have been very comfortable at the time of these inspections, all personnel maintained a safety harness during this inspection inside the reformer furnace, as shown in Figure 8.

Figure 8. Use of safety harness inside furnace

Filling of Catalyst and Adsorbent Materials

In parallel to systems that were still under construction or under cleaning, hand-over-systems whereby catalyst and adsorbent loading was required were also catered. These activities posed a completely different safety challenge – dealing with hazardous chemicals.

In order to minimize exposure to catalyst dust, all workers assigned to the catalyst handling and loading jobs, were specifically required to wear a dust mask in addition to the regular personal protection equipment (PPE), shown in Figure 9.

To improve on efficiency and avoid duplication of efforts; the same crew of workers was maintained for catalyst loading activities wherever possible.

Figure 9. Dust masks during catalyst loading

Additionally, catalyst and adsorbent loading meant lifting of heavy drums to elevated structures. In addition to safety coordination and skillful lifting, as shown in Figure 10, this required organization of working areas, providing the workers with sufficient space to execute the task properly, as well as barricading areas at grade, directly under the lifting radius to prevent people from working directly under a lifting activity.

Figure 10. Lifting of empty catalyst drums

Final Check of HOS against the P&ID

Following the line and equipment cleaning, the HOS was reassembled and a more thorough check was done to ensure that the system was ready for pressurizing and testing for tightness. During this check, subtle details of the piping are verified to ensure that the system is ready to be pressurized. This was one of the greatest strengths of the commissioning team on site.

When dealing with a facility of this magnitude, it is not uncommon that some items are missed during construction due to time constraints and emphasis on getting the job done. However, sufficient time and effort was scheduled for the final checks of each HOS. Some of the interesting findings included loose flanges, leftover temporary construction blinds, damaged gaskets, missing insulation etc, shown in Figures 11a. and 11b.

Finding these items upfront meant that the HOS was handed back to the construction department for further reinstatement. This minimized future issues whereby the system may have already been pressurized.

Figure 11a. Loose flange on PSV inlet

Figure 11b. Damaged gasket

Testing of the Control and Trip System

Another parallel and very critical item during pre-commissioning was plant-wide safety function testing. Prior to charging process gas into the system, the control philosophy and the trip system were verified at length.

For the testing of Safety Instrumented Functions (SIFs), a HART communicator was used to simulate inputs from the transmitters in the field. In the case of pressure transmitters and differential pressure type flow transmitters, a pressure calibrator was additionally used to verify the working of the transmitter. The testing of each SIF was carried out based on individual SIF Protocols prepared by the engineering team. These tests comprised of manually testing the final element first, followed by the testing of the transmitter and lastly the testing of the control logic implementation and verification of final element action.

One distinct Safety Instrumented System (SIS) was the flame detection system for the reformer burners. With fuel gas not available on site, this was done using a portable gas cylinder. A detailed Job Safety Analysis (JSA) was performed in preparation of the activity. Based on this, a flint lighter was to ignite a flame, Figure 12a, in the presence of a fire extinguisher and only 100% fire retardant clothing was permitted for this activity, Figure 12b.

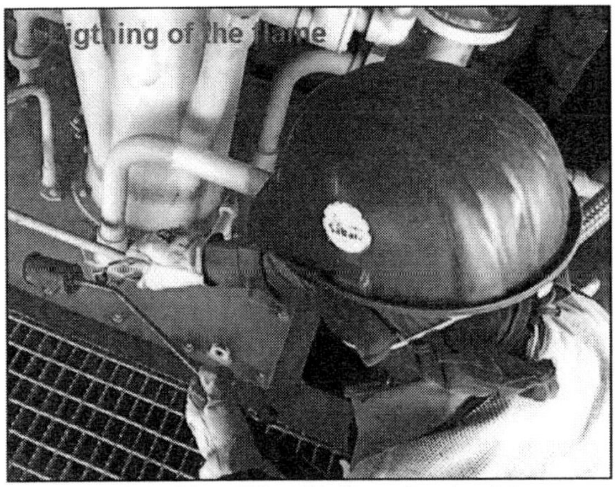

Figure 12a. Flint lighter for ignition

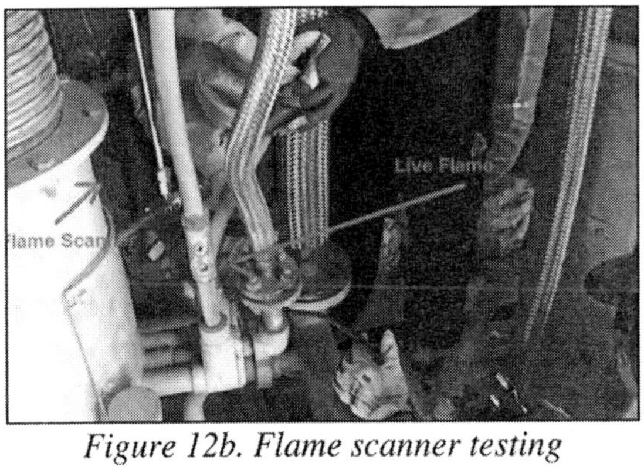

Figure 12b. Flame scanner testing

Once again, the motivation of the workforce and entire site team was that of achieving unparalleled safety success. If that meant taking some time, planning and only then executing a task, this was never compromised on.

Tightness Tests

Once the final reinstatement of a HOS was completed, the system was pressurized with low pressure compressed air, in most cases to nearly 20-30% of normal operating pressure, and the pressure drop of the system was monitored over time, to determine if the system was "tight" i.e. holding pressure.

During pre-commissioning, this was considered one of the more monotonous tasks – pressurizing each HOS with low pressure compressed air and searching for leakages in the system by using a soap solution bubble test, as shown in Figure 13.

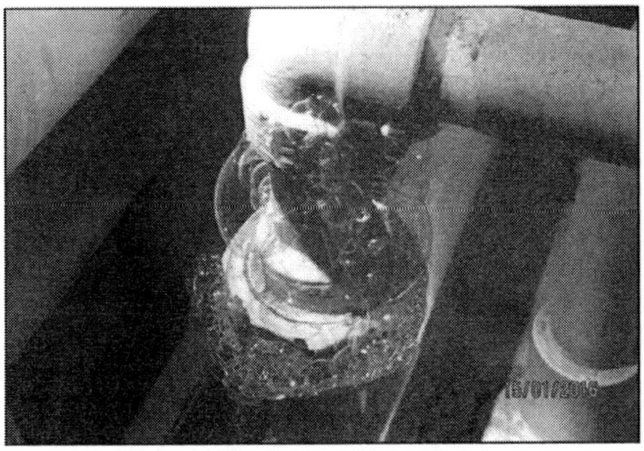

Figure 13. Soap Solution bubbles forming at gas leakages

However, in retrospect, this proved to be one of the most valuable investments of time and resources; especially after high pressure nitrogen was received later on and the systems were tested for leakages at higher pressures. The systems in the dual syngas generation trains were leak tested at normal operating pressure with high pressure nitrogen. However, due to the high operating pressure of the ammonia generation loop, these systems were only leak tested to about 20% of the normal operating pressure.

Once again, this was a prudent collaboration of several teams working together – first team working to find these leaks, then recording and maintaining tracking lists of these leaks, and lastly the team rectifying these leaks well in advance of actual pressurization with combustible, toxic process gases.

Vendor Packaged Units

For any plant, a particular area of concern, both technically as well as from a safety standpoint, is the tie-in of vendor packaged units to the balance of the plant. This plant had several vendor packaged units of this nature, ranging from simple centrifugal pumps and reciprocating compressors to the more complex refrigeration screw compressors.

Each of these packaged units was treated as its own hand-over-system. This meant that it was individually checked for construction against the P&ID, the lines were cleaned, and the control logic was verified. An important item of safely testing such units was the electrical isolation of the motor in the Motor Control Center (MCC) and temporary wrapping of the wiring with insulation tape, as shown in Figure 14.

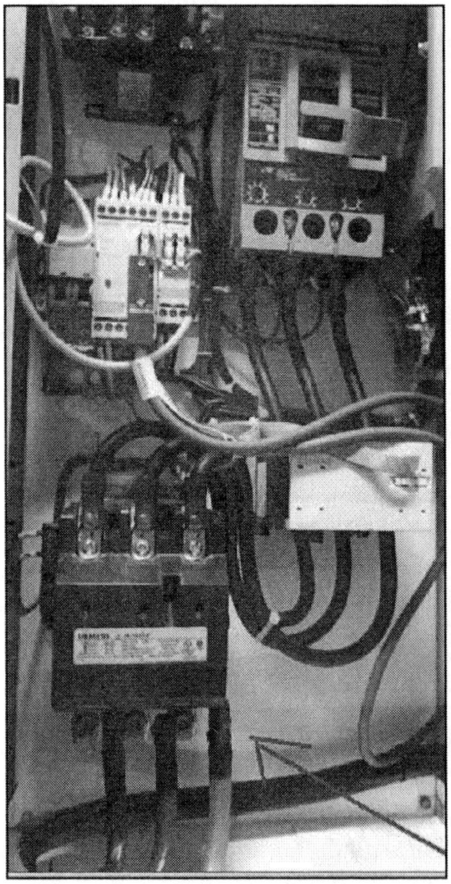

Figure 14. Electrical isolation of power supply

This isolation was required to allow testing the trip logic of the different compressors without having to actually run the machine. This allowed safe testing of the packaged units without risking the safety of the machines.

Commissioning

Commissioning and start-up of any plant involves take-over of utilities and start of nitrogen circulation within the process units. Prior to introduction of any flammable gases, the flare unit needs to be ready for commissioning and start-up.

This facility comprised of two individual flares – one for the dual syngas generation trains and a separate one for the ammonia production section. However, in preparation of charging the fuel gas header, the safety process and requirements were revisited. Several new requirements were mandated in addition to the existing rules and regulations, such as:

- 100% Flame Retardant Clothing (FRC)
- Personal gas detectors
- Usage of only non-sparking tools

The facility is currently in the middle of its commissioning and the experience during this phase is still being gathered. Safety of the process, plant and personnel continues to remain a top priority during these ongoing activities.

Flare Systems

The commissioning of the flares presented a new technical and safety challenge. A few pilots on the flare dedicated to the syngas section of the facility did not ignite upon introduction of pilot fuel gas. Since safely accessing and repairing these pilots from the platform was considered risky, despite the necessity to meet a variety of commissioning schedules, a scaffold had to be erected on top of the flare, approximately 60 meters (200 feet) tall, to inspect and troubleshoot the pilots, shown in Figure 15.

Figure 15. Scaffold for flare troubleshooting

Once again, a detailed JSA was performed and special emphasis was placed on the climatic conditions of Al Jubail, due to extreme windy conditions. Even at acceptable wind speeds, work on the flare was sometimes suspended to ensure that the safety of personnel and plant was not compromised.

Syngas Section

Following the modification of the syngas flare, the feed gas introduction to the plant was possible and the first reformer was fired up. Positive isolation was implemented by closing permanent blinds on the second syngas train to prevent unplanned pressurization.

Despite leak tests that were performed during pre-commissioning, some small leakages were detected via portable individual gas detectors on the fuel piping of individual burners.

Additionally, the amine wash process for removal of CO_2 posed a totally different challenge. A chemical suit was to be worn for the dilution and loading of the solvent into a storage tank.

The combination of young commissioning engineers and senior engineers provided a very proactive approach to the commissioning of the syngas train. The CO_2 compressor was initially left isolated to minimize the potential of carrying over dilute amine into the compressor in case the regenerator column had foaming and/or a sharp level increase. While doing so, the operations team was advised to monitor the CO_2 vent silencer closely.

The CO Compressor housing had CO leakages in spite of thorough leak checks and successful tightness tests. Fixing these leakages was new safety challenge given the toxic nature of CO, a.k.a. "silent killer". The compressor shelter was left open to dilute the atmosphere inside and the exhaust fans were switched ON. Additionally, an instrument air hose was used to continuously purge the area of activity and the flange being worked on. The number of personnel inside the compressor housing was minimized for the duration of this activity and each individual was carrying a personal CO detector with alarms.

One unit at a time, process gas was introduced to different sections within the syngas train. Slowly, several online analyzers were also commissioned and on spec production of H_2 and CO was confirmed.

Ammonia Section

Due to slight schedule delays, the commissioning activities of the ammonia section have not yet commenced. The planned activities include, but are not limited to, replacement purge of the nitrogen atmosphere with ammonia, commissioning of the compressor packages, Casale convertor, refrigeration unit and the ammonia tank.

Conclusions

Linde Engineering has engineered, delivered and constructed a new multi-product syngas and ammonia facility at the new SADARA Chemical Company in Al Jubail, KSA. Throughout the pre-commissioning and commissioning activities so far, safety has been a top priority for the entire team. By paying attention to delicate details, the entire project team successfully established a culture of interdependent safety at the site.

The entire workforce, comprising of Linde Engineering Division, Linde Gases Division and the construction contractor, has worked together as one team with a common goal of achieving excellence in safety by taking ownership and responsibility for themselves as well as for others. Adhering to all safety regulations, the site team has successfully achieved in excess of 10,000,000 man-hours without a Lost Time Incident.

The world class safety culture achieved by the team in this project can be attributed to a number of key success factors, such as:

- Valuable investment of time and resources in pre-commissioning during tightness tests and P&ID checks to minimize downstream complications
- Distribution of roles and responsibilities between the entire team resulting in ownership and an interdependent working relationship
- Functional implementation of a detail oriented permit to work system
- Comprehensive training of the workforce and efficient specialization of their working skills (like watchmen for confined space entry works)
- Planning, discussing, supporting and challenging each task to ensure that the safest practices are followed

Throughout the project, the entire team has truly believed and strived for long term sustainable safety practices to achieve the target of zero injuries. As milestones have been achieved, the entire team continues to work towards the execution of commissioning activities with the same diligence, and is committed to achieving the next safety landmark.

Acknowledgements

The authors gratefully acknowledge the contributions of the following individuals and organizations, without whom this paper would not have been possible:

- Linde Jubail Industrial Gases Factory
- Selas Linde North America
- Linde AG Engineering Division
- Linde Engineering India
- Kenneth Lamb
- Jamil Amir Baduwi
- Winfried Leclaire
- Bernd Klein
- Nirav Gabani
- Florian Darchinger

Increased Risks for Leakages in Aging HP Urea Gas Lines due to unexpected Failure Mode

In chemical plants the piping systems in general get less attention compared to the static and rotating equipment. This can be explained by the fact that in many organizations the responsibility for the said piping systems is unclear since they travel across battery limits, the piping systems are often difficult to inspect internally and have in general poor accessibility (piping racks). This is not justified, since most of the failures in chemical industry are reported in piping systems. This is also true for urea plants. Recently similar cases of leakages occurred in the high pressure process gas lines of aging urea plants (more than 30 years of operation). The failure mode in all cases are similar but not fully understood, but can be classified as a kind of stress corrosion cracking phenomena. In this paper the case histories will be discussed and a possible failure mode will be described. Furthermore, based on these experiences, it is advised to incorporate this failure mechanism in the RBI programs for high pressure urea synthesis gas lines. This also necessitates introducing appropriate inspection methods to detect and evaluate these defects.

Johan Thoelen and Vincent Duponchel
Yara International Brussels, Belgium

Alex Scheerder,
Stamicarbon BV, Netherlands

Introduction

In chemical plants the piping systems in general get less attention compared to the static and rotating equipment. This can be explained by the fact that in many organizations the responsibility for the said piping systems is unclear since they often travel across battery limits, the piping systems are often difficult to inspect internally (accessibility) and externally (insulation, tracing) and have in general poor accessibility (piping racks). This is not justified, since many of the failures in chemical industry do occur in piping systems. This is also true for urea plants. Recently, several similar cases of leakages and failures occurred in the high pressure process gas lines of aging urea plants (more than 30 years of operation). In this paper, three case histories will be discussed and a possible failure mode will be described.

Case History 1

The first case occurred in a 1975-tpd urea plant based on Stamicarbon stripping technology.

The urea plant was commissioned in 1981 and most of the high-pressure synthesis lines are still the original AISI316L-UG lines.

In December 2012 a first leakage occurred in the HP scrubber gas line, between the top dome and the bottom head. After one year of operation a second leak occurred in December 2013. In this case the leak was situated in the gas line from the urea reactor to the HP scrubber, near the urea reactor outlet nozzle.

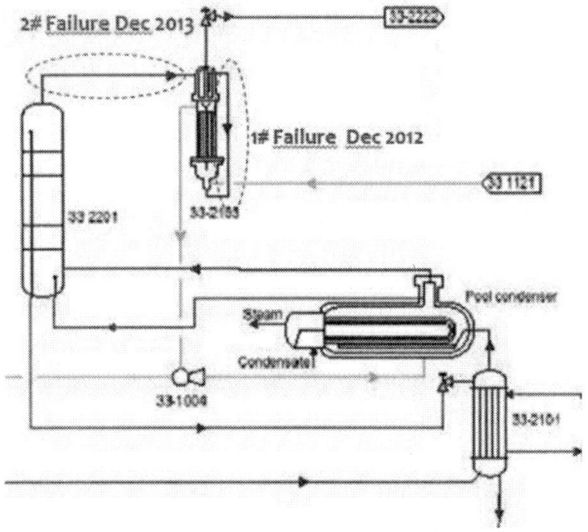

Figure 1 – Urea Synthesis loop indicating the leak areas

The first leak in 2012

In December 2012 a leak occurred in gas bypass line of the HP scrubber. The unit was stopped and the elbow was removed for failure analysis. The removed cracked elbow was fully inspected, inside and outside, by cutting it in two half pieces, see figure 2 and 3.

The second leak in 2013

The second leak was observed in the urea reactor outlet gas line in the third elbow, before the long pipe up to the HP scrubber on the 8th floor. Figures 4 and 5 show the location of the leak.

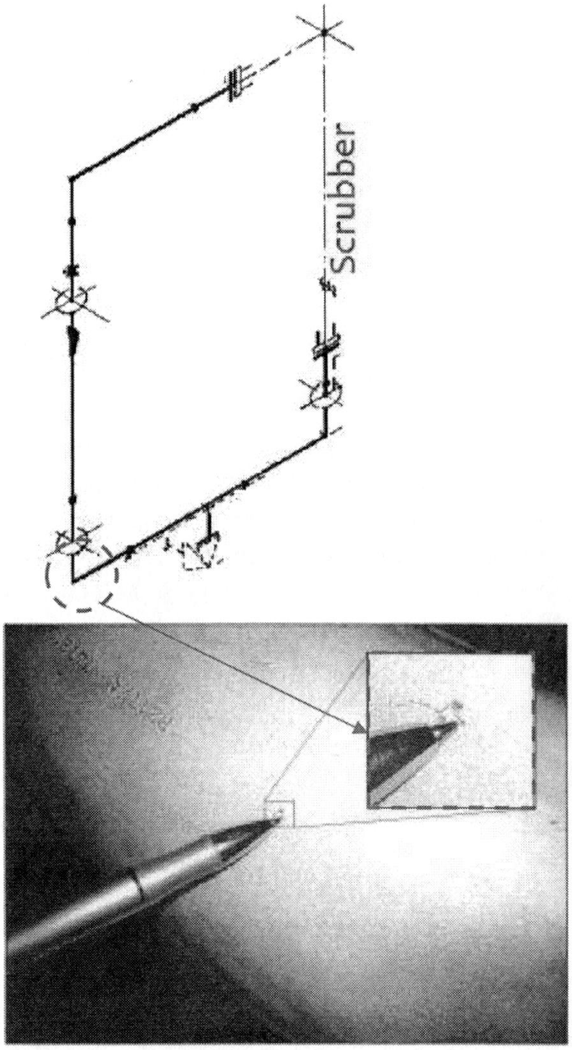

Figure 2 – HP scrubber gas line crack observed

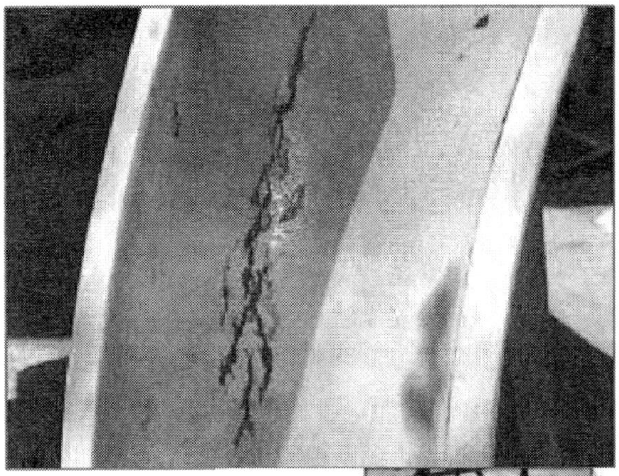

Figure 3 – Crack like indications inside the elbow of the HP Scrubber gas Line AISI316L UG, nominal diameter 150 mm (6in) - 1# leak

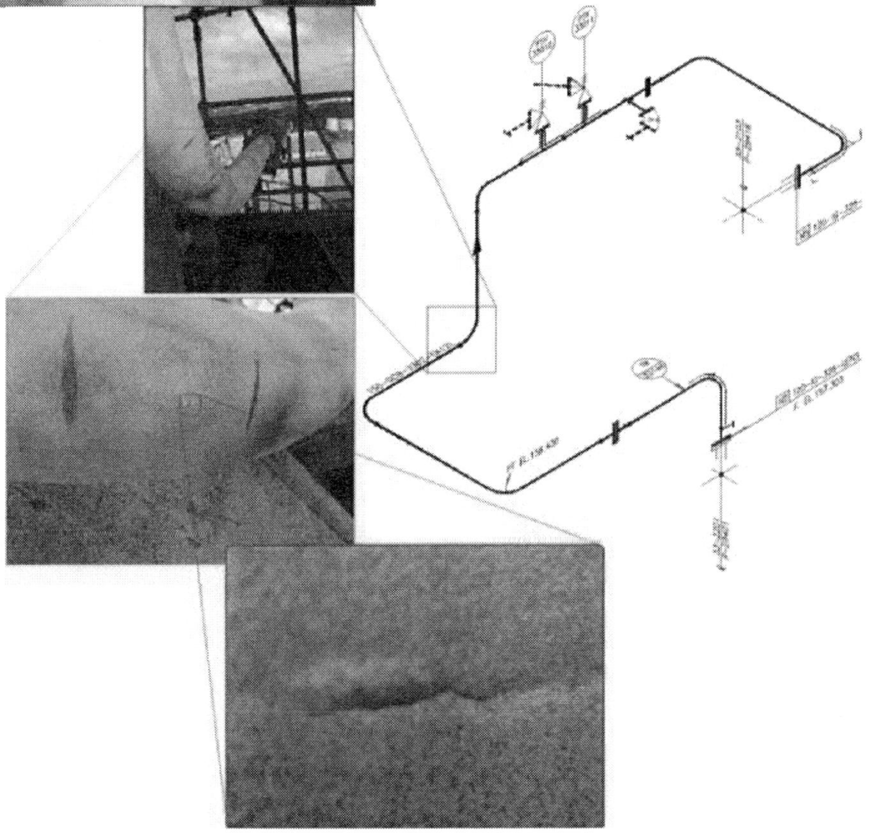

Figure 4 – Location of the second leak in 2013 – External crack after dye penetrant test

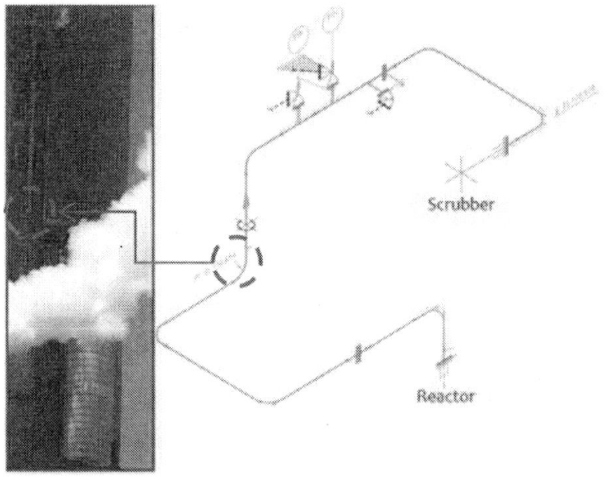

Figure 5 - Second leak in reactor outlet gas line

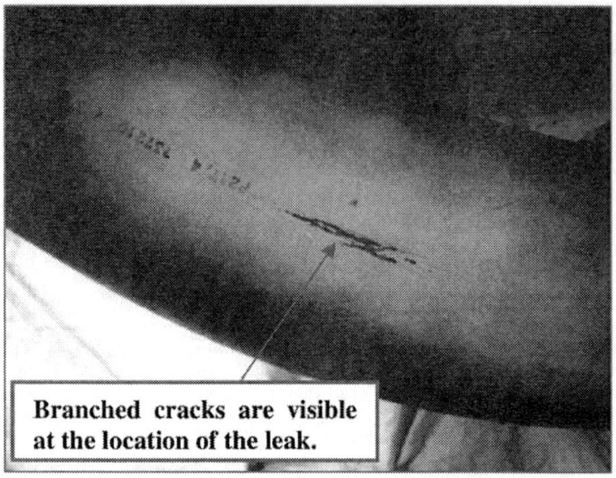

Figure 6 - Inspection by dye penetrant test at the external surface after removal of a boat sample

Examination of the removed cracked elbow pieces

Besides cracked elbows, also straight pipe segments were removed for metallographic examination. Visual inspection and dye penetrant testing confirmed that the cracks started at the process side. Figure 6 shows the leak area. Numerous branched cracks are visible, which is characteristic for stress corrosion cracking (SCC).

The majority of cracks were found in the elbow sections, and mainly located on the neutral line of the elbows, not in the inner or outer regions of the bend, see figure 7

During the inspection of all removed elbows, severe condensation corrosion was observed in the heat affected zones (HAZ) of the welds, see figure 8. Apparently condensation corrosion is more severe in these areas. However, there was no evidence that the AISI316L-UG material was sensitized in the HAZ.

Severe condensation corrosion was also found on the pipe surface close to the steam tracing tube. A possible explanation is that, over an extended period, the steam tracing temperature was not sufficient to keep the pipe wall above the dew point of the gas.

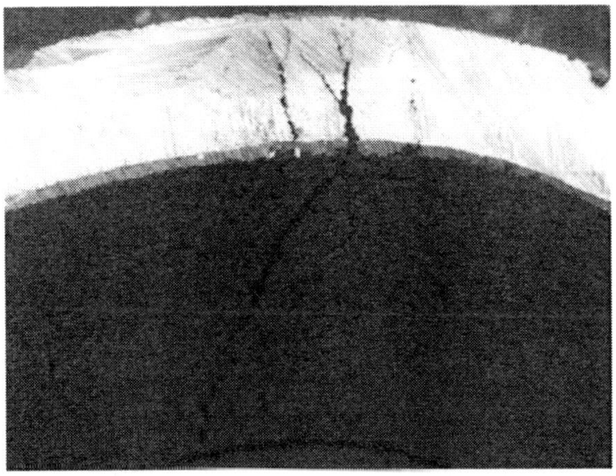

Figure 7 Stress corrosion cracks starting from inside

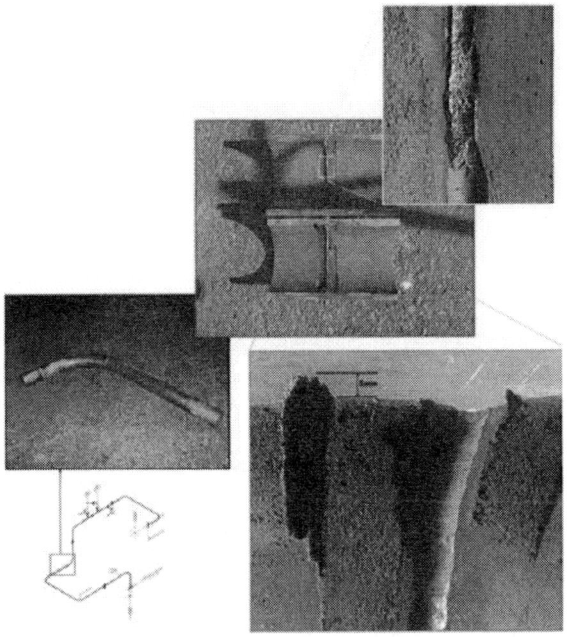

Figure 8: Intergranular corrosion near the welds

In general one can state that condensation of gases in the high-pressure gas pipelines cannot be completely avoided. Not even when a proper steam tracing is used (tracing temperature above dew point of the gasses). This typically will result in localized intergranular condensation corrosion (wall losses). However under these condensing conditions, stress corrosion cracks can apparently develop also. A metallographic examination was conducted to investigate which other factors play a role in this respect.

Metallographic examinations

Metallographic examination revealed that the bends are cracked predominately intergranular, see figure 9. This could indicate chloride stress corrosion cracking.

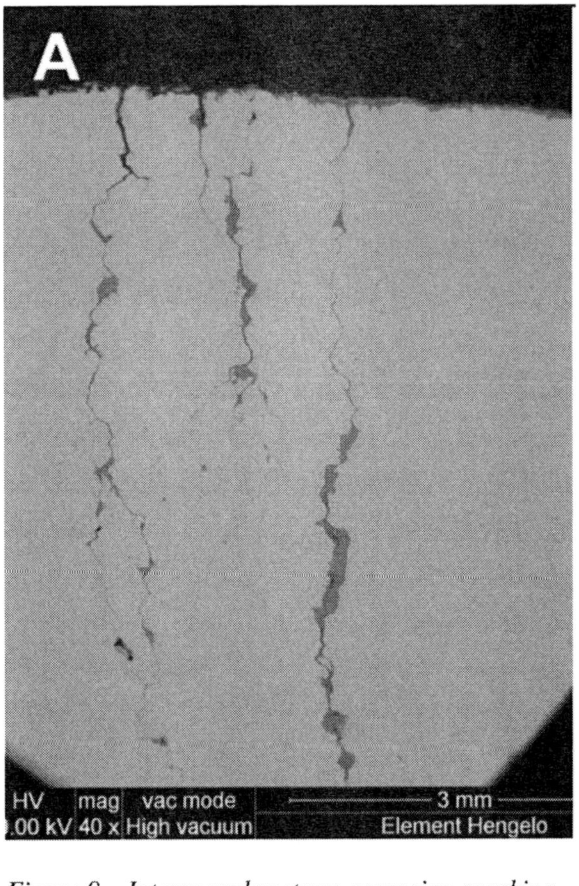

Figure 9 – Intergranular stress corrosion cracking

The presence of chlorides was confirmed using Energy Dispersive X-ray spectroscopy (EDX analysis), see Figure 10.

However, chloride-induced stress corrosion cracks are pre-dominantly trans-granular in nature, so a firm conclusion of the known chloride stress corrosion cracking phenomena cannot be drawn.

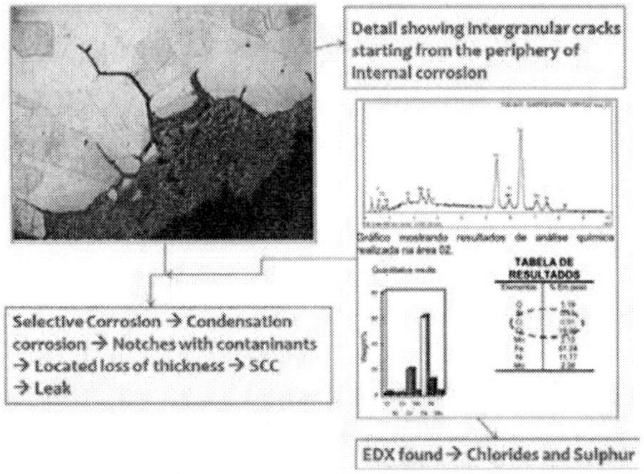

Figure 10 - Result of EDX analysis

Striking is also the presence of oxides along the cracks (see figure 9) and this points in the direction of slowly propagating cracks such as so-called Strain Induced Integranular Cracks (SI-IC). This is a failure mode which is observed in the liners in gas-phase areas of urea reactors.

In this failure mode the following conditions are a pre-requisite:

- The presence of an electrolyte as a result of condensation of ammonium carbamate in the gas phase.
- Plastic deformation of the material.

Other conditions may enhance this failure mode, such as:

- The presence of chloride and/or sulfur in the porous oxide layer.
- Inferior material (i.e. large grain size),
- Large number of pressure cycles

Presence of an electrolyte is proven by the presence of condensation corrosion. In the EDX analysis the presence of sulphur and chlorides are also confirmed.

The grain size number found in the two AISI316L-UG elbows was 2-3, see figure 11. This also indicates that both elbows were prone for this type of corrosion attack.

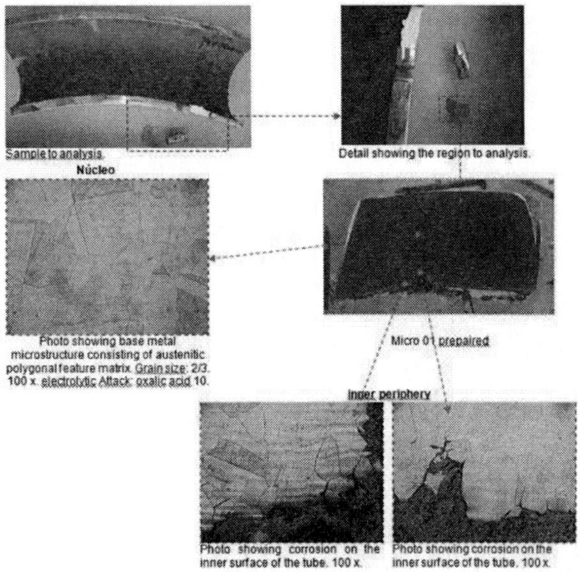

Figure 11: grain size investigation

No evidence was found in the microstructure for plastic deformation of the material (glide steps, for example). However, the cracked area showed some degree of deformation macroscopically (flattening, see Figure 12), which might indicate plastic deformation.

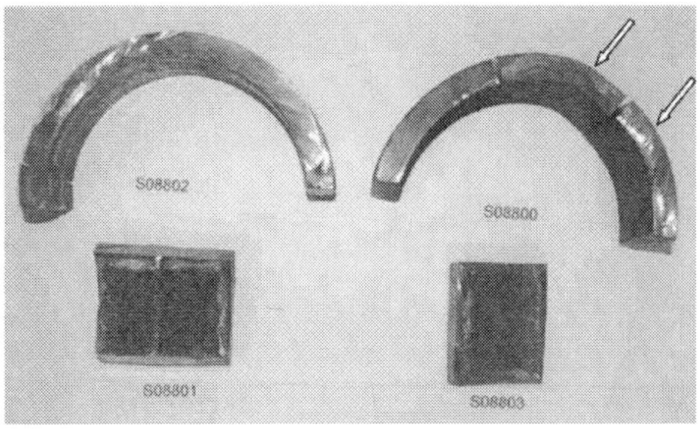

Figure 12 - Pipe flattened (deformed) at cracked area; might indicate plastic deformation

So, almost all pre-requisites are present for SI-IC, to occur.

Finally, the fact that these cracks started to leak after more than 30 years operation also points in the direction of SIIC. As already mentioned the crack propagation of SIIC is slow and progresses predominately during pressure cycles (start/stop).

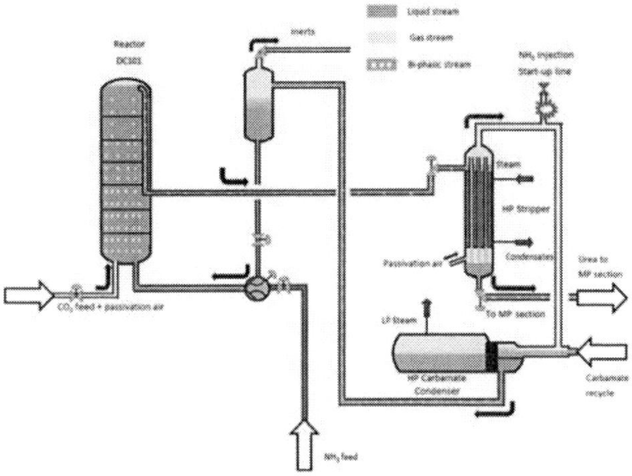

Figure 13 – Urea Synthesis loop indicating the leak location

The pipe ruptured in 2011 with major ammonia emission and leading to the temporary closure of the main road nearby and a local media coverage. Fortunately, no one was injured.

Case History 2

The second case occurred in a 2100-tpd urea plant based on Saipem technology. The urea plant was commissioned in 1979 and most of the high-pressure synthesis lines are still the original AISI316L-UG lines.

The failed pipe is used as a start-up line of urea plant. The line is only used during the filling phase of the reactor to inject ammonia. The pipe is 2", made of 316LUG material and has been traced with 5.5 bar steam.

Cause of the pipe failure

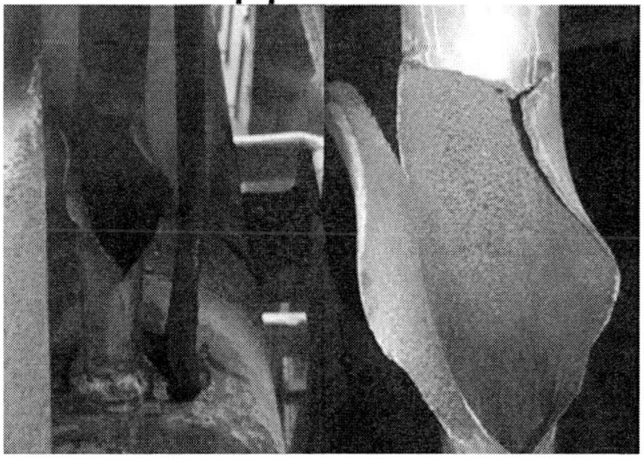

Figure14: findings after removal of the insulation

From visual inspection it became clear that the rupture is associated with large plastic deformation. This is clear evidence of a strain related rupture failure mode: remaining cross section of the pipe wall was not able to sustain the internal pressure anymore.

UT thickness readings made on failed pipe shows uniform loss of thickness around pipe circumference and a loss thickness profile along the line as illustrated in the figure 13 below. Taking the main gas line as a reference, the reduction in thickness increases with the distance z from the main process gas line until the center of failure position at z=200mm and then it decreases down to zero: at about 1000mm. Far from the main gas line the actual thickness is equal to the nominal thickness.

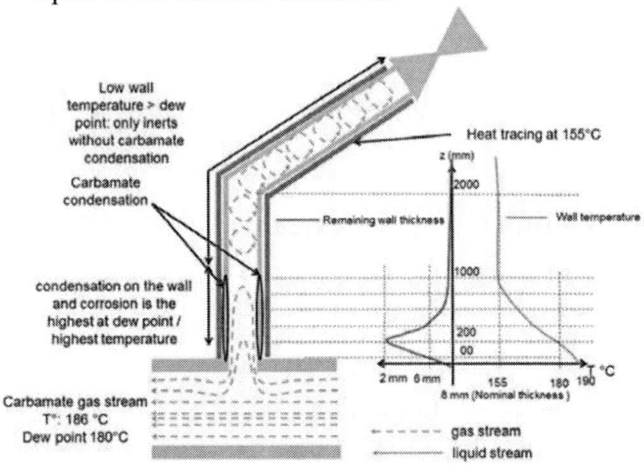

Figure 15: cause of the localised heavy corrosion

The main reason of this pipe failure is the condensation of carbamate on the pipe wall. The highest corrosion rate was observed where the condensation starts in the area of highest temperature.

The top part is unaffected for two reasons:

- Inerts build up in this area, reducing the condensation temperature
- Lower temperature in this part of the piping due to the "cooling effect" of the tracing

Case History 3

The third case occurred in a 1000 mtpd urea plant based on Stamicarbon technology. The urea plant was commissioned in 1969

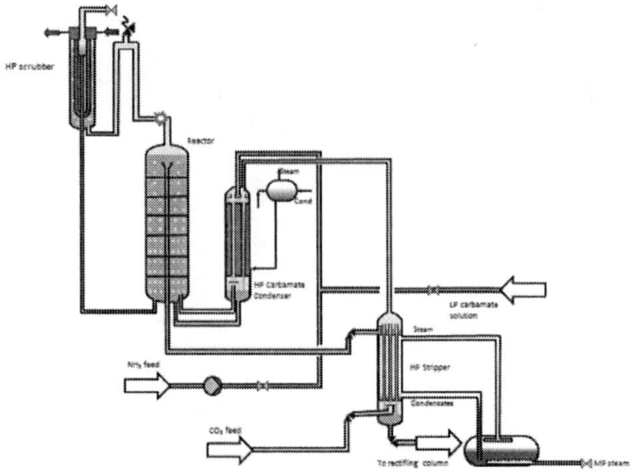

Figure 16 Urea Synthesis loop indicating the leak location

The failed line is the connection between the reactor and the scrubber. In 2011, most of this piping had been replaced in X2CrNiMoN 25-22-2 and only the spool piece, used to remove the top cover of the reactor, was re-used because it was recognised as easily accessible for inspection. This gas line was DN80 with 11.35 mm original thickness, insulated and traced with 9 barg steam. Due to its criticality and the experience from 2011, the pipe was part of the RBMI study and it was inspected 2 months before the failure with the following results:

- Thickness measurement showing no losses (11 UT measurement marked on figure 15 below)
- As well X-ray examinations have been done (2 locations "radio" in figure 15 below) without any findings.
- The only findings of the inspection was the flange face which has been re-machined during this TA

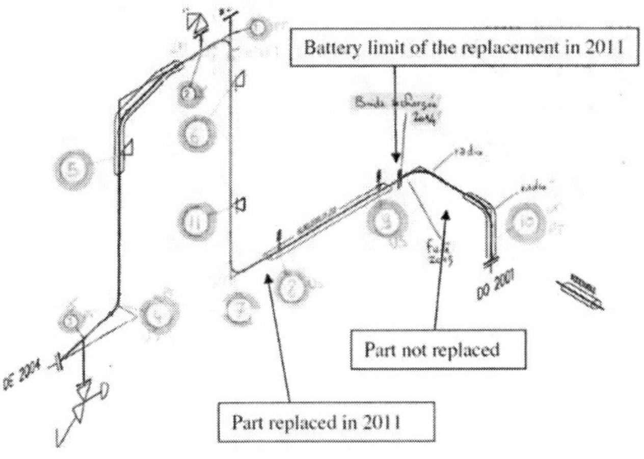

Figure 17: inspection plan of the failed pipe

The incident

On 09/02/2015, two ammonia sniffers, located on the top of the reactor and installed to detect a potential leak of one of the flanges, went into alarm and reached a maximum value of 300 ppm for short period. The plant personal decided to remove the insulation of the top of the reactor to search for a potential leak unsuccessfully. Then the insulation of the rest of the piping was removed to inspect the other flange and it showed a large amount of carbamate on the insulation. A small pinhole was finally found in the HAZ of the pipe. Immediately after this discovery the plant was stopped and drained.

Figure 18: findings after removal of the insulation

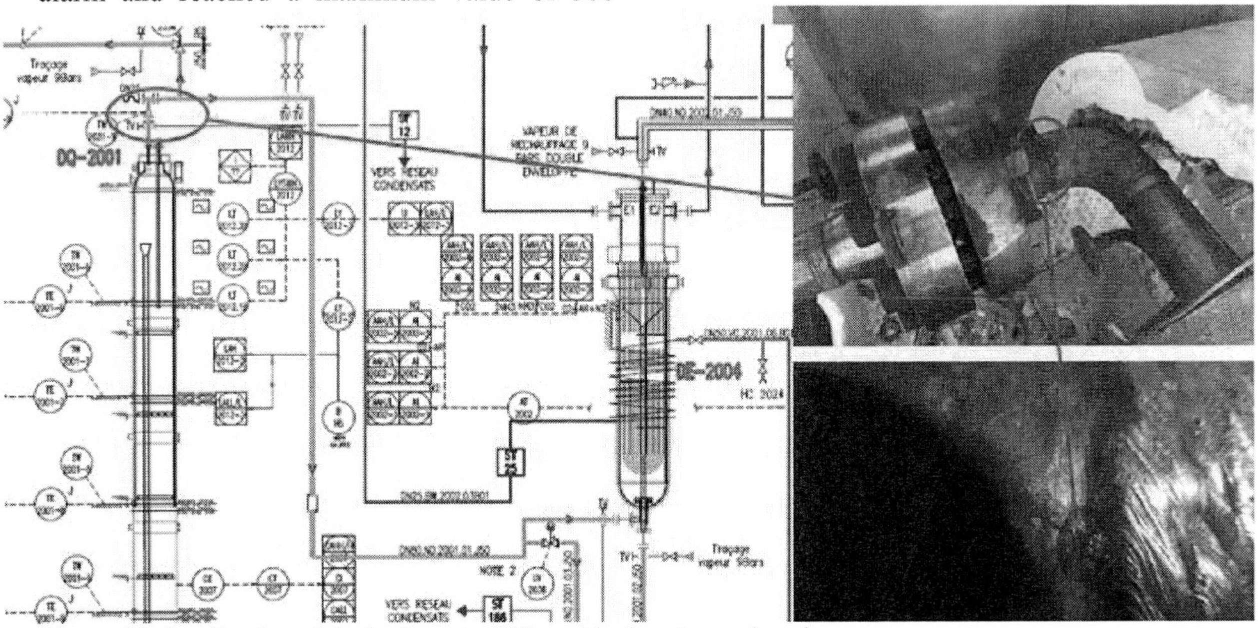

Figure 19: localisation of the leak on the PID and leak as it was found

When performing the repair by grinding off the leaking area, a lot of cracks were found in the neck flange as can be seen in figures 18 and 19.

Figure 20: Multiple cracks in the localisation of the leak

Moreover, when executing the X-ray examination some catastrophic figures in terms of thickness losses have been found (Some area with 90% of thickness losses)

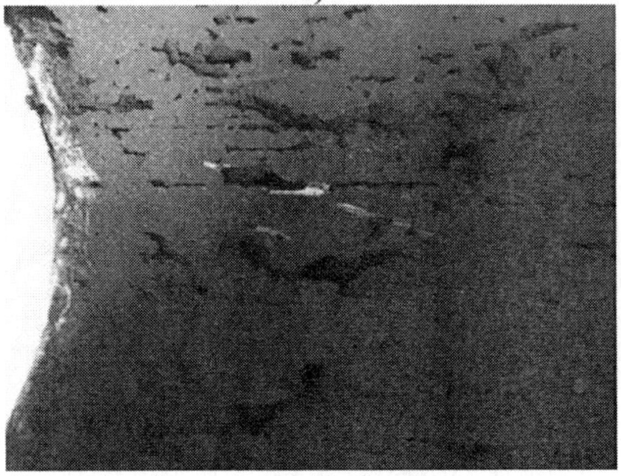

Figure 21: Corrosion inside the pipe

Cause of the pipe failure

Samples in the cracked area have been taken for microscopic examination and SEM analysis. The EDX mapping indicates that the crack is filled with oxides and that the parent material of the flange shows a typical composition for stainless steel type 316 Urea grade. Metallographic examinations show an oxide layer covering the entire inner surface of the flange in the sample. On several locations, intergranular cracks initiate from this oxide layer. Some of these cracks have penetrated through the entire wall of the flange. The samples exhibit a typical austenitic microstructure for the base material and heat affected zone. Both the heat affected zone as well as the base material exhibit an ASTM grain-size 0. As in previous cases the failure mode is pointing in the direction of SIIC triggered by condensation and large grain sizes in the material.

Figure 22: Microscopic examination of the leak area

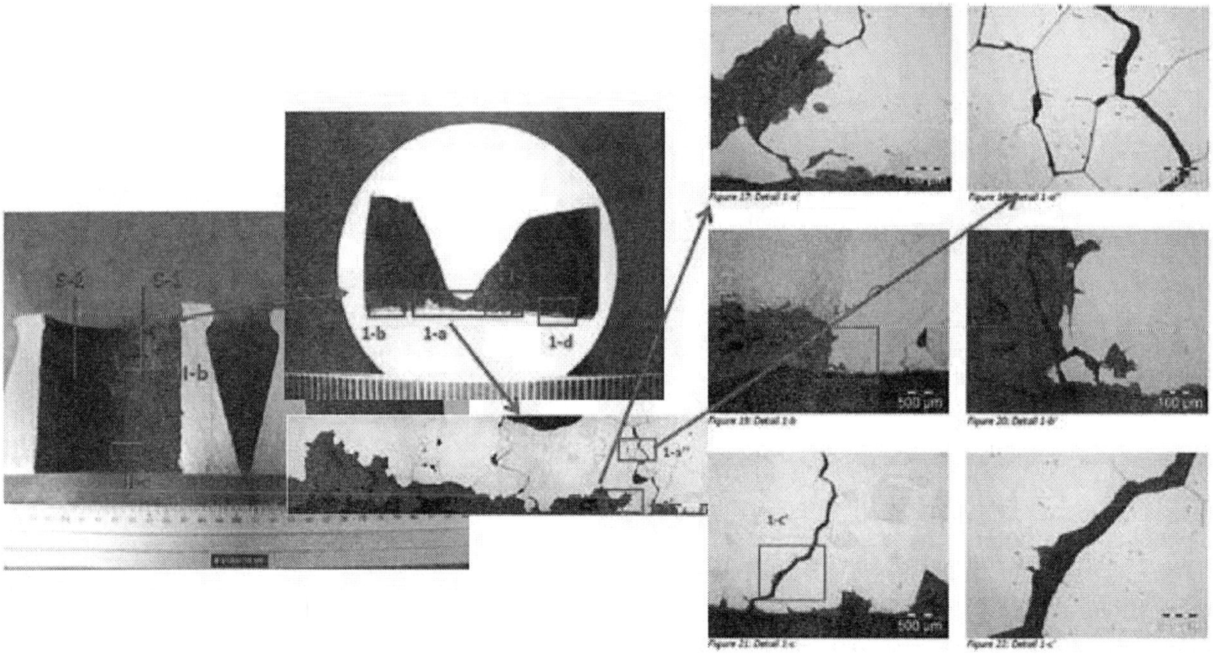

Figure 23: Microscopic examination of the leak area

Further findings after case 3

Immediately after case 3, an inspection/replacement program was initiated in the other gas line of the urea plant. This program led to the preventive replacement of the 35 year old gas piping from the HP stripper to the HPCC made of 316LUG. The dismantled gas line showed some spot corrosion and some heavy corrosion in the HAZ location.

Figure 24: condition of the dismantled pipe after 35 years of operation

Figure 25: condition of the dismantled pipe after 35 years of operation

In another plant, after 23 years of operation, some severe corrosion was found during the inspection, leading to the replacement of the piping at the next opportunity. The thickness loss was mainly caused by thermal bridge due to the pipe support at several spots.

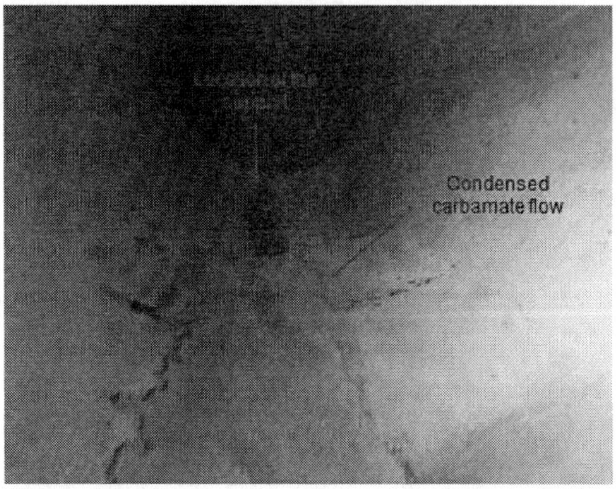

Figure 26: view of a gas line after 23 years with a cold spot

Figure 27: Another cold spot requiring early replacement of the gas pipe after 23 years

Conclusions

1. All reported failures occurred in AISI316L-UG HP gas lines of more than 30 years of age and are a result of stress corrosion cracking starting from the process side. Chlorides and sulfur play a role in this respect, but the appearance of the cracks does not point unambiguously in the direction of chloride-induced stress corrosion cracking. Contaminants like chlorides and or sulfides need to be avoided at the process side of urea plants, especially in older plants equipped with 316L UG piping.

2. In all reported cases the defect morphology points in the direction of so-called strain-induced intergranular cracking (SIIC), although no real evidence has been found for plastic deformation. All other prerequisites for this failure mode are present, such as the type of crack propagation (intergranular), the presence of oxides in the crack, the relatively large grain size of the cracked pipe pieces, and the rather slow crack propagation. The presence of chloride and sulphur increase the risk for this type of cracking but is not a prerequisite.

3. Strain-induced intergranular cracking (SIIC) is commonly observed in the areas of AISI316L-UG liners in aging urea reactors that are exposed to condensing conditions in the gas phase. Basically the same process conditions prevail in the top of the urea reactor as in the gas outlet lines.

4. This failure mode should be taken into account when assessing the structural integrity of AISI316L-UG HP gas lines, especially in older (>20 years) urea plants.

5. This damage mechanism shows the need for a different inspection program for AISI316L-UG HP gas lines. Condensation corrosion is more readily detectable by using ultrasonic wall thickness gauges. For detecting stress corrosion cracking, more sophisticated UT methods or other technologies (including visual internal inspection of gas lines using a video endoscope from flanged connections) should be applied.

6. Furthermore it is concluded that the maximum life time of AISI316L-UG HP gas lines is limited to approximately 30 years. When replacing these old HP pipes, it is advisable to install a more corrosion-resistant material such as a BC.05 (X2CrNiMo 25-22-2) or BE.06 (Safurex®). Safurex® is proven not sensitive to condensation corrosion, chloride-induced stress corrosion cracking or strain-induced intergranular cracking.

Recommendations

1. During TA, spot thickness measurement is not sufficient since it will not necessarily check the area with the lowest thickness. The specific program must include visual inspection. If visual inspection is not possible, preventive replacement of the piping must be performed after 30 years of operation.

2. Condensation of the process gas should be avoided as much as possible inside the pipes. Installation of proper insulation and tracing is a pre-requisite. Care must be taken to use the proper LP steam pressure to heat the pipes above the dew point of the process gas.

3. Gas pipe in urea plant must have an appropriate review during design phase (removal of unused nozzle, choice of steam tracing pressure, adapted support without cold bridge)

References

[1] Urea Synthesis Gas Lines Suffering from Leakages Due to Stress Corrosion Cracking. Ruben Wageck et al, Araucária Nitrogenados, Brazil, Alex Scheerder, Stamicarbon BV, The Netherlands.
13th Stamicarbon Urea Symposium, Rotterdam, May 2016.

Fire Protection and Detection for Critical Rotating Machines

There are vastly different approaches taken for the fire protection and/or detection of critical items of rotating equipment across the ammonia manufacturing industry. What is needed is a risk based philosophy rather than a prescriptive approach. This risk based approach will enable effective risk management of the exposure posed by loss of this critical equipment.

Ian Clarke
Zurich Insurance Company

Introduction

One of the largest risk exposures to any ammonia plant, and by definition the fertilizer production businesses that frequently depend on it, is damage to critical unspared items of large rotating equipment. Examples of such equipment are the ammonia and syngas compressors, and associated oil handling systems. There have been several losses in industry caused by mechanical failures and fires on these types of equipment, and on similar items in other related industries such as petrochemicals.

Insurance covers losses from these events, such as the property damage and associated business interruption (loss of profits). What it does not cover, however, is the ensuing loss of market share and/or key supplier positions with customers, because the loss has damaged your reputation in what is a very competitive market as a reliable supplier of product. Insurance company risk engineers travel the globe providing risk advice and recommendations to key customers, and Zurich is no different. However, internally, we have recognized a need for this advice to be more consistent.

There is an understandable reluctance in some quarters to provide water based protection systems over these types of machines, and in particular where they are steam or even gas turbine driven. What we have tried to do at Zurich is become more pragmatic about the risks and exposures, and provide a tailored risk based solution as opposed to the more typical "one size fits all" recommendation for sprinkler protection. This approach is valid for all types of large rotating equipment items but for the purposes of this paper we will restrict its application, and the examples, to the critical machines found on ammonia plants and downstream fertilizer businesses, as there will be many people in this group who probably have an interest in both.

The approach considers all of the factors that are critical in making a case for fire protection, fire detection, a combination of the two, or indeed none of them. With this in mind we take into account the property damage risk factors such as layout, duty, type of machine, existing fire protection and/or detection, credibility of manual response and other site and/or equipment specific details. We then also take into account the business criticality elements, such as the criticality of the individual machine and/or unit, replacement options and overall business resilience aspects such as any credible ability to import product as a replacement.

Using this approach we hope to work with our customers to deliver a more appropriate approach to this problem, and eliminate some of the contradictions and confrontations that can be endemic to the existing approach. We may not always agree on the outcome, but at Zurich we feel that this approach demonstrates a better understanding of your business and the constraints you work under.

Critical Rotating Equipment

Before we begin it is probably prudent to define critical rotating equipment, and what is in and out of scope. This paper is principally focusing on large, unspared rotating equipment items on ammonia or fertilizer plants, such as:

- Hydrogen and nitrogen compressors on hydrogen plants that feed an ammonia plant
- Startup circulation compressor on hydrogen plants
- Syngas compressors
- Refrigerant compressors
- Carbon dioxide compressors (urea plant)
- Air compressors (ammonia, nitric acid and subsequently ammonium nitrate plants)

The following items are considered to be out of scope:

- Gas and steam turbine generators on power plants, which are typically supplied with their own fire suppression systems within an enclosed cabinet
- Utility systems such as air (non-process uses), nitrogen or others which are typically supplied on an "n+1" basis

Risk Factors

To allow for consistent assessment, a specific assessment tool has been devised utilizing a number of defined Risk Factors. These were selected to cover the major risk areas.

It is suggested that the fire protection and/or detection philosophy for these machines should be risk based, taking into account a number of specific risk factors. These risk factors can be grouped under two general headings, for property damage (PD) and business interruption (BI) exposures. BI is an insurance term which is usually defined as the potential impact on a business, in terms of loss of gross profit, from a property damage incident. In almost all cases the associated BI costs will be several orders of magnitude higher than the actual property loss suffered from a serious incident on this type of equipment.

PD Risk Factors are listed as follows:

- Equipment duty
- Equipment value
- Layout of the machine and associated oil handling equipment
- Machine characteristics
- Machine protection systems
- Maintenance systems and efficacy
- Existing fire protection
- Existing fire detection
- Manual firefighting equipment and response capability
- Operating procedures and training

BI Risk Factors are as follows:

- Criticality of equipment (e.g. spared or unspared)
- Duality of supply (e.g. one or multiple trains and potential to make up lost production)
- Time for replacement, including any "guaranteed" arrangements with equipment suppliers
- Spares holding

- Credible alternative options (e.g. importing product, "group" options if a multi-site manufacturing organization)

PD Risk Factors

Equipment Duty

The material the equipment is processing is a contributing factor driving the need for any form of protection and/or detection. A compressor in ammonia or similar service obviously poses a greater hazard than one supplying carbon dioxide (as in a urea facility, for example). It is also a consideration in determining the type and extent of loss. If the material being processed is hydrogen (as an example) the primary hazard will be jet fire, unless the compressors are located within a confined structure (see following point under layout).

Equipment Value

The value of the equipment is important. Although insurance will pay for a replacement machine, deductibles are often substantial and these will be met by the owner/operator.

Machinery Layout

The layout of the equipment is critical. There are several aspects to consider. Firstly, whether there are multiple machines within the same compressor house and hence the potential for collateral damage to adjacent machines. For example, on some ammonia plants it is not unusual to have all the key compressors within a single structure. If they are arranged side by side (unusual, but possible) there is potential for a major event to impact adjacent machines.

Secondly, the layout of the machines themselves with reference to associated lubrication oil skids. Ideally the lubrication oil skids, which are a significant cause of industrial fires, will be located remote from the machine rather than directly beneath it. The compressor house itself may also have various construction features which will make it more resistant to a sustained fire.

Machine Characteristics

System design is an often overlooked aspect. Some compressors and similar machines are well designed, with minimal small bore lubrication oil pipes and inlet and outlet services entering from below, rather than encircling the machine with potential leak sources. Others are not so well designed.

Machine Protection Systems

Most equipment of this type comes with built in safeguards, including vibration monitoring systems and numerous trips for abnormal machine conditions. Where these do not come as standard they may often be installed separately or in addition to the existing safeguards. The most common manufacturer of these systems is Bently Nevada, but there are many other similar systems available. Obviously where these systems are installed and monitored the potential for a catastrophic failure and subsequent fire on the machine is less likely. Also, installation of these systems, if functioning correctly, suggests less process upsets and hence potential for releases from associated piping systems. Emergency stop systems should be located at appropriate places to ensure a shutdown can be activated a safe distance from the machine.

Maintenance Regimes and Efficacy of Protection Systems

The equipment will most likely have an array of automatic protection and monitoring systems, but these are of limited use if they are not correctly maintained and tested. In particular, overspeed testing on turbine driven equipment is particularly important. The lack of a good asset integrity culture will at some point compromise the machine safeguards and make an incident more likely.

Existing Fire Detection and Protection Systems

Where a retrofit of fire detection and protection systems is being considered, obviously the existing installations need to be evaluated. For the purpose of this exercise these will be defined as either automatically (sprinklers) or manually (deluge) activated systems installed over the equipment itself, and/or the lubrication oil systems alone. Where the systems are installed, the maintenance regime for them needs to be confirmed as adequate. Where systems are manually activated, the means of detecting a fire needs to also be evaluated, as does the emergency response capability (also below). The reliability of manual response capability needs to be considered as without this capability fire detection is of limited value.

Manual Fire Fighting Systems and Response Capability

The manual firefighting systems includes remotely or manually activated monitors or hydrants. The remotely controlled systems can be activated and controlled from either a control room and/or other remote locations. Hydrants are reliant upon the fire brigade or associated personnel setting up hoses to fight the fire. Some or all of these systems may include foam, either manually or automatically added.

The manual firefighting systems are obviously reliant upon an emergency response capability which needs to be based either on site or a short distance away. The capability and availability of the emergency response personnel are critical in evaluating the likely efficacy of a response to any major emergency. The training regimes, including tactical control training for emergency scene commanders, need to be adequate, as do the available manpower. Resources need to be adequate for the risk. If these resources and capabilities are not available on site then the municipal brigade and/or mutual aid facilities from nearby facilities needs to be evaluated.

In conjunction with the resources and training, the pre-fire planning of the relevant response units needs to be considered. Ideally emergency scenarios for the large rotating equipment items will have been evaluated and practiced as part of routine exercises.

It should also be noted that in many instances a local/public fire brigade response will not attempt to extinguish a fire on any equipment item that is handling hazardous process fluids unless there are people at risk. If the equipment fire only poses an equipment damage risk then the fire brigade will generally only attempt to contain the fire from spreading further and let the fire burn itself out. This is a particularly important consideration for rotating equipment located in rooms or enclosures.

Operating Procedures

The operating procedures for the equipment should be well established. The operating procedures should include standard operating procedures (SOPs) that contain start up and shut down procedures, and emergency operating procedures (EOPs) that cover rapid response actions in the event of unusual occurrences. These could be fire, loss of power, mechanical failure or others. Operators should be aware of the procedures and should have regular drills and practices to ensure they respond correctly in an emergency.

BI Risk Factors

Criticality of Equipment

It seems somewhat counter intuitive to have a risk factor for criticality of the equipment. However, in some instances there may be several machines providing the same duty, although for large compressors this is unusual due to cost. Where duplicate machines are provided the layout needs to be considered before giving too much credit for this aspect. As an example, where two machines are very close together and share a common lubrication oil system, a fire on this system could potentially damage both items.

Replacement Times

Typically replacement times for these large compressor units are 24-36 months, depending on the unit itself. However, there may be special arrangements whereby guaranteed replacement times have been arranged with a supplier. Also, where an operator has several similar sites, they may have decided that as a protection against loss they have purchased a warehouse spare item that could be used at any of their facilities in the event of a catastrophic loss. Note that this is not typical.

Spares Holding

Typically for critical rotating equipment items a complete set of spares except for the main casing itself would be the minimum accepted standard. All spares must be stored under appropriate conditions and the storage of rotors in particular needs to be to accepted industrial standards. The inspection and maintenance of spares is required on a regular basis to ensure they remain fit for use. In some cases a complete spare machine may be held, but note that this is not typical.

Credible Alternative Options

There are numerous other potential mitigating actions that can be considered here. Firstly, a large multinational with many operating sites may have the capability to transport product across borders with relative ease at acceptable cost and may consider this mitigating strategy as an adequate hedge against the loss of production from a single site.

Similarly, in some instances the intermediate products that typically are produced in plants with these types of machines (e.g. ammonia) can be imported from sister sites or from third party suppliers to continue the production of the end product (e.g. ammonium nitrate in the example given). Where this import capability is credible, and the quantities required could be sourced, this is an appropriate mitigating strategy.

Most importantly for all mitigating strategies involving the BI Risk Factors, it is important that these strategies are validated at least annually to ensure they are still credible. This should be undertaken using business continuity planning or an enterprise risk management philosophy. Without validation these options should be viewed or assumed with caution.

Calculation Tool

The methodology that we are proposing calculates a score using the risk factors that we have discussed above, and uses the resulting score to select an optimum fire protection and/or detection strategy. Note these are not absolute scores and have some overlap. It should also be pointed out that currently this spread sheet is a work in progress and needs to be validated by field use.

For each aspect, there are a range of scores from 1-5, 1-3 or 1-2, where each score represents a level of assessed risk following the guidance notes. The scores are multiplied together to give an overall score value. This value is then used to identify an appropriate course of fire prevention using the table of actions listed. To simplify this process, the attached spread sheet tool carries out the calculations delivering a final overall score for each assessment.

Note that some of the risk factors have been combined together to a single aspect, to reduce the complexity of the tool. As an example, there is a single score for business interruption. Also, the higher the score, the greater the level of assessed risk.

Two examples have been shown at the end of the paper. The first example (Figure 1) is a syngas compressor in medium hazard duty with generally good layout and below average protection features. The tool demonstrates a solution involving fire detection on the lubrication unit only, assuming a good manual firefighting response.

The second example (Figure 2) is a carbon dioxide compressor in low hazard duty with generally poor layout and protection features. The tool demonstrates a solution involving fire detection on the entire unit, again assuming a good manual firefighting response.

Two versions of the spreadsheet tool have been developed. The one shared here is for compressors, as this is the more common critical equipment item in ammonia based industries. There is a second tool for pumps, which is slightly different and outside the scope of this paper.

Guidance

The methodology presented above is a guideline to enable the risks associated with a loss of critical rotating equipment to be fully understood. It is not intended to be viewed as a "standard", and cannot hope to consider with 100% accuracy all possible eventualities in industry. Facility operators can utilize the framework to decide upon an appropriate level of fire protection and/or detection, but, as always, these solutions need to be consistent with the overall risk appetite of the operator.

Conclusion

Our objective is to achieve a risk based philosophy for the determination of fire protection and/or detection of critical machines, rather than a prescriptive approach. In our opinion the prescriptive approach will not be an adequate solution for all.

To achieve this objective a risk-based decision tool is recommended, and an example of one has been presented.

ASPECT	COMPRESSORS	DESCRIPTION
EQUIPMENT DUTY	3	Score 1-5. High Hazard (ethylene, propylene, propane) - 5, Medium Hazard (methane, hydrogen, ammonia) - 3, Low Hazard (air, carbon dioxide) - 1
EQUIPMENT VALUE	2	Score 1-2. Low (Less than US$25M) - 1, High (More than US$25M) - 2
LAYOUT	1	Score 1-5. Good (separated from other machines, lubrication oil system out from underneath the machine) - 1, Poor (next to other machines, lubrication oil system beneath the machine) - 5
MACHINE CHARACTERISTICS	2	Score 1-2. Good (modern design and systems, lubrication oil piping and feed/product lines well arranged) - 1, Poor (old design and systems, lubrication oil piping and feed/product lines poorly arranged) - 2
MACHINE PROTECTIVE SYSTEMS	3	Score 1-3. Good (hardware controls limit overspeed and vibration, independent fixed on line vibration monitoring (such as Bentley Nevada) fitted) - 1, Poor (limited hardware controls and no independent fixed on line vibration monitoring) - 3
MAINTENANCE	1	Score 1-2. Good (detailed PM schedule meets OEM requirements) - 1, Poor (no detailed PM schedule and no input from OEM) - 2
OPERATIONS PROCEDURES	1	Score 1-2. Good (SOPs and EOPs established and practiced, operators well trained) - 1, Poor (SOPs and EOPs not established or practiced, operators not trained) - 2
BUSINESS INTERRUPTION	5	Score 1-5. Good (spare machine (unusual), duplicate trains, agreements on supply to reduce replacement times, good spares holding, credible alternative manufacturing capability) - 1, Poor (single train, standard replacement time, below average spares holding and limited alternative supply capability) - 5
TOTAL SCORE	**180**	
>3000	Automatic Sprinklers on unit and associated lube oil systems	
1800-3000	Manually activated deluge on entire unit and associated lube oil systems with good fire detection coverage	
1500-2000	Automatic sprinklers on lubrication oil systems only with good fire detection coverage on the compressor	
1000-1600	Manually activated deluge on lubrication oil systems only with good fire detection coverage on the entire unit	
700-1200	Automatic sprinklers on lubrication oil systems only with nothing on the compressor	
400-800	Manually activated deluge on lubrication oil systems only with good fire detection coverage but nothing on the compressor	
200-500	Good fire detection coverage only on the entire unit. NOTE assumes good manual fire fighting response and capabilities	
100-300	Good fire detection coverage on the lubrication oil systems only. NOTE assumes good manual fire fighting response and capabilities	
<100	No fire protection or detection. NOTE assumes good manual fire fighting response and capabilities	

Figure 1. Syngas Compressor

ASPECT	COMPRESSORS	DESCRIPTION
EQUIPMENT DUTY	1	Score 1-5. High Hazard (ethylene, propylene, propane) - 5, Medium Hazard (methane, hydrogen, ammonia) - 3, Low Hazard (air, carbon dioxide) - 1
EQUIPMENT VALUE	1	Score 1-2. Low (Less than US$25M) - 1, High (More than US$25M) - 2
LAYOUT	5	Score 1-5. Good (separated from other machines, lubrication oil system out from underneath the machine) - 1, Poor (next to other machines, lubrication oil system beneath the machine) - 5
MACHINE CHARACTERISTICS	2	Score 1-2. Good (modern design and systems, lubrication oil piping and feed/product lines well arranged) - 1, Poor (old design and systems, lubrication oil piping and feed/product lines poorly arranged) - 2
MACHINE PROTECTIVE SYSTEMS	3	Score 1-3. Good (hardware controls limit overspeed and vibration, independent fixed on line vibration monitoring (such as Bentley Nevada) fitted) - 1, Poor (limited hardware controls and no independent fixed on line vibration monitoring) - 3
MAINTENANCE	1	Score 1-2. Good (detailed PM schedule meets OEM requirements) - 1, Poor (no detailed PM schedule and no input from OEM) - 2
OPERATIONS PROCEDURES	2	Score 1-2. Good (SOPs and EOPs established and practiced, operators well trained) - 1, Poor (SOPs and EOPs not established or practiced, operators not trained) - 2
BUSINESS INTERRUPTION	5	Score 1-5. Good (spare machine (unusual), duplicate trains, agreements on supply to reduce replacement times, good spares holding, credible alternative manufacturing capability) - 1, Poor (single train, standard replacement time, below average spares holding and limited alternative supply capability) - 5
TOTAL SCORE	**300**	
>3000	Automatic Sprinklers on unit and associated lube oil systems	
1800-3000	Manually activated deluge on entire unit and associated lube oil systems with good fire detection coverage	
1500-2000	Automatic sprinklers on lubrication oil systems only with good fire detection coverage on the compressor	
1000-1600	Manually activated deluge on lubrication oil systems only with good fire detection coverage on the entire unit	
700-1200	Automatic sprinklers on lubrication oil systems only with nothing on the compressor	
400-800	Manually activated deluge on lubrication oil systems only with good fire detection coverage but nothing on the compressor	
200-500	Good fire detection coverage only on the entire unit. NOTE assumes good manual fire fighting response and capabilities	
100-300	Good fire detection coverage on the lubrication oil systems only. NOTE assumes good manual fire fighting response and capabilities	
<100	No fire protection or detection. NOTE assumes good manual fire fighting response and capabilities	

Figure 2. Carbon Dioxide Compressor

Improving Down-fired SMR Flue Gas Flow Uniformity Using the New StaBlox^{TM} Reformer Tunnel System

For decades, down-fired SMR furnaces have utilized flue gas tunnels or "coffins" along the radiant section floor to collect flue gas flow as uniformly as possible. These tunnels utilize standard refractory shapes and installations which constrain the flue gas to non-uniform flow which has been correlated to non-uniform catalyst-tube temperatures and accelerated tube aging. In addition, these tunnels have been less than reliable in structural integrity and in some instances have leaned onto catalyst tubes creating additional stresses on the tubes in a hot condition. The ability to design and construct tunnels using new highly-engineered refractory shapes is new to this industry and could be the answer to greater structural reliability and achieving near-perfect flue gas flow among and into the tunnels resulting in improved catalyst tube reliability and longevity.

Joe Price and Tanner Howell
BD Energy Systems, Houston, Texas, U.S.A

Joe Quintiliani
Blasch Precision Ceramics, Albany, New York, U.S.A

Introduction

Catalyst Tube Failure

Top-fired reformer arch burners heat catalyst-filled tubes for steam methane reforming. The catalyst tubes are heated by both radiation and convection from the flue gas, which flows downward through the radiant box. Catalyst tubes typically are designed for a service life of 100,000 hours. In reality, tube longevity varies. Some tubes remain in service for 20 years, while others age more rapidly, failing far ahead of schedule. Sometimes reformers develop regions where tubes degrade more quickly, seemingly without cause until a closer analysis is performed.

Catalyst tube inspection reports indicate that excessively high catalyst tube temperatures are correlated positively to accelerated catalyst tube aging and premature tube failure. Even when the average tube temperature is within design limits, non-uniform flue gas velocities cause some tubes to be heated more than others. Higher than average tube temperatures have been correlated to radiant box regions with greater than average flue gas velocities.

Flue Gas Flow

Hot flue gas flows from the arch burners into tunnels on the floor directly below each burner row. Flue gas enters the tunnel through wall

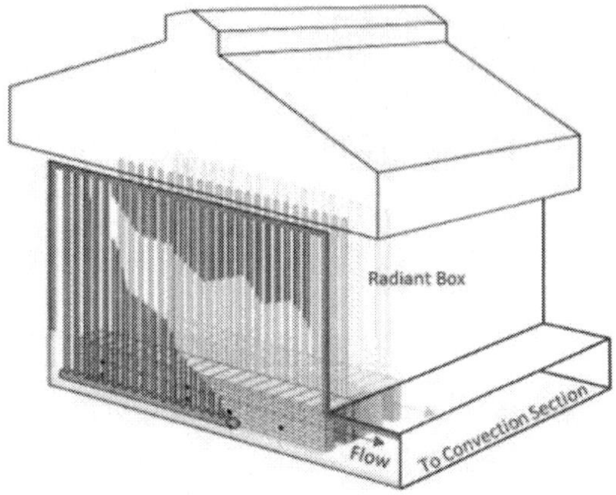

Figure 1 - Radiant Box Flue Gas Flow

openings distributed along the tunnel length and exits the open end of the tunnel to the Convection Section for additional heat recovery (Figure 1). Ideally, radiant box flue gas flows vertically downward with uniform velocity throughout the box resulting in uniform catalyst tube heating, temperature, and good flame patterns. The ideal total flow into each tunnel is proportional to the firing rate of their respective burner rows and the ideal incremental flue gas flow rate into each tunnel is uniform along the length of that tunnel. Typically, burner outer-row firing is lower than inner-row firing because the outer row of burners must heat only one row of tubes, whereas inner rows heat tubes on either side of these burner rows. As a result, outer-row burners produce less flue gas than inner-row burners (Figure 2).

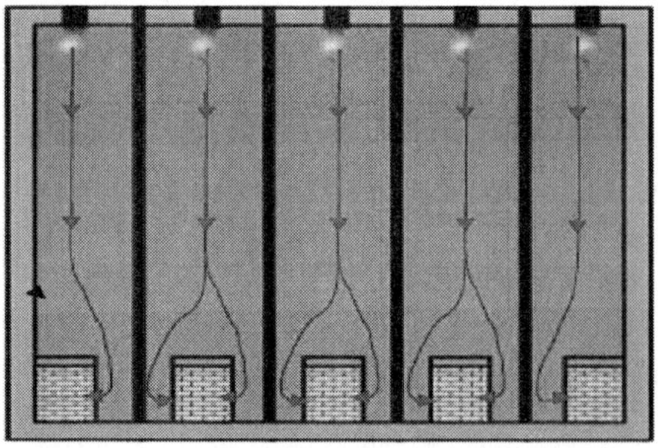

Figure 2 - Flue gas Flow into Radiant Box Tunnels

Therefore, proportionally less flue gas must be transported by the outer tunnels than by the inner tunnels. For example, if an outer burner row were designed to fire at 65% of the rate of an inner burner row, then the ideal outer tunnel flue gas load would be 65% of that of an inner tunnel. The ideal flow ratio between outer and inner tunnels is achieved by designing the outer tunnels with less tunnel-wall open-area and a smaller cross-section compared to inner tunnels.

Uniform flow into a tunnel along the length of the tunnel requires less open area near the tunnel open end than near the tunnel closed end (Figure 3). Flue gas is drawn into the tunnel wall opening by the differential static pressure across the wall which is proportional to the flue gas velocity inside the tunnel. Near the closed end of the tunnel where the in-tunnel flue gas velocity is very low (e.g., 0 ft/s) the differential static pressure is also low, so more open area is required to draw in flue gas. Toward the open end of the tunnel, where the in-tunnel flue gas velocity is high, the differential static pressure is also high, so less open area is required to draw in flue gas. Ideal tunnel wall open area decreases gradually from the tunnel closed end to the tunnel open end. Due to certain physical constraints (see below), ideal open-area distribution is not achievable for the conventional tunnel system design.

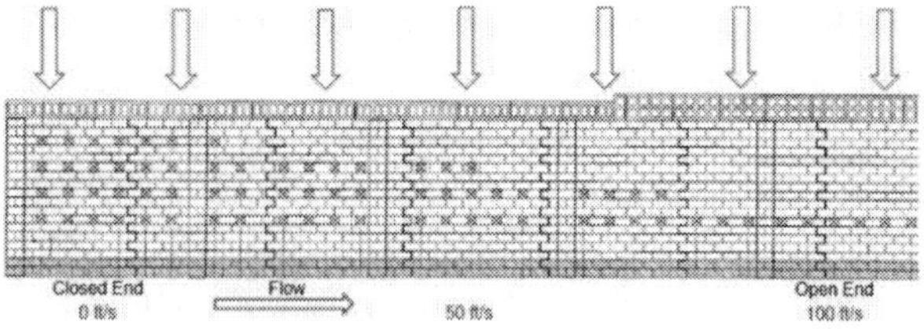

Figure 3 - Ideal In-Tunnel Flue Gas Velocity

Conventional Tunnel System

Physical Characteristics

Conventional brick and mortar tunnels are typically 4 to 10 feet (1220 to 3050mm) high, 2 to 3 feet (600 to 900mm) wide, and 40 to 100 feet (12200 to 30500mm) long. Nominal brick dimensions are 9"x3"x6" (230x75x150mm) (LxHxW). Buttresses extend from the outer wall surface for support and stability. Discrete expansion gaps ('expansion joints') are intended to accommodate thermal expansion to avoid wall distortion ('snaking'). Tunnel-wall openings are created by removing ½ bricks (4½"x3") (115x75mm) from the tunnel wall. Columns of openings are arranged at least 1½ to 2 brick lengths apart (center-to-center) and are distributed along the length of the tunnel to control flue gas flow into the tunnel (Figure 4). These brick tunnels historically have been a significant source of reliability concerns and tunnel section failure or even full tunnel collapse are common. The causes of these failures can be classified into three main sources: Mass, Material Selection, and Expansion Management.

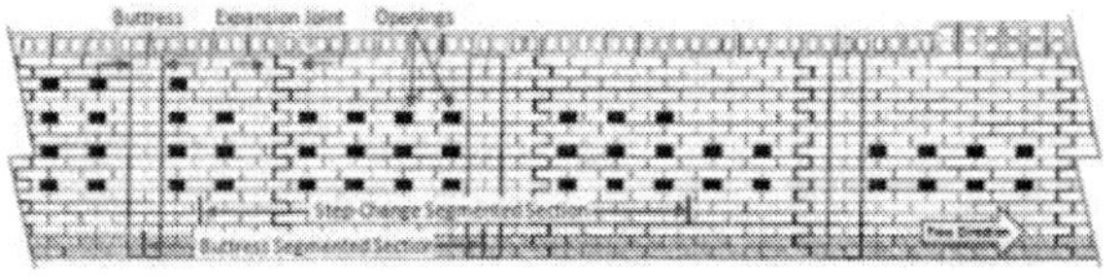

Figure 4 - Conventional Tunnel Wall Sketch

Mass

Mass is an important thermal load factor, but thermal load failures often are perceived incorrectly as static load failures. For example, after tunnel slabs are broken or collapsed (Figure 5) it may be believed that a middle-of-the-slab

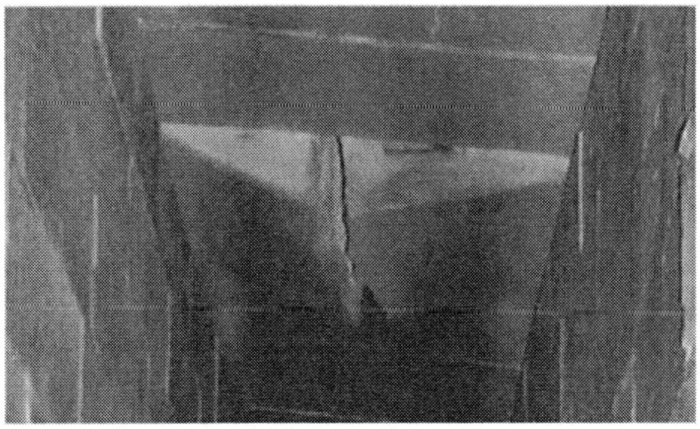
Figure 5 - Tunnel Slab Failure

crack is correlated to the ratio of component span to material thickness. After the replacement slab thickness is increased, however, the result is an even worse failure than before. This is because the slab failure is not a result of static load but rather, the component fails due to higher stresses caused by its excessive mass. Hand calculations coupled with computer simulation clearly prove that the static load alone imparts very little stress on the slab and will not result in failure. Consider, for example, a tunnel slab measuring 9" (230mm) wide, 9" (230mm) tall, and 42" (1067mm) long analyzed at a 1900°F (1038°C) constant service temperature with no external forces acting upon it except its own weight (Figure 6). The result is a 10 psi (69 kPa) maximum stress. Most of the engineered refractory material suppliers characterize the Hot Modulus of Rupture (HMOR) of their products and supply a slab material option having a sufficiently high HMOR safety factor for the static loads. Based on the Finite Element Analysis (FEA) results compared to the published HMOR it can be concluded that most slab failures are not a result of static load alone and therefore must be the result of thermal state stresses.

The component can also fail as a result of thermal stress. Thermal stress is not limited to instances of large upsets, but occurs during normal operation as a result of any differential temperature. Thermal stress failure results when the thermal expansion from one area of a component differs from that of another area, resulting in a stress greater than the material yield strength. If the radiant section temperature differs from the temperature inside the tunnels, even for a short period of time, thermal stress potentially is present. Consider an identical slab installed on a tunnel as before, but this time with the slab upper surface temperature increased to 1910°F (1043°C) (Fig 7).

The slab has no external forces acting upon it other than its own weight. A differential temperature of 10°F (5.6°C) across the slab results in a 1500 psi (10342 kPa) maximum stress, which is well above the HMOR of many refractory materials. In a situation where a great percentage of tunnel slabs fail during the same operating period without a wall collapse, thermal stress is the most likely failure mode (Figure 8).

Excessive mass in these tunnels not only causes individual component thermal stress, but also may affect the entire reformer radiant section due to the large weight of refractory materials. All of this weight rests on a base layer of insulating refractory on the floor. A tunnel cross section with 6" (150mm) wide bricks, 96" (2440mm) high tunnel walls, and a solid 9" (230mm) thick slab represents an 11.6 psi (80 kPa) load on the supporting insulating refractory layer. Published data using ASTM testing suggests that at these reformer furnace temperatures the insulating refractory layer will deform a full 1% under those loads in 100 hours. The insulating refractory layer deformation will translate in two ways: the deformation will result in an unstable wall, and will reduce the overall insulating value. Either of these can result in tunnel failure (Figure 9).

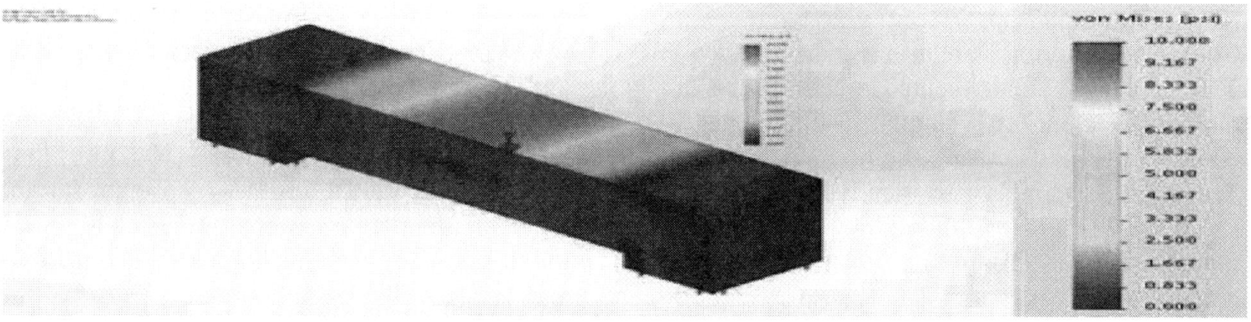

Figure 6 – FEA of Tunnel Slab at Constant 1900°F

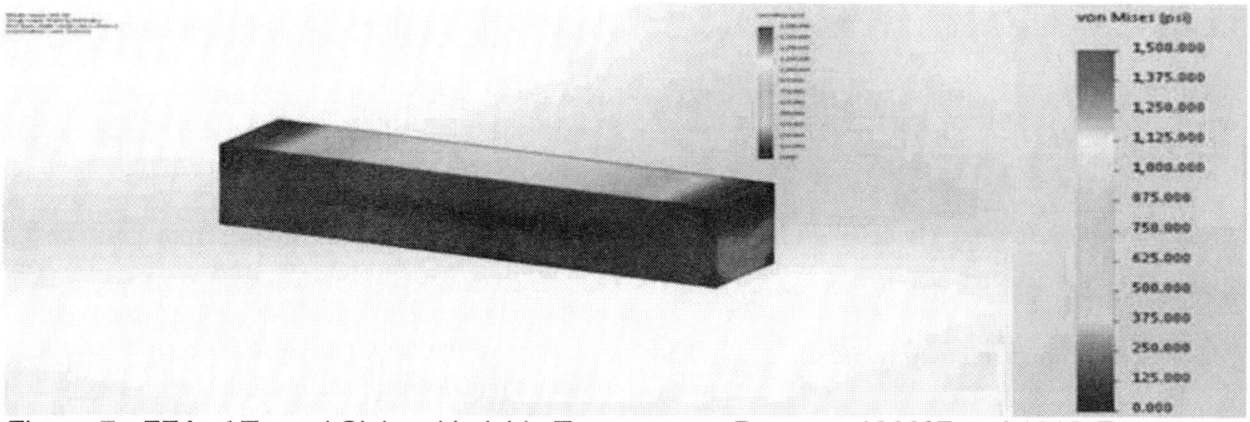

Figure 7 - FEA of Tunnel Slab at Variable Temperature Between 1900ºF and 1910°F

Figure 8 - Tunnel Slab Thermal Stress Failure Mode

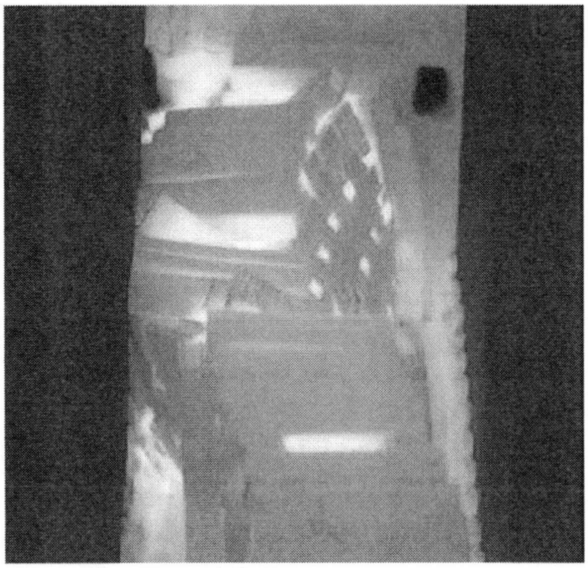

Figure 9 – Tunnel Failure

The effects of temperature and tunnel mass are not limited to the furnace internals, but can cause supporting furnace structure deformation, leading to an uneven furnace floor. Conventional tunnel design utilizes a mortared joint, effectively turning the large number of small bricks into a small number of large wall sections. These wall sections move as a single body and cannot accommodate any major dimensional change in the furnace floor, meaning that deformation of a supporting furnace structure will result in tunnel collapse.

Material Selection

In order to address this thermal stress build-up, sufficiently strong material selection is essential. Two primary material properties, Creep Resistance and the HMOR, are very important. Creep occurs when a material slowly but permanently deforms under long term high stress exposure at elevated temperature below the material yield strength. For refractory material, this commonly is referred to as Refractoriness Under Load (RUL). This property comes into play in two distinct areas: first in the slabs, and second on the unsupported brick sections that span the openings in the wall. These unsupported sections will inevitably sag over time and as they do, tensile stress develops on the lower surface and causes the parts to crack (Figure 10).

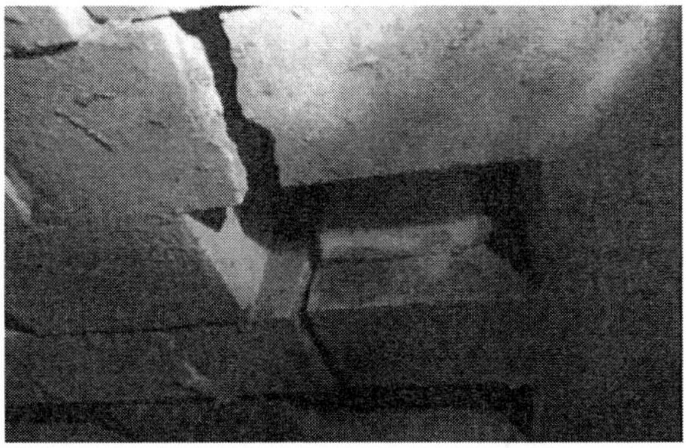

Figure 10 – Unsupported Brick Failure

Slab deformation has a secondary effect on the tunnel walls. The slabs are designed to transfer all their weight in the vertical direction, which compliments the strength and structure of the wall. In a slab, creep will result in center span "sagging" and will change the interaction force between the slab and the tunnel walls resulting in a moment which eventually will lead to failure as this non-vertical force pushes apart the tunnel walls (Figure 9). Creep can be characterized with ASTM standard testing. ASTM tests on Super Duty Brick have published results of a 7.86% deflection at 2,600°F (1427°C).

HMOR also is an important material property. It is critical to note that knowing only the ambient MOR at room temperature is not sufficient for proper tunnel system design. As refractories are exposed to high temperatures they begin to lose strength over time (aging). This non-linear process will diminish over time. ASTM tests expose components to 2500°F (1371°C) environments for 100 hours and then measure the MOR. The test result is called the HMOR because it indicates material strength at elevated temperatures. To put this in perspective, ASTM tests on Super Duty Brick have published results of 1100 to 1500 psi (7584 to 10342 kPa) ambient MOR, but this decreases to a HMOR of approximately 600 psi (4137 kPa) at 2,500°F (1371°C). Proper material selection should include confirmation that the MOR at the furnace service and excursion temperatures has a sufficient safety factor compared to the associated static load stresses.

Expansion Management

One of the most important design aspects of flue gas tunnels is thermal expansion accommodation along the length of the tunnel. A conventional tunnel design includes an expansion gap every 6 to 10 feet (1830 to 3050mm) of tunnel length. Refractory mortar is used to bond all of the components between gaps into a wall section which acts as a single structure. Unfortunately, furnace temperature distribution is not uniform and differential thermal expansion can occur across a given wall section creating thermal stress much like a tunnel slab. These wall section stresses can cause failure of the wall section.

To illustrate the differential temperature stress levels from the top to the bottom of a fully mortared wall section, consider a 10 foot (3050mm) fully mortared wall section, which can be treated as a single body for the purposes of this analysis. Apply 1925°F (1052°C) and 1900°F (1038°C) temperatures to the top and bottom of the wall segment, respectively, with a uniform temperature distribution in between. The analysis also included simulated gravity and tunnel slab weight, with no other external forces (Figure 11). Under these conditions, the system stresses exceed the 500 psi (3447 kPa) standard refractory mortar HMOR. Because the mortar joints are the weakest points in the wall, they will crack to alleviate the stress. As more cracking occurs, the mortared wall sections become smaller, and stresses become lower for any one section.

Properly accommodating thermal expansion is, perhaps, the most difficult thermal application design aspect. Conventional tunnel slabs and bases have differing material and design. Many tunnels are designed with low density insulating refractory or fiber board insulation in the "base" area of the tunnels. The tunnel slab

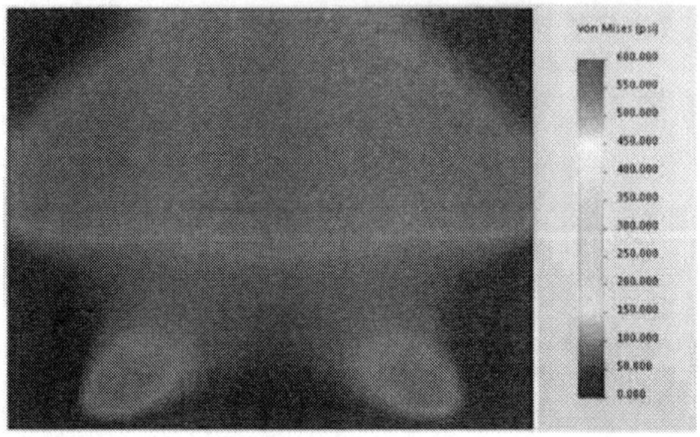

Figure 11 – FEA Results for a Fully Mortared 10' Tunnel Wall Section with Variable

can expand by as much as 0.375" (9.5mm) pushing apart the tunnel walls, whereas the base insulation will not impart any expansion forces on the tunnel walls. The resulting non-vertical force increases the likelihood of wall buckling and collapse.

During turnarounds, tunnel wall inspection has in some cases shown an alternating lateral movement. This more commonly is known as "snaking" and is the result of the overall tunnel attempting to expand more than the built in allowance. This movement will crack the mortar, separate walls from slabs, and can separate walls from bases, all of which can lead to failure. Proper material selection and installation procedures are important to prevent snaking. Many materials will increase in size when re-heated, increasing variability and making thermal expansion management more difficult. Because the Coefficient of Thermal Expansion (CTE) for refractory components is nonlinear, it fully must be characterized and understood to ensure that proper expansion joints are created.

Design Technique

Open-Area Distribution Design

To distribute open area along the tunnel-wall length, the tunnel is segmented into discrete lengths and the average ideal open area for each length segment is calculated. Open area then is distributed to approximate the calculated ideal distribution. Conventional tunnel lengths are segmented evenly into approximately 4 sections. Alternatively, they may be segmented at the buttresses or at the open-area step changes (Figure 12). Conventional open area is added in nominal 13½ in^2 (8700 mm^2) increments (4½"x3") (114x75mm) and for structural reasons cannot be located at buttresses or at expansion joints. To compensate for the lack of open area at buttresses and expansion joints, open area is increased elsewhere in the wall section. The result is non-uniform incremental flow into the tunnel.

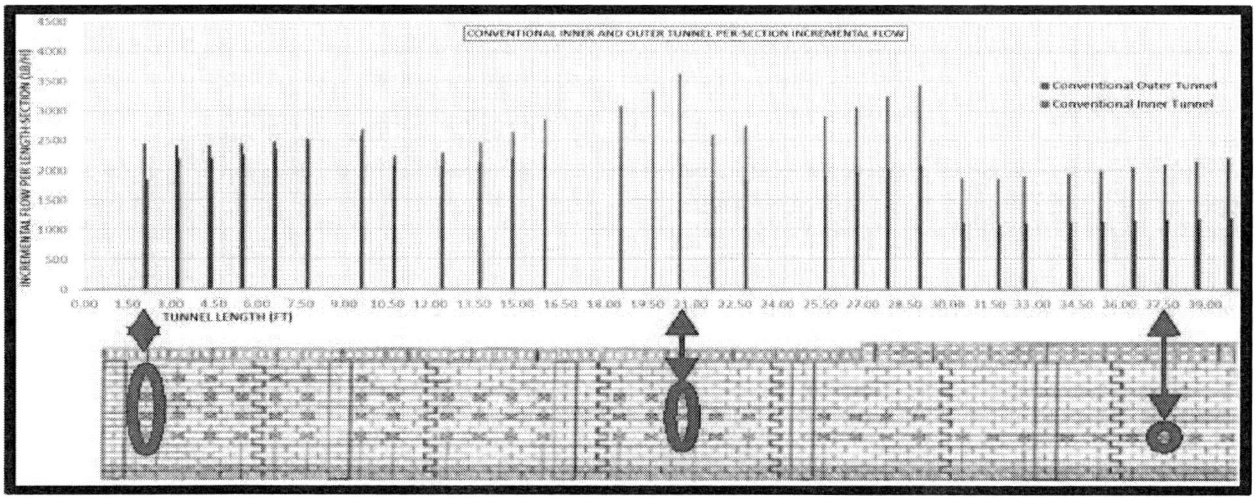

Figure 12 – Incremental Flow per Open-Area Column for Convention Inner and Outer Tunnels

Evaluation of Conventional Design

A conventional 40 foot (12190mm) long tunnel system is evaluated (Figure 12). Calculated total flow through each outer-tunnel is approximately 77% of the inner-tunnel flow, but this ratio is not constant along the tunnel length. For example, at each column of holes, the ratio of one outer-tunnel flow to one inner-tunnel flow (the 'flow ratio') is calculated. The resulting average flow ratio is 77.4% ±55.1%, with a standard deviation of 21.3%. Additionally, the 'high-flow' columns tend to be grouped together. For the 6 columns closest to the closed end, for example, the average outer-tunnel incremental flow is 110% that of the inner tunnels. For the 6 columns closest to the open end, however, the average outer-tunnel incremental flow is 56% that of the inner tunnels.

Also, incremental flow is non-uniform along the length of each inner or outer tunnel. Following any open-area step change, the incremental flow increases with each successive column of holes (Figure 12). As in-tunnel velocity increases along the tunnel length, incremental flow through each successive column increases until an open-area-per-column step-change reduces the open area and, consequently, the incremental flow. After the step change, incremental flow increases, again. The areas near buttresses and/or expansion joints are 'no flow' regions.

Therefore, at the tunnel closed end (Figure 12), some flue gas generated above the inner tunnels moves laterally toward the outer tunnels and toward the radiant-box center. At the tunnel open end, however, some flue gas generated above the outer tunnels moves laterally toward the inner tunnels and toward the middle of the outer tunnel. Half-way between the tunnel ends, flue gas generated above the outer tunnels moves toward the radiant-box center.

StaBlox™ Interlocking Tunnel System

Physical Characteristics

The StaBlox™ system utilizes 5 core tunnel construction components. The construction begins with a base component that spans the bottom of the tunnel and mates to the side wall blocks. The side wall blocks are 18 inches (457mm) long by 9 inches (230mm) high and each block contains two 4½" (114mm) diameter openings. An orifice insert with an engineered inside diameter is installed in each opening (Figure 13). In addition to orifice inserts, these openings can hold ancillary parts such as tie-rods inside the tunnels which connect the two tunnel walls and replace buttresses as an additional support structure. These tie-rods can be easily removed and replaced during outages to allow easy access down the tunnel interior for inspection or repairs (Figure 14).

Figure 13 – StaBloxTM Base, Insert, and Block Components

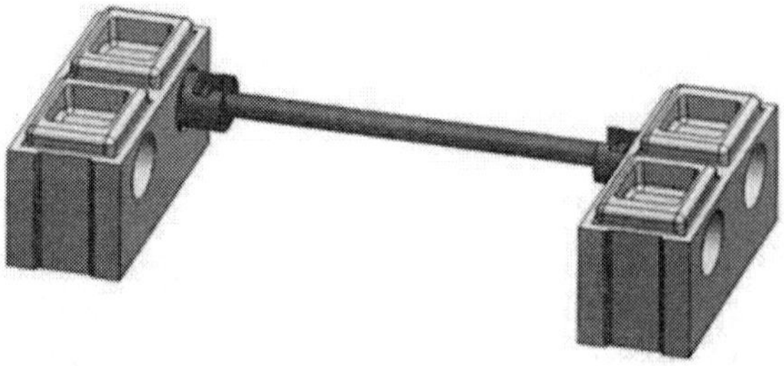

Figure 14 – Support Tie Rod

At the top of the tunnel is a light weight slab (Figure 15) that mates to the side walls (Figure 14). In order to allow for a completely mortar-free construction, the slabs also employ a shiplap joint to prevent any gas bypass through the tunnel slabs. Every component of this system is designed to accommodate its own thermal expansion. In order to eliminate any gas bypass through the expansion joints a fiber gasketing system is employed.

Figure 15 – Thin Walled Slab (Underside)

Mass

As previously mentioned, temperature differential stresses across a body can result in failure. The most direct way to reduce the thermal stresses to below the refractory component yield strength is to decrease the refractory component wall thickness (Figure 15) to allow the wall temperature to equilibrate more quickly and eliminate temperature differential stresses. The wall thickness should be as thin as possible without sacrificing the overall stability of the tunnel system. Because the tunnel system is self-supporting, component wall-thickness reduction decreases the overall system weight. With the StaBlox™ design, tunnel system weight can be reduced by up to 60%. This weight reduction provides added benefits when selecting a suitable base insulating material. The bricks that have insulating value sufficient to keep the furnace floor and supports from deforming are too weak to support the tunnel system, while the bricks with the higher strengths conduct too much heat. As mentioned before, a conventional tunnel cross section with 6" (150mm) wide bricks, 96" (2438mm) high tunnel walls, and 9" (230mm) thick solid slabs, will result in an 11.6 psi (80 kPa) load on the supporting refractory layer and a 1% deformation within the first 100 hours of use.

Decreasing the overall weight of the tunnel system by 60% will translate into a lower psi load, and result in a deformation to the supporting refractory layer one order of magnitude less. The tunnel system design also utilizes a "base" component that distributes the wall load over an area 5 times larger than the conventional design. A light weight design coupled with a larger base component results in a 1.4 psi (10 kPa) load on the insulating material. This makes possible the use of highly insulating materials which would improve structural furnace support reliability and improve overall system reliability.

Material Selection

Because a reduced creep will prolong tunnel system life and prevent premature failures, selected materials for tunnel system components should have the highest creep resistance reasonably available. ASTM tests on Super Duty Brick have published results of a 7.86% deflection at 2,600°F (1427°C). An ideal candidate to replace Super Duty Brick is Mullite bonded Alumina refractory material, which has published results of 1.11% deflection at 3,000°F (1650°C).

In addition to creep resistance, a high HMOR also is critical for these tunnels. Comparing the same mullite bonded alumina to super duty fire brick results in a similar difference in material strength properties (Figure 16).

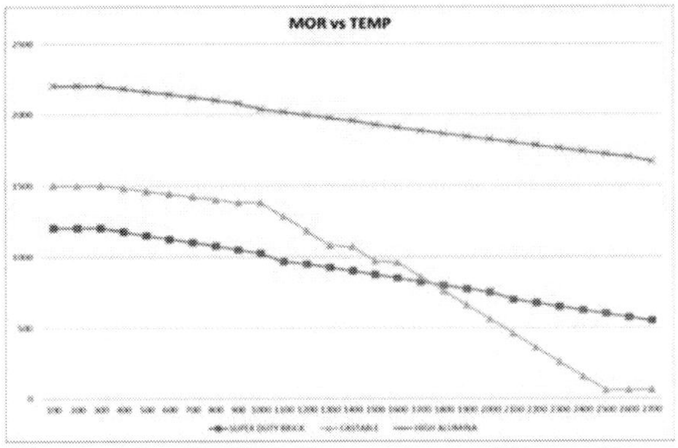

Figure 16 – HMOR Comparison of Super Duty Brick, Mortar, and Mullite Bonded Alumina

Super duty brick has a <600 psi (4137 kPa) HMOR, high grade castable can have an almost 500 psi (3447 kPa) HMOR, and mullite bonded alumina has a 1700 psi (11720 kPa) HMOR. A fully characterized CTE, higher HMOR, and increased creep resistance will improve the overall tunnel system reliability.

Expansion Management

Reducing the component wall thicknesses is just one way to relieve thermal stress; it is also possible to reduce the tunnel system 'section' sizes to minimize single section differential temperatures. Ideally, the tunnel system sections should be only as large as the individual building components. In order for this to be accomplished every block must accommodate its own thermal expansion and the entire system must be mortar free, but for stability reasons, must be completely interconnected. With thin walled parts it is feasible to increase wall component size. An ideal size for these parts is 18" (457mm) long and 9" (230mm) tall. An analysis of a 10 foot (3050mm) wall section composed of these blocks (Figure 17), under operating conditions and external forces identical to what was analyzed previously (Figure 11), results in a peak stress of less than 65 psi (448 kPa).

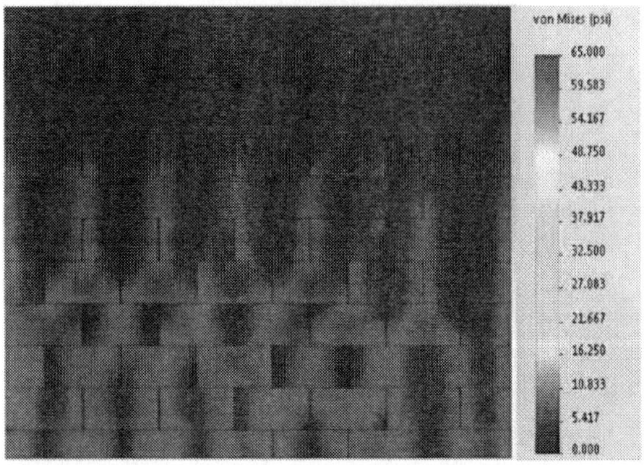

Figure 17 – FEA of a Non-Mortared 10' Tunnel Wall Section with Variable Temperature

This distributed expansion requires precision formed refractory components or an installation procedure that measures and accommodates the variability in every single component. The CTE is very important in the analysis of uniformly distributed expansion allowances. For the tunnel system materials, this value should not be assumed to be a linear function. A fully characterized CTE is important to ensuring properly managed expansion and is critical for single component expansion management. Proper thermal expansion management also requires a base component of the same material and having dimensions similar to the slab component. This ensures that the tunnel expands and contracts uniformly on both the top and bottom of the wall, maintaining the overall structure and reducing stresses that may cause buckling.

Even if the thermal expansion is properly managed, buckling still may occur as a result of delayed combustion or an uneven furnace floor. Additional cross beam supports at predetermined locations are recommended to arrest any buckling (Figure 18).

Figure 18 – Tunnel System with Support Tie Rod

These supports replace the buttresses used in conventional tunnels and reinforce the tunnel structure by reducing the unsupported tunnel length per Euler's column equation:

$$F = \frac{\pi^2 EI}{(KL)^2}$$

Where:
F = allowable load
E = Modulus of elasticity
I = Moment of inertia
L = Length of column
K = Column effective length factor

Buckling can also be arrested using a tight tolerance mating feature in the wall components such that the rotation of a block in relation to the block below it results in direct contact. In order to dislodge this tight tolerance mating, an amount of stress sufficient to break the block wall is required. By contrast, a traditional tongue and groove design will be effective in preventing lateral movement, but will not arrest buckling because the rotation of a block in relation to the block below it will separate the tongue from the groove, allowing a full system collapse.

Design Technique

Open-Area Distribution Design

The StaBlox™ design method is to segment the tunnel length into ½ block (i.e., 9" (230mm)) sections, one column of openings per section. For example, a 75 foot (22860mm) tunnel-wall length is segmented into 100 sections with openings in every course (row). Open area is distributed along the tunnel length by specifying the orifice-insert internal diameter for each opening, rather than by specifying the per-section opening quantity (Figure 19).

Figure 19 – StaBlox™ Interlocking Blocks with Varied Orifice Insert Diameters

Evaluation of StaBlox™ Design

The Figure 12 conventional tunnel system is redesigned using Blasch technology (Figure 20). The goal is to achieve 65% flow through each outer tunnel compared to the flow through each inner tunnel and to achieve near uniform incremental flow along each tunnel length. The inner and outer walls are segmented into 9 inch sections, one column of openings per section. Buttresses are not necessary and expansion gaps are evenly distributed among the blocks. Required open area per column is calculated and appropriate orifice insert diameters are selected for each opening. Each orifice size is illustrated with a unique color.

For each column, more than one orifice size may be specified so the average orifice diameter per column gradually decreases closer to the tunnel open end. The outer tunnel requires very little orifice diameter change compared to the inner tunnel because the outer tunnel closed end flue gas velocity is very low and the outer-tunnel closed end acts somewhat like a manifold, whereas the inner-tunnel closed end develops high-velocity flow almost immediately. To be conservative, the minimum orifice-size step-change is arbitrarily specified well above manufacturing limitations. More finely tuned diameters are feasible.

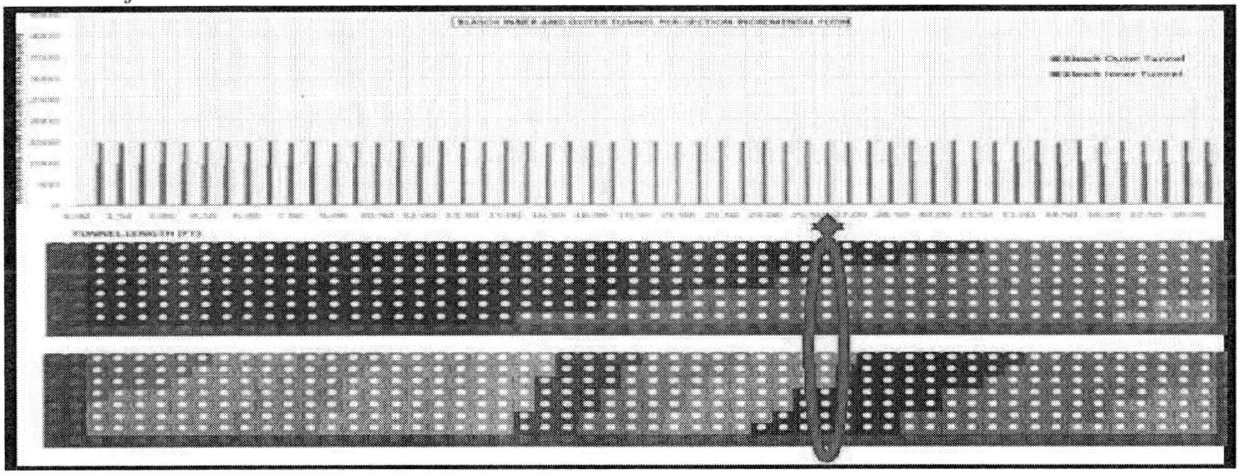

Figure 20 – Blasch Tunnel Evaluation

Calculated total flow through each outer-tunnel is approximately 65.64% of per-inner-tunnel flow, and this ratio is approximately constant along the tunnel length. At each column of holes, for example, the ratio of one outer-tunnel flow to one inner-tunnel flow (the 'flow ratio') is calculated. The resulting average flow ratio is

65.64% +/- 1.8%, with a standard deviation of 0.7%. Also, neither high-flow nor low-flow regions are grouped together. For the 6 columns closest to the closed end, for example, the average outer-tunnel incremental flow is 65.62% of the inner tunnels. For the 6 columns closest to the open end, the average outer-tunnel incremental flow is 65.60% of the inner tunnels.

Additional Advantages

Reduced Pressure Drop

StaBlox™ tunnels walls are roughly one half the width of conventional tunnels when the width of the buttresses is taken in account. Therefore, the StaBlox™ interior width of the tunnel tunnel can be increased by approximately 11" (280mm) without increasing the tunnel footprint. This increased interior width results in approximately 1/3 greater cross-sectional flow area and significantly reduces pressure drop across the tunnel system.

Process Fléxibility

An increase in capacity results in increased flue gas flow. This results in the existing tunnel wall open-area distribution not performing optimally. For a conventional tunnel, open-area distribution redesign may require partial or complete reconstruction, incurring material and labor costs and possibly lengthening the turnaround schedule. For a StaBlox™ tunnel system, however, open-area redesign requires only the removal, addition, or relocating of some of the orifice inserts (a relatively quick and inexpensive procedure), re-using the wall tunnel structure in its entirety.

Reduced Heat Storage

During heat up or cool down it is important to evenly heat or cool the catalyst tubes to avoid accelerated tube aging. The tunnels absorb heat during heat-up and release heat during cool down, causing non-uniform heating or cooling of individual tubes. During a trip, for example, while the catalyst-tube upper ends are cooling more quickly than normal, the tunnels radiate heat to the catalyst tube lower ends. Because StaBlox™ tunnel mass is up to 60% less than conventional tunnel mass, StaBlox™ tunnels absorb and release 60% less heat, reducing the degree and period of uneven cooling.

Conclusions

The new tunnel system introduced in this article provides for a more stable structure and more uniform flue gas flow into the tunnels resulting in more uniform flue gas flow around the bottom of the catalyst tubes for more even heating. These two factors increase both the reliability and safety of the radiant section of downfired primary reformers.

Improving Combustion Efficiency, Productivity, Reliability, and Safety in Primary Reformers Utilizing Multiplexed Laser Based Monitoring and Diagnostic Systems

Operators are often challenged to maintain well-balanced, uniform combustion throughout primary reformers and large fired heaters in order to optimize efficiency, reliability, production yields, and safety and to reduce emissions. While most reformers and large furnaces have several individual sensors available to monitor temperature, O_2, and CO at the furnace exit, typically such sensors are not available for use directly inside the firebox. This paper describes a novel but well-proven wavelength-multiplexed, laser-based monitoring and diagnostic technology that measures temperature, oxygen, and carbon monoxide directly in the radiant zone of large primary reformers.

It allows operators to tune burner controls to achieve fuel savings, increase production yields, lower excess O_2, monitor CO, reduce NO_x and CO_2 emissions, and extend reformer tube life through well-balanced temperature and oxygen profiles across the entire envelope of the reformer.

Andrew D. Sappey
Greg Heilbrunn
Zolo Technologies, Inc.
331 S. 104th St.
Suite 100
Louisville, CO 80027

Introduction

Primary reformers use a significant amount of energy to convert a feedstock of CH_4 and steam into CO and H_2 via an endothermic catalytic reaction. Further reactions can convert these products into NH_3 for fertilizer production and methanol, or the hydrogen may be used directly in refining or other processes. The initial conversion process takes place in tubes filled with catalyst and heated externally inside a furnace. The tubes and catalyst are designed to operate in a narrow temperature range for optimum yield. If temperatures outside this range occur, either the reaction is incomplete leading to poor yield at low temperature, or the tubes and catalyst can be damaged at high temperature. Balanced combustion in the furnace can have a large impact on many of these factors and therefore the profitability of the plants. A typical reformer furnace has sensors installed in different locations for the purpose of monitoring the combustion process. However, sensors available for use directly inside the combustion zone are lacking. Historically, the dearth of information about combustion zone conditions made it very difficult to optimize temperature, O_2, and CO profiles to realize process improvements. Imbalances in temperature, O_2, and CO across the furnace can adversely affect hydrogen production as well as combustion efficiency. Tube leaks caused by thermal hot spots can reduce plant productivity and ultimately result in an unscheduled outage for repairs.

Until recently, there existed no means to measure simultaneously all of the key combustion constituents directly inside the radiant zone of the furnace. This paper addresses the benefits of *in-situ,* real-time combustion monitoring to maximize production yields, reduce energy usage, improve asset reliability and safety, and reduce emissions.

Overview and Architecture of Multiplexed TDLAS Laser Systems

Multiplexed Tunable Diode Laser Absorption Spectroscopy (TDLAS) systems currently provide *in-situ* combustion measurements in multiple industrial applications including coal-fired power plants, chemical and steel plants, and refineries with over 80 installations worldwide. Multiplexed TDLAS is well proven, accurate and reliable and TDLAS is designated as a best practice by API for large fired heater applications.

Multiplexed TDLAS combustion diagnostic systems simultaneously measure the average temperature, O_2, and CO in real-time, directly in the radiant zone along multiple lines of sight. The system requires no insertion probes, no sensitive electronics are located near the furnace and no in-field calibration is required. These real-time measurements with two-dimensional resolution provide a more accurate and reliable representation of the furnace conditions than traditional single-point thermocouples, tube temperature measurements, ZrO_2 sensors and traditional TDLAS as illustrated in Figure 1 which shows an actual O_2 distribution in a furnace. As can be seen, the distribution is quite inhomogeneous, which can lead to measurement sampling errors using single point sensors or even single path-averaged TDLAS. Multiplexed, multipath TDLAS systems monitor a two-dimensional field and therefore provide much more accurate information to optimize the furnace. (See Figure 1).

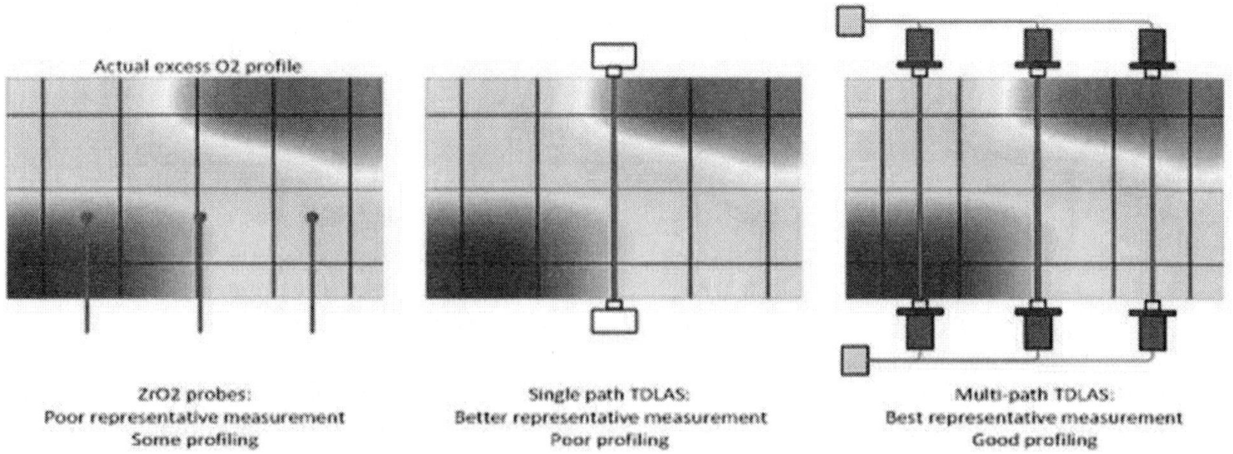

Figure 1. Sampling an inhomogeneous combustion zone with point sensors, single path TDLAS, and multipath TDLAS

Multiplexed TDLAS systems (Figure 2) combine multiple lasers onto a single optical fiber; generally one laser is required to measure each species. In the case of the most advanced multiplexed TDLAS system, pairs of laser heads incorporating automatic alignment are installed in strategic locations along the furnace walls to send (pitch) and receive (catch) the laser signals along a line of sight. The combined light is sequentially transmitted by an optical switch to each of the paths where a pitch head collimates and transmits the laser light across the furnace. Light is collected by a catch head and routed back to the control rack via optical fiber. In the control rack the combined laser light is separated so that the light absorbed by each constituent is measured to determine the average concentration of O_2, CO and temperature along each laser path. Multiple paths are arranged to provide combustion information corresponding to the combustion control parameters for each zone in the furnace.

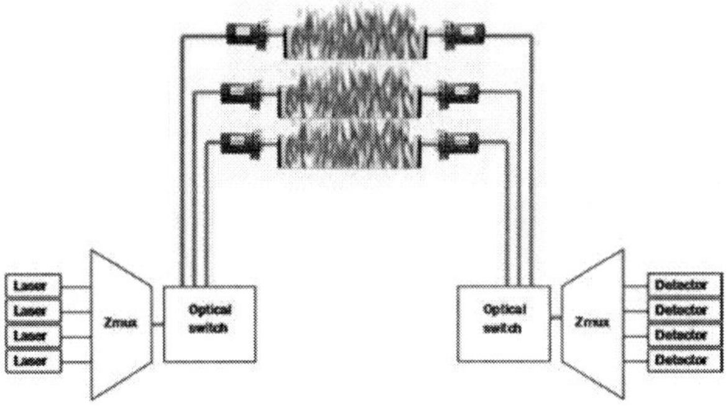

Figure 2. Fiber-optic system architecture

The control rack for this system is NEC Class 1, Div 2 and ATEX compliant and can be located away from the furnace on account of the fiber-coupled nature of the system architecture. The control rack houses all of the sensitive electronics such as the lasers, detectors and computer. The heads mounted to the side of the furnace do not require any cooling or purge air and automatically maintain alignment of the laser beam across the furnace to maximize the received signal thereby achieving high reliability and accuracy.

The optimal layout for measurement paths depends upon the specific burner configuration and controls, process tube/pass configuration and spacing, and the process and combustion monitoring objectives for the specific furnace. A typical layout to provide high resolution for an SMR is shown in Figure 3 with paths located at an elevation just above the tunnels. The configuration includes a rectilinear grid as well as multiple diagonal paths which allows cells consisting of groups of four burners to be defined and measured. This level of resolution has been demonstrated but is generally considered to be more than is required to obtain the majority of the benefit provided by the system. In this case, simpler is generally better.

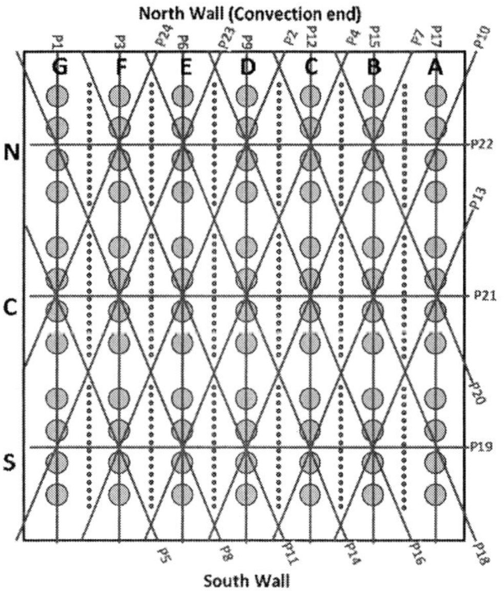

Figure 3. Plan view of high-resolution path layout for an SMR. The yellow circles are burners; the black dots are tubes; and the red lines are laser paths

Figure 4 shows a lower resolution path layout for a different reformer and it incorporates only a rectilinear grid which has proven sufficient to balance the furnace in the east-west and north-south dimensions and derive the majority of the economic benefits of the system. In addition, a single path is typically located at the exit of the radiant section to provide a more representative measurement of overall excess O_2 than a single point O_2 sensor.

A standard system can accommodate up to 24 measurement paths from a single control rack. This provides a cost-effective means to measure multiple paths in one furnace for high spatial resolution or in multiple adjacent furnaces at lower resolution using a single TDLAS system. The system interface provides real-time balance information using each measurement path across the SMR furnace. Measurement times range from 3-10 seconds/path, depending on automatic path alignment requirements, with a complete profile of the furnace updated typically within two minutes depending on the number of paths. Operators can use information from the system interface to make adjustments to the air/fuel ratio of the burners to improve the temperature and oxygen balance in the furnace. The real-time measurements can also be integrated directly into a combustion control system to achieve the greatest benefits available with automated control. An example of the interface is shown as Figure 5 which displays path-averaged concentrations and temperature along each line of sight.

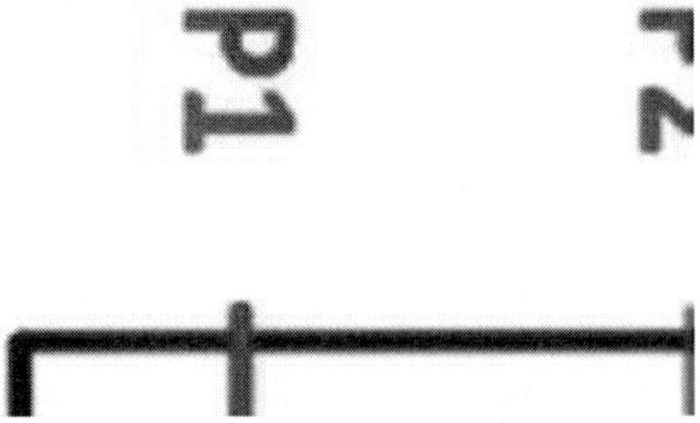

Figure 4. Plan view of lower resolution, rectilinear path layout for an SMR. Typically 8-14 paths are required for a simpler installation.

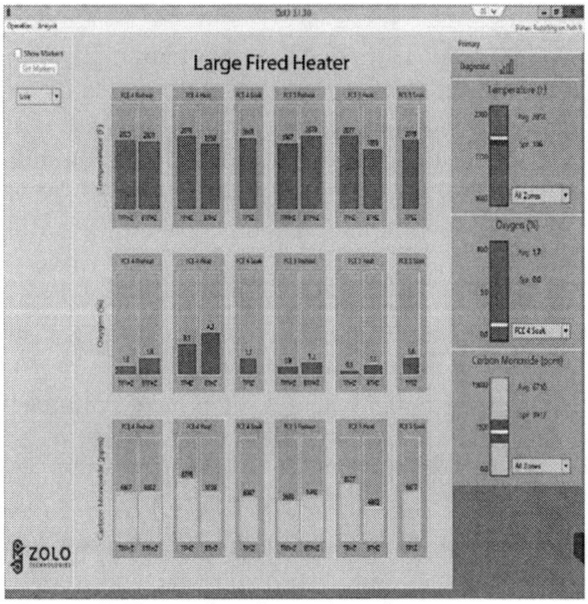

Figure 5. Example of user interface

Results

Accuracy of the measurements

In addition to extensive calibration of absorption constants and system verification using calibrated gas mixtures in the laboratory, the field data collected on a daily basis also confirms the accuracy of the multiplexed TDLAS measurements. For instance, Figure 6 shows temperature and O_2 data on a single path for a hydrogen plant reformer. The periodic nature of the signal results from the pressure swing adsorber that separates the product hydrogen from the effluent gases. The temperature variation is a result of the fuel pressure changing due to the operation of the PSA. Additionally, when the fuel flow varies, the excess O_2 varies as well.

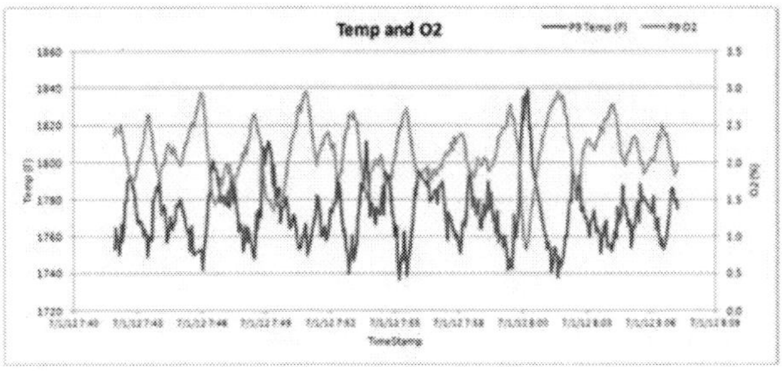

Figure 6. Plot of TDLAS measured temperature (red) and O_2 (green) concentration versus time

Figure 7 shows TDLAS measured temperature versus furnace thermocouples. The absolute temperatures are offset slightly due the difference in measurement location (thermocouples are downstream and therefore in a cooler zone), but the correlation is obvious.

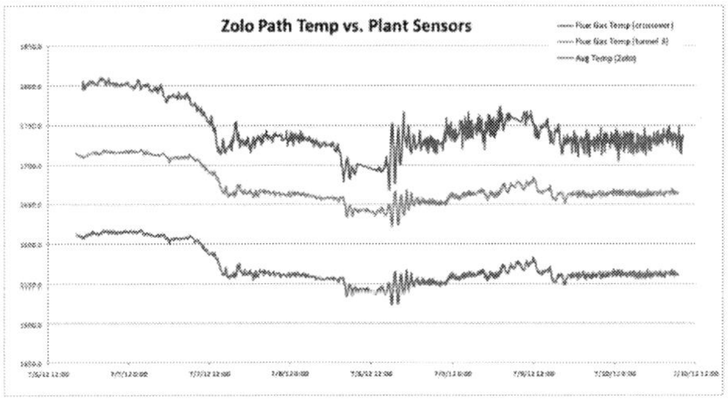

Figure 7. TDLAS temperature (blue) versus downstream thermocouples (orange and red)

Figure 8 shows the TDLAS O_2 measurement (green) compared to two ZrO_2 probes (light and dark blue). Note the very high correlation between the TDLAS measured O_2 and the ZrO_2 probes.

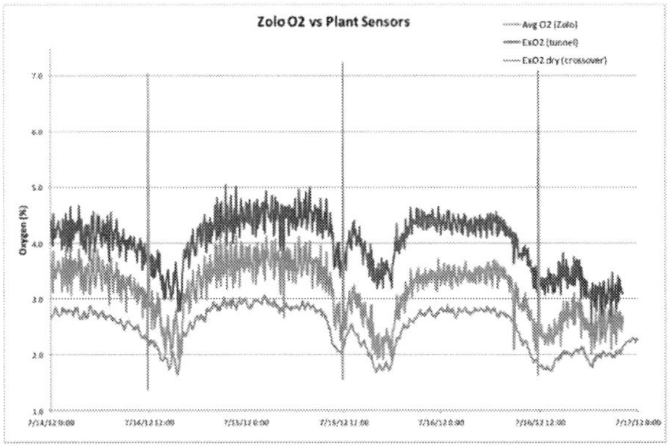

*Figure 8. TDLAS measured O_2 versus ZrO_2probes.*The field data shown in Figures 6-8 strongly support the claim that TDLAS measures accurately precisely because the measured data correlates so well with known sensors (thermocouples and ZrO_2 probes) and with itself.

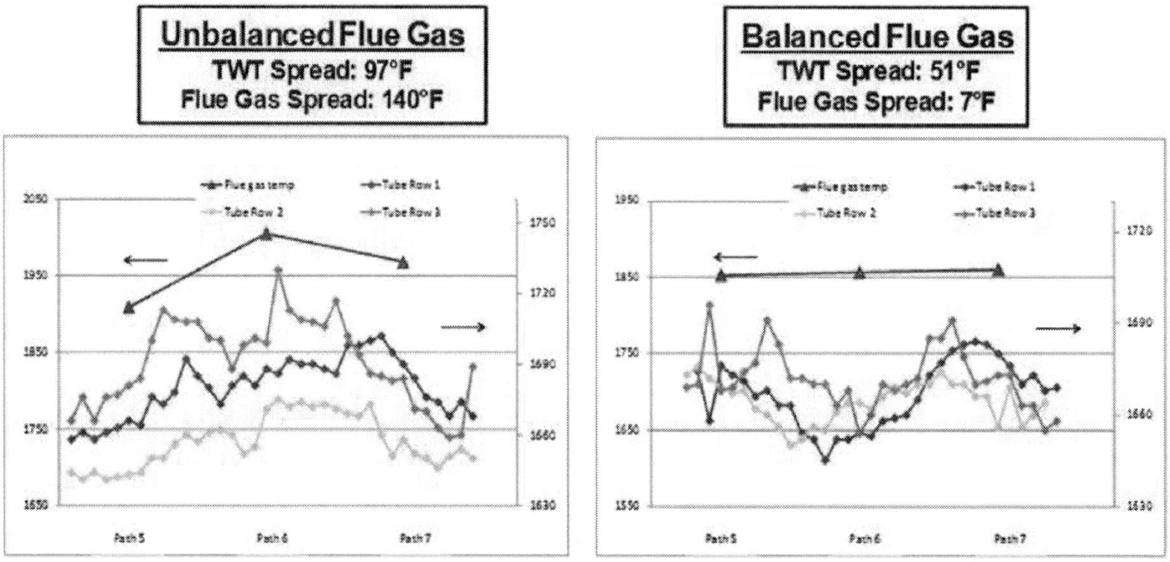

Figure 9. Tube wall temperature profiles for an SMR before and after balancing using TDLAS data

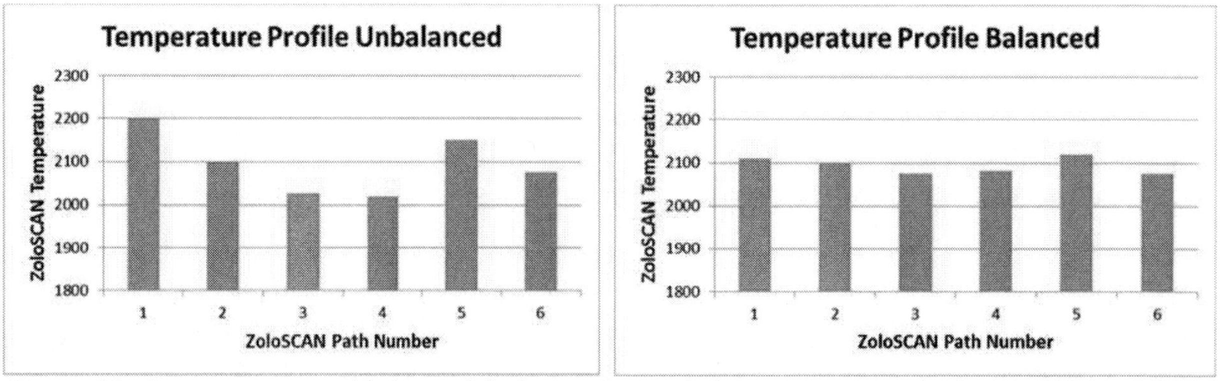

Figure 10. Temperature profile by path before and after balancing using multiplexed TDLAS data

Benefits of reduced temperature spread

Full-scale commercial multiplexed TDLAS systems have been installed on operating Steam Methane Reformers (SMR's) in Hydrogen and Ammonia production plants in an attempt to balance the flue gas profiles by changing (biasing) the burner settings and therefore influencing the tube wall temperature (TWT) profile. Figure 9 above shows actual SMR plant data demonstrating a ~50 F (10 C) decrease in the spread of TWT across the reformer using the information from the multiplexed TDLAS system to balance flue gas temperature (Figure 11). The aim was to reduce "hot spots" and therefore reduce the number of "at risk" tubes. Metal fatigue failures are adequately modeled using the Arrhenius expression in which failure is exponential in temperature (Larson-Miller equation). Typically, a 5% increase in absolute temperature will increase the rate of creep (and eventual failure) by a factor of ten. A 20 F (6.667 C) increase in temperature is projected to reduce tube life by a factor of two times. Therefore, reduction of tube failures through well balanced reformers and furnaces can result in increased intervals between unplanned shutdowns. More uniform tube temperatures also allow a larger majority of the tubes to operate within the optimal temperature range thereby improving process yield and efficiency. Because of the conflicting requirements of lower temperature to

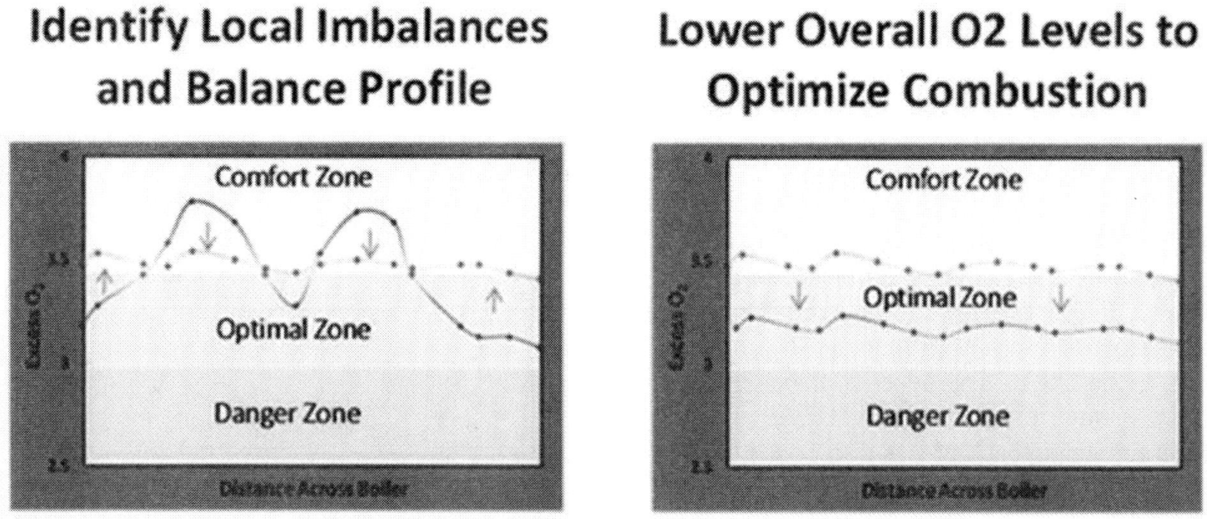

Figure 11. By identifying and eliminating inhomogeneity in the O_2 distribution, increases in process efficiency and reduction of pollutants such as NO_x result

avoid tube failures and higher temperatures to promote reaction, the optimal operational temperature is a highly non-linear function and encompasses a very narrow temperature range. A more uniform temperature distribution also reduces thermal NOx production which is non-linearly and directly proportional to temperature.

Benefits of a uniform O_2 distribution

As shown in Figure 11, multiplexed TDLAS technology can be used to identify localized areas of high (or low) O_2. Adjustments to the burners' air/fuel ratio can be made to create a more balanced O_2 profile. With a more balanced

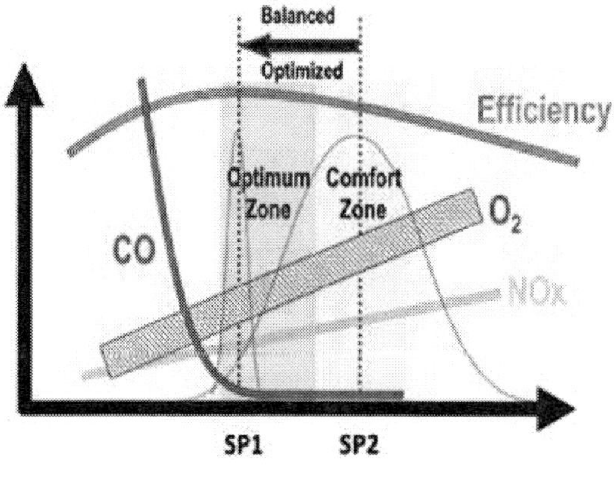

profile, the overall excess O_2 levels can be safely lowered without the risk of high CO. Since less air (O_2) is used, less nitrogen is introduced into the furnace leading to reduced energy consumption (since less cold nitrogen must be heated to combustion temperature) and lower NO_x production (less nitrogen means less NO_x). At plants with high fuel costs (i.e. EU), a small improvement in fuel efficiency can result in significant fuel savings. Furthermore, firing less fuel at lower excess O_2 levels also reduces NO_x and CO_2 emissions (Figure 13). A reduction in excess O_2 levels of only 1% can result in savings of 0.4% in fuel costs. Depending upon the burners and the level of excess O_2, NO_x levels can be reduced by 5-20%.

Figure 12. Process efficiency, CO, and NO_x plotted as a function of O_2 concentration

Return on Investment - Primary Reformer Savings with Multiplexed TDLAS Technology

The benefits of real-time combustion monitoring and control in primary reformers can be significant. Figure 13 shows the potential benefits for a primary reformer producing 100 mmscfd of H_2 in the USA. The benefits of better combustion using a multiplexed TDLAS system have a simple payback of less than 1 year.

Summary of ZoloSCAN Value/Savings		
Reformer H2 Output Capacity	100	MMscfd
Base Efficiency (feed+fuel)	375	Btu/scf
Fuel Cost	$2.70	$/mmbtu
Process Effiency Improvement		
	2.8	Btu/scf
Value of Process Efficiency	$275,940	$/yr
Combustion Efficiency Improvement		
	2.3	Btu/scf
Value of Combustion Efficiency	$221,738	per Year
Tube Life Extension	$0	per Year
Catalyst Life Extention	$0	per Year
Safety	$0	per Year
Total Annual Savings	**$497,678**	**per Year**

Figure 13. Return On Invested Capital

Primary Reformer Combustion Monitoring and Diagnostic Technology Comparisons

Table 1: Combustion Monitoring

	Traditional Sensors	Traditional TDLAS	Multiplexed TDLAS
Proven, Reliable & Accurate	Proven, but prone to operator or calibration errors	Yes	Yes
Maintenance	High	Low	Low
In-situ measurements directly in furnace	Limited by high temperatures. Extractive sampling can	Yes	Yes

	provide erroneous O_2 & CO measurements.		
Auto-alignment of laser beam	No	No Relies on divergent beam to maintain alignment which lowers sensitivity.	Yes Actively aligned collimated beam for maximum power, sensitivity and measurement reliability.
Simultaneously measures multiple constituents from a single sensor	No	No Single constituent (O_2 or CO) for each sensor.	Yes Temp, O_2, and CO measured simultaneously from a single measurement
Multiple measurement paths from a single sensor	No	No Only 1 constituent per sensor.	Yes Up to 24 paths from a single sensor to measure in multiple areas.
Multiple furnace measurements from a single sensor	No	No	Yes A single multiplexed TDLAS system can service multiple adjacent furnaces with up to 24 total paths
Maximum measurement temperature	1500°F (815°C)	2750°F (1500°C)	3500°F (1925°C)
Periodic, in-field calibration required	Yes	No	No
Representative measurement of combustion area	Poor Single point only.	Fair Average concentration across a single path.	Best Average concentration across multiple paths.
Combustion balance information	No	No	Yes Multiple paths provide balancing information across furnace

Table 1 compares types of TDLAS systems available for monitoring combustion processes in large furnaces.

Conclusion

Multiplexed TDLAS systems clearly provide substantially increased capabilities that can lead directly to efficiency improvements and cost savings; whereas, lesser TDLAS systems typically address only potential safety issues and regulatory-driven emissions monitoring – both important issues but not readily monetized. Multiplexed TDLAS combustion monitoring systems can assist primary reformers and large fired heater operators to improve productivity, process yields, process efficiency, energy efficiency, and equipment life through real-time combustion monitoring and diagnostics. These systems provide temperature, O_2, and CO measurements directly in the furnace which can be used to balance the combustion flue gas profiles. Excess O_2 can be reduced safely since CO is monitored in real time with spatial resolution providing comfort that minimum CO limits will not be exceeded as a result of the optimization process. The ability to measure and control furnace temperature and O_2 gas profiles in a narrow and precise range across the full area of the furnace can achieve improved equipment productivity, reliability, emissions reduction, and safety.

Safe Cleaning of Scraper Arms in Urea Plant

The cleaning of scraper arms inside a prill tower is a great safety concern in Urea Plant as it is a high risk activity with potential severe consequences. Several accidents have occurred during the manual chipping of Urea deposition on scrapper arms all over the World. Entering the scraper floor is dangerous even with all safety protections like steel or wooden movable structures as well as personnel protective equipment (PPE). Urea lumps can fall with high velocity due to the height and even small lumps can cause severe injuries to the personnel working inside. Some industries are using movable steel structures under which workers carry out cleaning. Sometimes moving the steel canopy inside the scraper floor is an issue. At Nagarjuna Fertilizers and Chemicals Limited (NFCL), India, scraper arms are cleaned without entering into prill tower scraper floor area. This paper provides insight into how Nagarjuna Fertilizers and Chemicals Limited (NFCL), India has achieved both improved Safety and Reliability by implementing these modifications..

Kumar K R R
Nagarjuna Fertilizers and Chemicals Limited

Anand G V S
Nagarjuna Fertilizers and Chemicals Limited

Introduction

NFCL follows a very safe practice of washing the urea deposition on scraper arms with process condensate generated in Urea Plant. It requires no grinding or hammering operations and does not involve entering the prill tower scraper floor area. This process modification of washing the scraper arms with process condensate has zero impact on the environment with full recovery of urea accumulated. Very few plants in the world have implemented this modification. To date, many plants are following the practice of entering the prill tower to conduct cleaning, exposing persons to serious hazards. This paper describes the full process scheme, modification details, procedures being followed while washing of arms and possible preventive measures for reducing the severity of deposition. Overall this modification has improved plant safety and operational control.

NFCL has very good safety records and administer all important systems like Process Safety Management System, Responsible Care system 14001:2013, ISO 9001:2008, ISO 14001:2004, OHSAS 18001.

Background

Nagarjuna Fertilizers and Chemicals Limited, under the flagship of Nagarjuna Group, operates a large, modern, integrated Ammonia-Urea complex with an annual capacity of 1.56 Million MT (1.71 Million ST) of urea, laid out in two streams. The complex is located at Kakinada on the East-Coast of India.

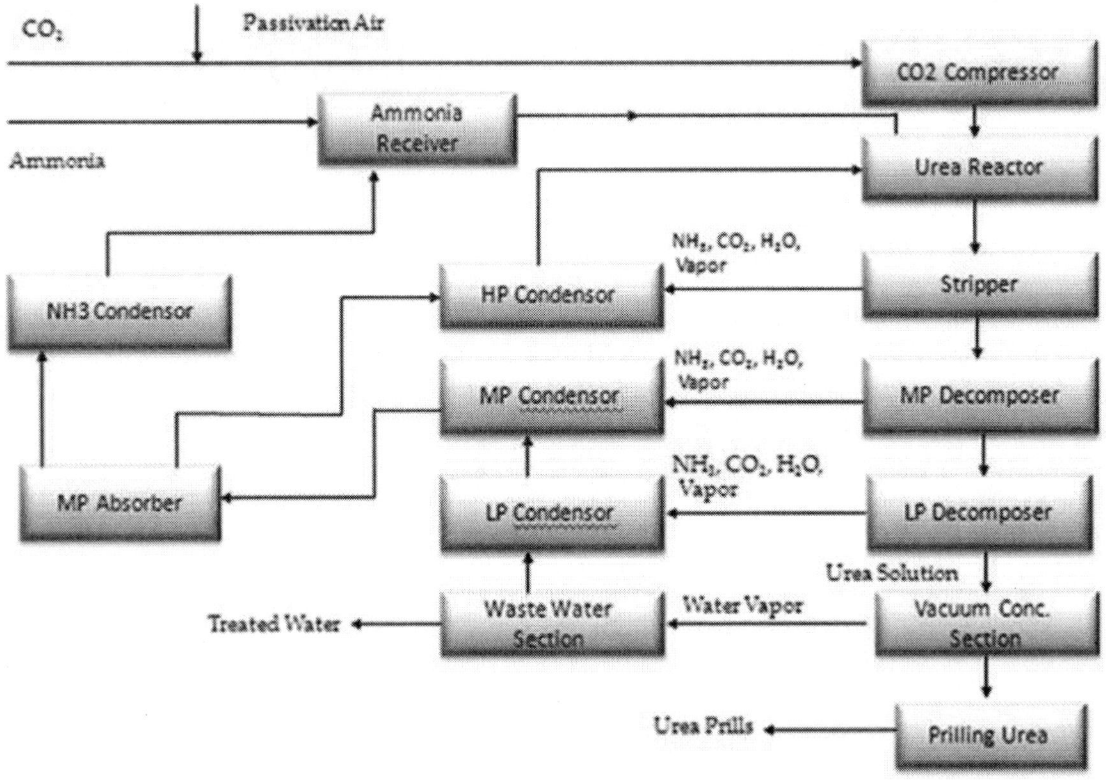

Figure 1. Typical Urea process

This complex is surrounded by an excellent green belt of more than 283 hectares (700 acres) with flora and fauna, which is 70% of total plant area. The ammonia plants are based on the Haldor Topsoe's (HTAS) Steam Reforming process and the urea plants are based on Snamprogetti's ammonia self-stripping process. The typical Urea process flow diagram is shown in Figure 1. The main modifications are in the Prilling section and Waste Water section in the above diagram.

Main Features

The following are the typical features of the Prilling and Waste Water sections at NFCL. The schematic diagram of these sections is given in Figure 2.

Prilling Section

The Urea melt solution from vacuum section is pumped to a rotating prilling bucket which is located at 72 meters elevation at the top of the prill tower. The urea melt solution comes out of the prill bucket in the form of fine droplets. These droplets fall down the prill tower and pass counter-current to a cold air flow from the bottom louvers of prill tower which causes solidification. The air at the top of the tower, containing low levels of urea dust is washed with purified process condensate in the de-dusting system before it exhausts from the top of the prill tower.

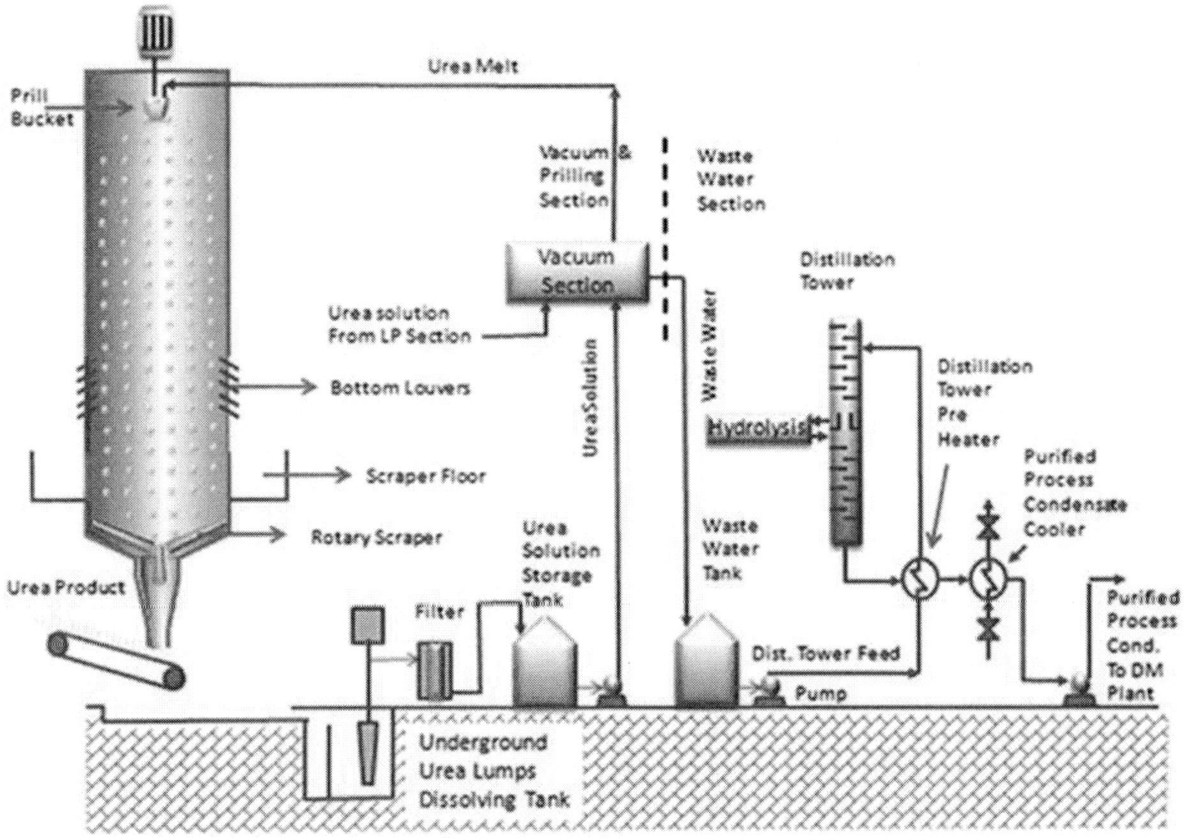

Figure 2. Before Modification – Schematic Diagram

The prill tower has a conical bottom floor and the urea prills falling on this floor are sent to a belt conveyer with the help of a rotary scraper.

An underground tank is used to dissolve urea lumps and to collect urea solution from all drains. This tank is equipped with stirrer and submerged pumps to recover solution.

This solution is pumped to a Urea solution storage tank. The urea solution is diverted to this tank in the event of shut downs or back end problems.

Waste Water Section:

The condensed vapors from the vacuum section containing urea, ammonia and carbon dioxide in the form of waste water are collected in waste water tank. The waste water is pumped by feed pump into the distillation tower after being preheated in the feed effluent exchangers. After being treated in the distillation column and hydrolyzer (urea is converted to ammonia and carbon dioxide), the process condensate (purified water) leaves from the bottom of distillation tower. The purified process condensate is cooled first in the distillation tower preheater and then a cooling water exchanger (process condensate cooler). Finally it is sent to de-mineralized water plant for necessary recycling.

Before modification - Methodology

Urea gets deposited on the rotary scraper arms, due to continuous falling of urea prills on it. Over a period of time, this urea accumulation forms a heap on the scraper arms and gets hardened. This urea heap will grow bigger in size, leading to scraper motor load amperage fluctuation and may even touch the over load amperage at times. In addition to this problem, when urea lumps fall from the top ceiling of prill tower, the lumps obstruct the path of rotation of the scrapper and a small moving heap will form at the blades of the scraper. Due to this, scraper arms may be distorted to touch the floor a few places. This problem sometimes causes damage to the scraper floor also. Hence, cleaning of scrapper arms is a must as heaps become bigger in size.

Problems before modification

Earlier, before implementing this modification, after taking shut down, people used to go inside the prill tower and physically hammer these urea lumps on the scraper arms. While doing physical cleaning, minor injuries and near misses occurred at times. There are a number of safety risks associated with performing this operation as mentioned below:

- **Lumps fall:** This is the most dangerous aspect as urea lumps fall from a height of 72 meters from the top ceiling of the prill tower. This is a major safety concern.
- **Difficult for standing:** Working or even standing on the floor to perform the cleaning operation is difficult as the floor is conical in shape and can lead to risks of injury due to slips, trips or falls.
- **Injuries due to physical breaking:** People will have injuries while hitting with hammer. Sound propagation within the scraper floor due to hitting will enable more loose lumps to fall. The potential for injuries exists while using hammers to break up the urea lumps.
- **Shifting of Material:** Sometimes urea lumps (after physical breaking) will be large and heavy. Careful shifting of the material is important, otherwise manual handling it may lead to injuries, especially as the entry door is relatively high compared to the scraper floor.

The above physical cleaning was time consuming and it was difficult to ensure uniform cleaning on both arms.

Root causes for deposition

There are a number of causes for deposition. Few are mentioned below:

A. Frequent over flow of bucket
B. Bucket design problem
C. High prill temperature
D. More dust/fines in prills
E. Improper louvers operation at bottom and top etc.

Though the above problems can be addressed, it is not possible to completely eliminate the deposition problem. By addressing these problems, frequency of cleaning will be reduced, however the dangerous occurrences cannot be ruled out as cleaning of scrapper arms is a must. Details about how these problems were addressed have been covered in this paper below.

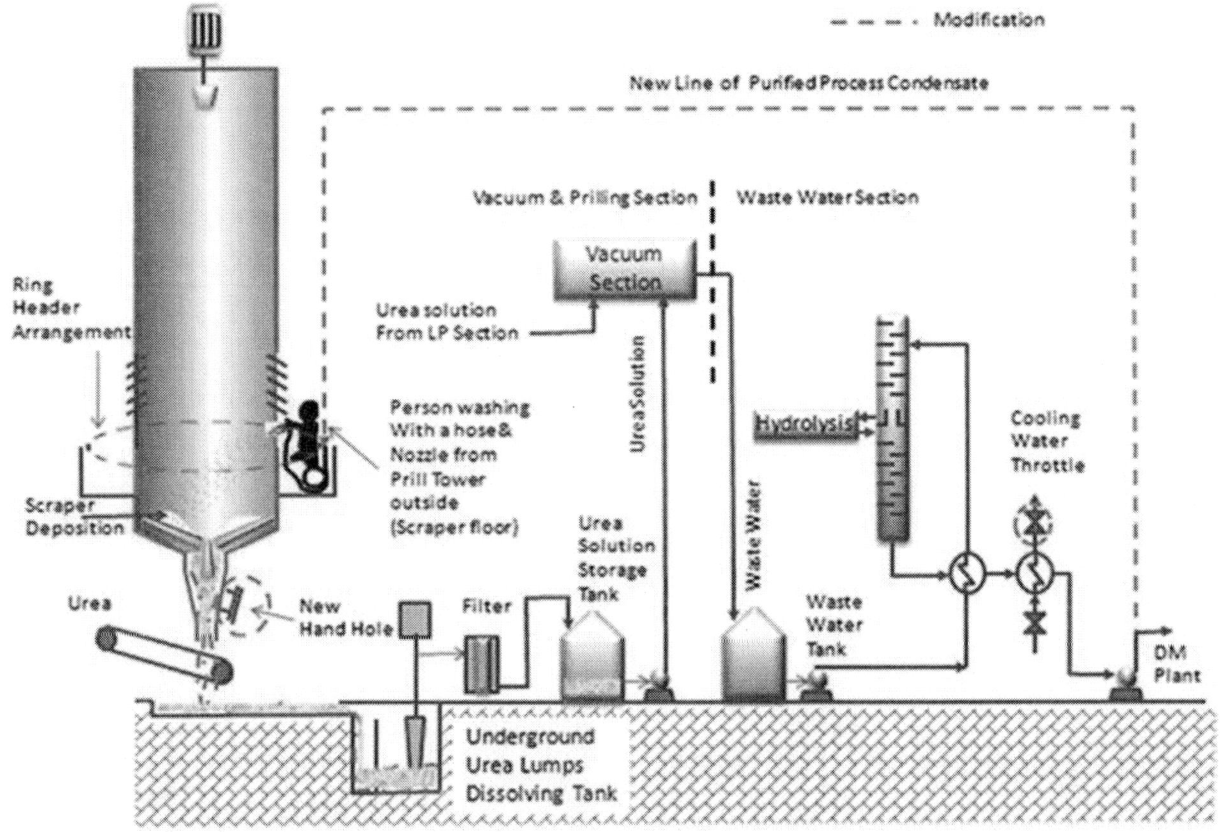

Figure 3. After Modification - Scheme for Urea Washing on Scraper Arms

After modification & Change in methodology

After implementing the new scheme, we are now using purified condensate from the waste water section for cleaning. The modifications implemented are as follows:

A. New process condensate line to scraper floor with a ring header arrangement.

B. Provision of rectangular hand hole for scraper chute.

Details of modifications are as follows:

A. New process condensate line to scraper floor with ring header arrangement:

A new line has been laid from purified process condensate line going to prill tower top for the de-dusting system as a makeup (Major quantity goes to DM plant). Each prill tower is 22 meters (72 feet) diameter. A ring header of 3” (7.62 cm) has been provided at the scraper floor with 10 tappings for washing with process condensate.

Figure 4. Window for scraper arms washing

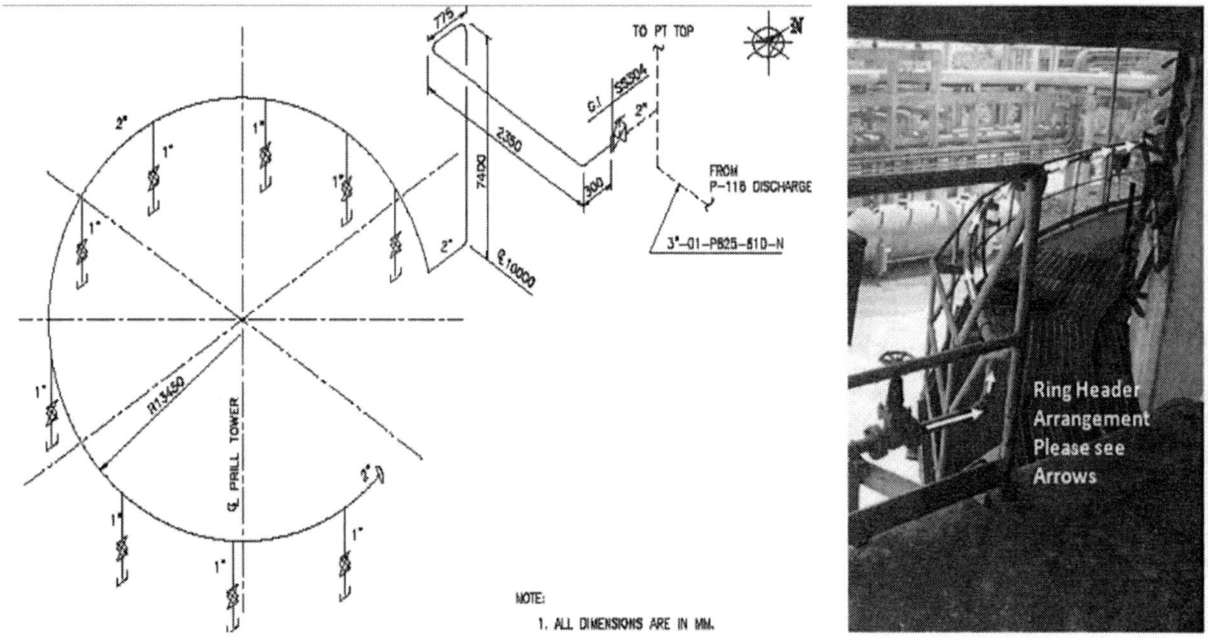

Figure 5. Ring header around the prill tower drawing and photo of actual implementation

The operator can project the water spray on to the scraper arms from any convenient location around the prill tower at scraper floor. Refer to Figure 5 for details of ring header arrangement and photo. Sufficient distance has been made at the bottom louvers of the prill tower for the washing operation. Refer to Figure 4.

B. Provision of a rectangular hand hole for scraper chute

A rectangular hand hole was made on the scraper chute at the bottom half (upper side). While washing the urea heaps on the scraper arms, large pieces break off and fall on the scraper floor. In addition, urea lumps which fall from bucket floor ceiling also accumulate on the scraper floor and are removed through washing only. These larger size lumps may choke the scraper chute at times. Hence, this hand hole is provided and is used for removing un-dissolved urea lumps. Refer to Figure 6 and Figure 7.

The purpose of the hand hole is to check visually and to remove urea lumps by using tools like crow bar etc. This operation eliminates the hazard of people entering the scraper floor for removal of urea lumps.

Figure 6. Chute modification

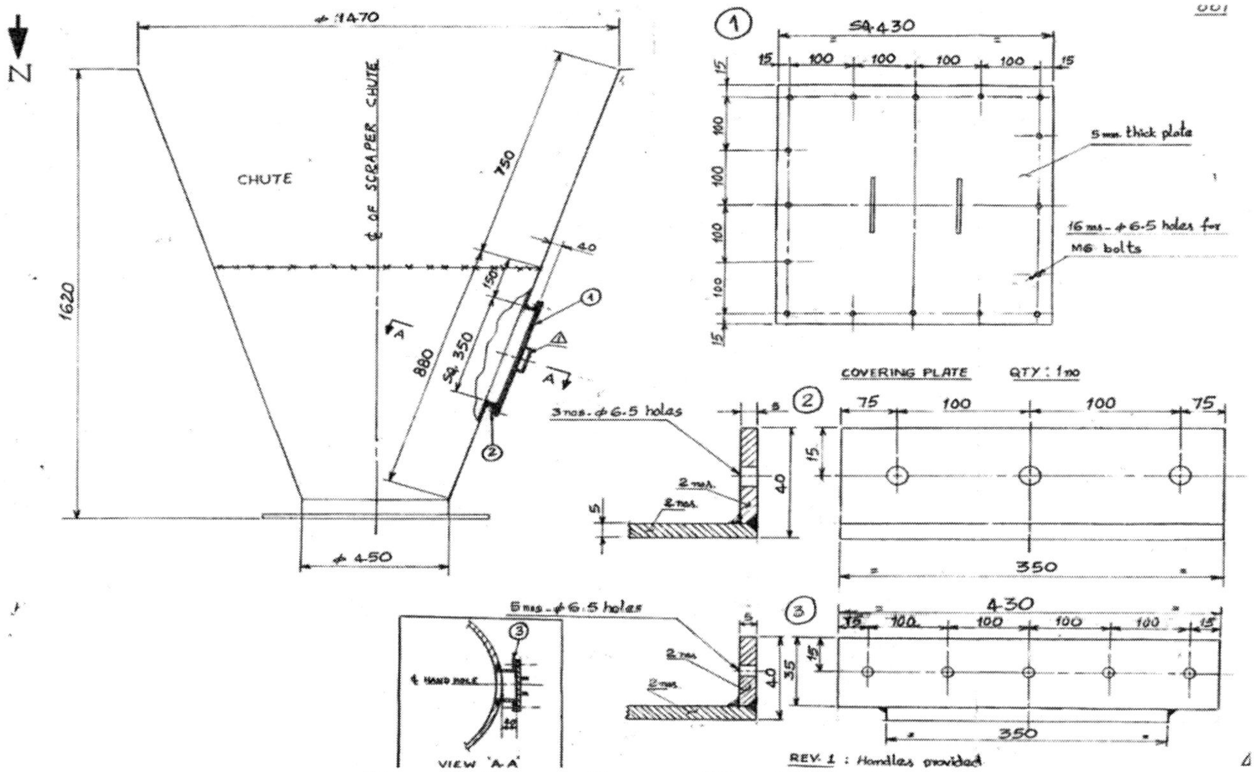

Figure 7. Scraper Chute modification details

Procedure for Washing of the Scraper arms:

The entire washing operation can be performed while the main plant is in operation. Before starting this washing operation, if required, plant load will be reduced to minimum possible load which is around 40%. In so doing, 99.7% urea solution produced will be less in quantity. The urea solution produced can be diverted to the urea solution storage tank. The scraper and conveyors are de-energized during this operation for safety and the motor is covered with polythene sheet for protection from water spray.

The process condensate temperature is raised to around 50°C (122 °F) by throttling the cooling water outlet valve of the process condensate cooler. Refer to the encircled valve in Figure 3.

This way the operator can perform the operation with comfort and due to increase in temperature, the urea dissolves at a faster rate.

Before starting the washing operation, once urea solution is diverted from the prill bucket to urea solution storage tank, the scraper is run for some time to ensure that the urea bed on the scraper floor is removed as much as possible. This will ensure that time is not wasted in dissolving the urea prills on the scraper floor while pushing of small lumps to the center of scraper chute will be easy.

During this operation, the (hand hole) rectangular opening provided with flanges in the scraper chute is kept open. Most of the washed condensate comes out from this rectangular opening.

The dilute urea solution generated during the washing of the scraper arms goes to an underground trench which leads to an underground urea lumps dissolving tank. From this tank the solution is pumped to the urea solution storage tank. This tank has a volume of 350 m3 (12360 ft^3) capacity.

The urea solution with 99.7% concentration and the dilute urea solution generated during the washing, both go to urea solution Tank. The washing operation takes around 3 hours. This time is sufficient to wash out the deposition, if frequency of washing of scraper arms is planned in a proper manner. Once the washing is completed, the hand hole will be boxed up. The scraper and conveyors are energized again and the motor polythene cover is removed. The conveyors and scraper are run for some time to ensure that all are running normal.

Urea melt can be lined up to prill bucket immediately as the vacuum section is in operation during the scrapper arms washing activity. This modification was also implemented in Urea Plant II and is running successfully for the past 18 years.

Benefits realized:

The above modifications significantly improved the safety of the plant operation.

1) No personnel ever injured after implementing this modification.

2) This modification enabled us to do washing of scraper arms while plant is in operation.

3) No personnel need to enter the scraper floor to conduct cleaning, even during turnarounds.

Addressing the root causes

As discussed above, ways to reduce the deposition problems were also identified and addressed at NFCL. Some of the points are mentioned below:

A. **Frequent over flow of Bucket:** There are many causes for over flow of bucket.

   - **Periodic Cleaning**: Periodic cleaning of the bucket will help in controlling fines rather than cleaning after over flow of bucket. On average, every 15 days, prill buckets are cleaned. Whenever urea solution recovery is done, bucket holes choked more frequently. This is due to carry over of small size impurities along with the solution even though strainers are used. Hence, whenever urea solution recovery is there from urea solution tank, prill bucket is being cleaned every week. Also, the urea solution tank is cleaned during every plant turn around to ensure that clean solution is being recovered.
   - **Ensuring thorough cleaning:** After dropping the bucket, it is cleaned with a pin while keeping a lighting lamp inside. In so doing, choked holes can be seen clearly from outside and are cleaned effectively.
   - **Tuning of Level controller to maintain low level:** Proper tuning of second vacuum vapor-liquid separator holder level control valve helps in preventing frequent overflow of the prill bucket. An output lock has been provided for level control valve so that the valve does not open to its maximum extent thereby overflowing the bucket.

B. **Bucket design:** Ensure that the plant load shall match with the rotation speed (rpm) of the bucket operation. Running the bucket at more than design condition may produce more fines, which will aggravate the deposition on scraper arms and ceiling. At NFCL, the prill buckets are suitable for various loads and also for summer conditions. Though they are procured for maintaining good quality of the product, utilization of the proper bucket as per the plant load will also help in reducing deposition.

C. **High prill temperature:** High prill temperature due to ambient conditions also will enhance deposition on scraper arms. The adjustment of the bottom and top louvers of the prill tower is important as per plant operation and weather conditions. NFCL has a wet de-dusting system at the top of the prill tower and it is in continuous operation. When there is flow restriction near the nozzles and air path, a shutdown of the de-dusting system is taken. Nozzles, air path and sump are cleaned and the de-dusting system is put back in operation.

D. **More dust/fines in prills:** Operating the prill bucket at optimum speed corresponding to plant load helps in limiting the generation of fines and in turn reduces urea deposition on scrapper arms. Hence at NFCL, it is ensured that fines are always minimum at the prill tower bottom by analyzing the fines by the operation personnel themselves (not dependent on laboratory analysis alone) and taking appropriate operational changes within the shift.

E. **Improper Operation of the bottom and top Louvers:** It is important to adjust the louvers according to the plant load to minimize the dust in urea depending on the change in wind direction. With the change in ambient conditions from day to night, the louvers are adjusted accordingly.

A number of problems were discussed above. After addressing the above problems, the deposition on scraper arms has reduced and cleaning frequency has also come down. Nowadays at NFCL, scraper arms are being cleaned once in 6 months (on an average) or depending on the plant condition. However, each plant has to identify the plant specific problems and suitable solutions to be implemented to reduce the severity of deposition.

Conclusions

The following conclusions are offered:

1. Washing of scraper arms without entering prill tower will avoid dangerous occurrences.

2. Washing with purified process condensate is very safe and easy with proper planning.

3. This modification does not cause any effect on the environment as the urea solution generated during scraper arms washing is recycled back to the system.

4. A plant shutdown is not required and only plant load needs to be reduced during the operation.

5. The root cause for the problems are identified and addressed to reduce the cleaning operation.

6. Latest technologies and latest versions of prill buckets are to be reviewed, procured and installed in order to minimize fines and dust.

7. Maintaining proper condition of the prill bucket will help in alleviating the problem.

Approaches to Better Hazard Recognition During a PHA

One issue in the minds of many Process Hazards Analysis (PHA) teams is the concern that they will "miss scenarios" – that is, potential hazards at their site will go unrecognized and unaddressed by the PHA team, only to be discovered when an unwanted incident occurs.

Many PHA teams focus only on the immediate information in front of them – typically, past PHAs and existing Process Safety Information (PSI) such as piping and instrumentation diagrams. It is often easy to overlook some simple steps that can ensure teams are considering incidents that have happened at other sites within the ammonia and fertilizer industry.

This paper discusses basic steps that PHA facilitators, operators, and team members can consider to provide assurance that they are increasing the number of major scenarios caught during their PHA studies. This paper will also discuss information sources that could lead to better hazard recognition and thus more complete risk management. It also describes the concept of a Hazard Checklist that can be used to improve hazard recognition and improve study quality while reducing study duration.

Peter G. Hereña
Baker Engineering and Risk Consultants, Inc.®, Chicago Office

Introduction

Process Hazards Analysis (PHA) is a required element for industry operators that are required to follow the Occupational Safety and Health Administration (OSHA) Process Safety Management (PSM) regulation 29 CFR 1910.119. It is a good practice even for those not subject to the regulation. A PHA can be applied using a variety of methods, from a comprehensive Quantitative Risk Analysis (QRA), to the Hazard and Operability Analysis (HazOP), to a simple "What-If?" method. They all share the same components of identifying how failures can result in process material releases, the possible consequences of the releases, and possible safeguards that would prevent or mitigate the releases.

The PHA is completed by a team of individuals with diverse technical backgrounds – the key individuals are persons generally knowledgeable in the day-to-day operations of the unit (typically an operator), knowledgeable in the design of the process (typically an engineer), knowledgeable of the mechanical integrity ("maintenance guy"), and knowledgeable in the PHA method (called a "facilitator" or "leader").

Most PHAs are qualitative studies that provide rough estimates of risk, which can drive significant recommendations for management discussion. These studies typically last from a few days to several weeks, depending on the complexity of the process. An effective PHA is a powerful tool to prevent hazards while they are in a theoretical stage. If the team does not have sufficient guidance, however, a PHA can lead to frustrations over study inconsistency, duration, and impractical recommendations.

Understanding the Hazards

PHA facilitators and teams tend to start with the site Process Flow Diagram (PFD) or Piping and Instrumentation Diagram (P&ID), separate the system into nodes, and start at Node 1. Although this is certainly an expedient way to start the discussion, one drawback is that it can tunnel-vision the team by focusing them only on expected hazard scenarios based on a limited amount of experience. Even a team that boasts 30 years of experience with their site pales in comparison to thousands of years of combined collective experience of other sites. Additionally, because most sites thankfully experience very few major hazards, they could tend to downplay the very real hazard potential that has happened to similar sites.

The greatest enemy of knowledge is not ignorance; it is the illusion of knowledge.
– Daniel J. Boorstin

One way to break this mentality is for a team member - typically the facilitator or process engineer - to spend some time reviewing the kinds of hazards that have happened in industry or other sites within the company. Does this take time? Yes, but experience shows the up-front cost of conducting research is rewarded by time savings of a reduced PHA length because the team spends less time "spinning its wheels" on what are credible potential hazards.

Accessing this type of industry information may seem like a burdensome process, but there may be several avenues of information that can be mined for helpful data, including:

- Published books and literature
- Safety-oriented websites
- Online incident searches
- Intra-company resources.

Published Books and Literature

There are many books in print that describe accident scenarios, the factors that led to them, and lessons learned. Some books describe not only the conditions but the aftermath of the incident with high levels of detail, such as *City on Fire* by Bill Minutaglio, which details the Texas City 1947 Ammonium Nitrate (AN) explosion. A different resource that details hundreds of anecdotal events are the Trevor Kletz books that include *What Went Wrong*, *Still Going Wrong*, and *Computer Control and Human Error* (ref 3, 4, 5). While all of the listed incidents are not specifically fertilizer-related, the types of events described and the errors that led to them are universally applicable. Concerns such as labelling, maintenance, and human error affect every industry. A key resource for both ammonia-industry safety professionals and PHA facilitators are the proceedings of the *Ammonia Plant Safety and Related Facilities Symposium* published each year by AIChE, with compilations in 2016. This is a solid compendium of industrial memory that is still valid to this day.

Some periodicals are worthy of review to identify industry, process, or equipment specific hazards. Two examples are Chemical Engineering Progress, published by AIChE, and Process Safety Progress, published by the Center for Chemical Process Safety (CCPS). Back issues of safety-related articles are searchable through their respective websites or in academic libraries. One example of a recent article from March 2015 is "Process Safety Progress", which details the explosion potential of mixing ammonium sulfate and sodium hypochlorite, which was related to an incident that occurred in Ontario in April 2014. This is a relatively rare combination of chemicals for many plants, but might be applicable to discuss in a "Mixing" deviation at some water treatment facilities.

Lees' Loss Prevention in the Process Industries contains good summaries of the hazards present in many different types of processes and equipment, including a description of Stress Corrosion Cracking in ammonia service, storage of refrigerated or pressurized ammonia, and the factors leading to AN detonations. Lees' third volume describes several case studies for consideration.

Available industrial standards or recommended practices for mitigation or control of hazards are published by many organizations, including the National Fire Protection Association (NFPA), International Society of Automation (ISA), and American Society of Mechanical Engineers (ASME). The American Petroleum Institute (API), originally established for petrochemical industries, has many publications that are considered good practices for a wide array of industries. One good example is *API Recommended Practice 556 on Instrumented, Control, and Protective Systems for Gas Fired Heaters*, which identifies common fired heater concerns and countermeasures. Even to this day, it is surprising to hear from some operators and engineers that the communication of fired heater hazards remains incomplete to all levels.

Safety-oriented Websites

There are several valuable online resources that provide incident details for better understanding of the hazard potential in certain processes. The most popular source of information is from the U.S. Chemical Safety Board (CSB), which investigates process incidents, publishes reports of major findings and produces high-quality summary videos of the events. While the reports can provide valuable insight, the videos are effective when presenting to groups, in order to increase awareness of hazard development. CCPS, for example, has a deep foundation of webinars, presentations and E-learning courses, many of which are available at no or low cost. A recent search brought up a presentation on an ammonia release at Millard Refrigerated Services in August 2010 that was a consequence from hydraulic shock due to an uncontrolled reset of an ammonia refrigeration unit.

Online Incident Searches

Another good resource is the list of industrial disasters found on the Wikipedia website. It lists hundreds of publicly-known incidents from a wide range of industries, and includes reference links for the reader to corroborate the information from the website. There are over two dozen incidents listed for ammonium nitrate alone. Alternatively, an interested party can do web searches specific to a piece of equipment or process. It is sometimes helpful to connect known processes with either a location or company name in order to find public records of incidents. BakerRisk® recently utilized this method for a study that sought to identify release consequences due to hydrofluoric acid release; over thirty incidents were discovered with reports from local newspapers.

Intra-company Resources

Intra-company resources present a unique avenue of investigation that is not available to a company with a single location. If such a practice has not already been established, a multi-site company should establish a means of communicating incidents and high injury potential near-misses between sites. This will broaden the awareness of the hazard and also enable other sites to correct deficient practices. It is also important to note that, in the U.S., if OSHA has identified a critical deficiency at one site, it is more likely to bring attention to the company's other sites for that same or other deficiencies. Therefore, it pays to communicate areas of concern.

Larger companies might have subject matter experts specifically dedicated to some types of processes or equipment and engineering standards in place where common hazards and operability concerns are addressed.

Multi-site or not, the PHA team is required to review past incidents that have happened onsite and confirm that the incident potential has been either captured or at least considered in the study. Such a search should extend to similar locations or units. For example, an incident that occurs in poly phosphate loading at one site could also apply to ammonia at another site.

Developing a Hazard Checklist

The knowledge of the process engineer, operator, or facilitator, backed up by some of the resources noted above, should result in a tolerably complete high-level understanding of the types of hazards that can be expected. While this is a good basis, the PHA facilitator can utilize this research to go one step further by creating a Hazard Checklist of the major anticipated scenarios for each unit or equipment.

Note: a Hazard Checklist does not replace team brainstorming that occurs during the PHA. Rather, it establishes checks to ensure that some minimum expected hazards are discussed. A checklist does not attempt to exhaustively determine causes, quantify the consequence of the hazards, risk rank the scenario, or determine safeguards. That task is the responsibility of the PHA team to carry out together.

An example list of a Hazard Checklist for a centrifugal pump is listed in Table 1 for a general service pump, but the list could be augmented to consider process specific hazards.

Condition	Potential Hazard
Loss of suction flow	• Pump vibration, damage, seal failure • Loss of process operation
Loss of discharge flow	• Pump vibration, damage, seal failure • Possible piping overpressure • Loss of process operation
Loss of lubrication or seal	• Possible seal failure
Abnormal feed temperature	• Possible accelerated corrosion or wear
Failure of seal	• Possible external release
Startup without feed	• Pump vibration, damage, seal failure
Startup overloaded	• Possible motor damage
Shaft break	• Loss of process operation

Table 1: Example Centrifugal Pump Hazard Checklist

The PHA team should address each PHA node according to the guidelines of the method (HazOP, Layers of Protection Analysis, What-If?, etc...) being used for the PHA. When the team analysis of a node is nearing completion, the facilitator can review the checklist to ensure the hazards in the checklist were considered by the team. If a particular hazard was not brought up for discussion, the team should consider whether that hazard is valid, and if so, add the scenario for further consideration. The Checklist serves only as a backdrop to a good team actively identifying hazards.

The checklist is not meant to identify the consequence that *will* occur, but to bring into discussion what has happened per industry examples under similar circumstances. There may be many reasons the consequence is not credible for a specific case. In the pump example above, the pump configuration, piping design, and process fluid properties all have important impact on the credibility of a hazard.

This checklist should only apply to equipment and processes with a long history of operation, so that failure modes have been identified through millions of hours of collective operating time. For a unique piece of equipment, a checklist may not be practical, so it may be more efficient to start without a reference. Requiring a completely novel type of technology is relatively rare, particularly in the ammonia industry, so this checklist concept can be broadly applied.

The good news is that, although it could take some effort to develop a Hazard Checklist, once it has been developed for a single piece of equipment or process unit, it should be applicable for any similar piece of equipment.

Hazard Checklists for PHA Preparation

Before anything else, preparation is the key to success.

– Alexander Graham Bell

The traditional method of preparing for a PHA is to assemble in physical or electronic form all Process Safety Information (PSI) that is available and within scope of the study. The collected information is then brought to the meeting or made accessible in the meeting and can be consulted as needed. Digging into the PSI can be a task that significantly slows down or completely derails the overriding PHA. The following are just some of the reasons PSI digging can lengthen study time:

- The team loses focus or forgets discussions on the PHA study while the PSI is retrieved
- More frequent breaks or early termination to find or read PSI
- Discovering that the specific piece of information is not available
- Unclear, erroneous, or ambiguous documentation that necessitates more review time.

This can be similar to an "open book" exam at school – one in which any resource can be used to help with a test. A good student knows that studying is still required for the exam. Why? Because no matter how many references are available, if the student must spend precious test time lumbering through the reference material, there will be no time left to answer all of the questions! The analogy is very true for a PHA. If an operator or engineer does not review or is not already aware of the PSI in advance of the study, it will significantly lengthen study time and likely undermine the quality of the work.

This begs the question: "How is it possible to 'study' for the PHA?" Wouldn't there be an infinite number of possible options that would take forever to prepare? If the checklist method is used, the answer is "no." The reviewer can use the checklist to focus on specific pieces of information that will most likely come up. This provides a massive hedge against being surprised during the study.

Ideally, a Hazard Checklist can include another pair of columns added to help focus the reviewer on what PSI to review:

1. Possible safeguards – this lists the potential safeguards that *might* be credited for the possible hazards described in the checklist. The reviewer should not perform a check for whether or not it is effective; that is the function of the PHA discussion.
2. Relevant PSI – This is a specific fact to check that should be available on an equipment specification or drawing. It should support either establishing a consequence or establishing the validity of a safeguard. It should not be a general piece of PSI like "PSV-111 datasheet." It is assumed that if the equipment is within scope, the datasheet is needed. Something more appropriate would be "What is the set point pressure of PSV-111?" or "Is PSV-111 designed for blocked flow case?" The answer to those questions would help confirm that specific PSV as a safeguard.

As an example, for the condition of Loss of Pump Discharge flow from Table 1 (see previous page), the following could help the reviewer in tracking down specific information prior to the PHA (refer to Table 2 below).

Condition	Loss of discharge flow
Potential Hazard	• Pump vibration, damage, seal failure • Possible piping overpressure • Loss of process operation
Potential Safeguards	• Pump dual/tandem/special seals • Discharge PSV • High discharge or low flow shutdown • Spillback valve • Loss of flow alarms • High pressure alarms • Operating procedures
Relevant PSI	• Pump seal type (or API plan) • PSV set pressure • Is PSV designed for blocked flow case? • Trip set pressure • Alarm set point • Alarm priority, if an alarm rationalization has been performed

Table 2: Expanded Hazard Checklist for the Condition of "Loss of Discharge Flow"

Hazard Checklists to Anticipate Consequence

Prediction is very difficult, especially if it's about the future.

– Source Unattributed

One additional benefit of using a Hazard Checklist is that it starts the process of understanding the potential consequence that can happen from a release. Much time can be spent discussing whether or not a scenario is credible and, if so, what the consequence might be. Sometimes it is because the scenario is complex, but more often, it is because the team is basing the hazard potential by "gut feel," which is driven by emotion. It is a challenge to move two people into agreement when they are both operating solely on intuition.

One way to reduce the "gut feel" is to make consequence and credibility decisions more fact-based. If a hazard is well understood for a common piece of equipment, a single key piece of information could swiftly end the discussion. Take, for example, the following discussion record that can occur at a PHA:

Facilitator: "So the cause is blocked pump discharge. What is the consequence?"

Engineer: "It could split the piping because of high pressure."

Operator: "We've deadheaded the pump before and it never did that."

Engineer: "Maybe we just got lucky and we didn't reach the pressure in time. It's a high head pump with a lot of power behind it. It's hard to imagine it not causing piping damage if left unattended."

This discussion could continue for some time, followed by a check of the PSI, which could take additional time and allow the other team members look on unproductively until a determination is made.

Now, imagine the scenario if the engineer had gone through the Hazard Checklist prior to the meeting. The duration of the discussion can be significantly reduced and the accuracy improved due to more fact-based dialogue.

Facilitator: "So the cause is blocked pump discharge. What is the consequence?"

Engineer: "The piping design pressure is 285 psig at 130 deg F. I checked the pump deadhead pressure from the pump Head-Flow curve and it is 245 psig. Pipe rupture is not plausible, though seal damage could still occur."

Facilitator: "Operator, do those design conditions sound right?"

Operator: "Yes, normal operation is about 200 psig and 95 deg F."

Facilitator: "Have you experienced deadheading with this pump before?"

Operator: "Yes, sometimes it causes the seal to weep, but we never saw piping gasket failures."

Advanced Hazard Checklist

The checklist can be used on an equipment-by-equipment basis for common pieces of equipment, and relatively little variance would be needed for them:

- Pumps
- Compressors
- Vessels
- Heat exchangers
- Air coolers
- Storage Tanks
- Storage Spheres.

Some complex scenarios would almost certainly be raised and would warrant some specific review and data collection before the PHA:

- Fired Heater rich/lean firing conditions
- Rate of tube temperature increase for fired heaters
- Loss of Forced Draft or Induced Draft fan for steam reformers
- Reactor runaway conditions for Methanators
- Ammonia vapor release distances for storage tanks and spheres, or ammonia-containing piping.

In some cases, PSI is not available or cannot be found before the study. If the hazard is critical to the process, one way or another, that PSI must be updated. Such deficiencies are valuable to identify at the beginning of the study so that a recommendation can be made to update the PSI and the issue can be resolved without unnecessary waste of team time.

Timeline for Review

How many things, too, are looked upon as quite impossible, until they have been actually effected?

– Pliny the Elder

The reader may be concerned by the considerable amount of forethought that the Hazard Checklist places on one or more members of the PHA team. The payback is a much shorter PHA with improved accuracy, and especially more appropriate recommendations. When weighed against the resources wasted in chasing after an unneeded recommendation, the effort more than pays for itself.

In order to develop a Hazard Checklist, the reader is advised to start with basic types of equipment, research the most common failure scenarios, and identify only the most crucial, specific data needed to characterize the hazard. Once the value of the checklist has proven itself, it can be expanded in scope or merged with those prepared by other subject matter experts within the organization.

The actual time needed to develop a beginning checklist depends on the complexity of the equipment and process, and the starting date of the PHA and accessibility of site PSI. The checklist should be completed about 4-8 weeks in advance of the PHA study, which provides some time for research of critical data. In the last few weeks prior to the study, approximately 15-30 minutes of data collection per node might be needed, although that time depends on the complexity of the system and the quality of the starting PSI source.

Conclusion

The concept of Process Hazards Analysis has been a part of industry for many decades. As PHA requirements and complexity have increased, so has the time needed to complete a study. Teams often report that their major frustration with the process is the length of time required and slow pace of the study, while managers report that their major frustration is inconsistency between studies. This inconsistency is exacerbated when different sites, facilitators and teams are used, even though the process is fundamentally similar.

Background hazard research through literature, websites, internet searches, and subject matter experts is valuable for adding to a team's understanding of the types of hazards that industry has already experienced. This will help the PHA team in PHA quality.

A Hazard Checklist can significantly reduce the time needed to perform a study by developing a in advance a list of commonly expected failure modes for typical equipment, and then providing overall guidance to an engineer, operator, or facilitator to look for specific data in PSI that will reduce the guesswork involved that produces so many frustrations.

The Hazard Checklist does not replace a good PHA team, and it does not qualify alone as a PHA method. It serves as a backup to ensure common hazards and scenarios have been discussed. It also frames the discussion to identify key data that is expected to be critical to effective discussion.

Although this method does take some preparation time for an individual, it does have many benefits, including:

- Reducing team meeting duration
- Increasing technical accuracy
- Improving personnel ownership of the PHA quality
- Levelling consistency between sites or units
- Minimizing missed hazards
- Reducing spurious recommendations
- Identifying important PSI gaps.

As operating companies and industry strive for more consistent, fact-based safety studies, a Hazard Checklist is the next step in the evolution of PHAs and one that, if applied correctly, should lead to reduced study time and improved quality.

References

1. City on Fire: The Explosion that Devastated a Texas Town and Ignited a Historic Legal Battle (February 15, 2014) by Bill Minutaglio
2. What Went Wrong?, Fifth Edition: Case Histories of Process Plant Disasters and How They Could Have Been Avoided (Butterworth-Heinemann/IChemE, July7, 2009, 5th Edition) by Trevor Kletz
3. Still Going Wrong!: Case Histories of Process Plant Disasters and How They Could Have Been Avoided (1st Edition, October 2003) by Trevor Kletz
4. Computer Control and Human Error (October 2, 1995) by Trevor Kletz
5. API Recommended Practice 556 Instrumentation, Control, and Protective Systems for Gas Fired Heaters, (2nd Edition, April 1, 2011 American Petroleum Institute)
6. Lees' Loss Prevention in the Process Industries: Hazard Identification, Assessment and Control (3 Volumes), (4th Edition published in August 2012)

Benchmarking – An Important Milestone in the Journey towards PSM Excellence

Cindie Pridy
Sr. Project Manager Global Benchmarking
Phillip Townsend Associates

Introduction

Chemical production and the associated processing industries dealing with hazardous and toxic materials have a significant commercial and societal responsibility to eliminate and/or mitigate the loss of primary containment or fires related to process materials while also ensuring that adequate safety systems are in place. Process Safety Management systems differ from personnel safety management in that process safety focuses directly on the processes, equipment and technology involved in causing an incident. The broader application of process safety management decision making and leadership behavior needs to address loss of containment events of greater and lesser consequences, safety system challenges and failures that could have led to failure, and operating discipline and management behaviors that ensure adequate layers of protection are being maintained. Process Safety Management is focused on a risk based strategy and systematic implementation of processes to prevent process safety incidents.

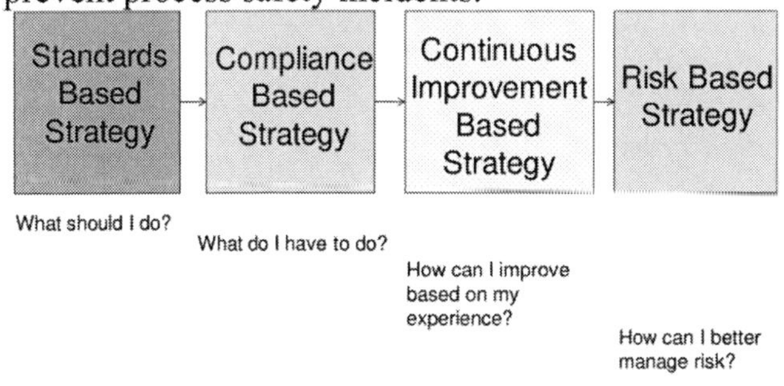

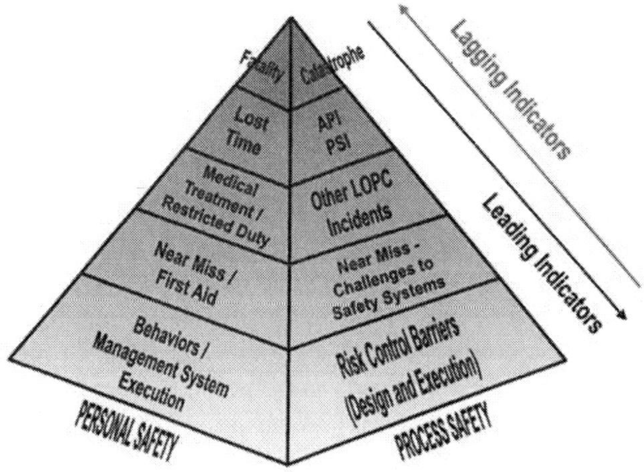

Figure 1: Personal Safety vs. Process Safety Indicator Pyramid

The focus shifts from the lagging indicators that track incidents to leading metrics that are used in a proactive manner to prevent incidents and continuously improve as shown in Figure 1.

The vision and recommendation of industry-wide risk based process safety originated as a result of the December 3, 1984 Bhopal incident killing more than 3000 people due to toxic Methyl Iso Cyanide gas release. The Centre for Chemical Process Safety (CCPS®) was established by AIChE and has identified guidelines for a risk based safety process (CCPS®, 2007). The 20 process safety elements identified form the Four Pillars of Risk Based Process Safety (CCPS®):

- Commitment to Process Safety
- Understanding Hazards and Risk
- Managing Risk
- Systematic Learning from Experience.

In 2009, the CCPS® and Phillip Townsend Associates, Inc. (PTAI) entered into an agreement to develop a benchmarking program for Process Safety Management Systems (PSMS).

PTAI and CCPS® six elements were selected for the benchmarking survey: Process Safety Culture, Compliance with Standards, Hazard Identification and Risk Analysis, Asset Integrity and Reliability, Management of Change, and Measurement and Metrics (CCPS®, 2007). The key goal was to seek an understanding of the relationship between process safety outcomes and management practices.

This paper discusses the details of the Process Safety Management benchmark, the value proposition to be gained, and specifically the experience of developing an understanding of the current status of their existing Process Safety Management system, and the practices that will improve and sustain Process Safety Management performance. The objective is to take Process Safety Management systems and practices to the top quartile performance through a benchmarking process of industry peers relative to Process Safety Management practices. This paper focuses on the most important decisions and behavior changes that leadership can make to spread process safety awareness throughout the organization, close the gap to the top performers, develop central and site specific action plans for improvement, and raise the internal level of all organizational units to the that of the top internal performer.

Methodology

Benchmarking identifies how well a company is doing compared to its peers in the industry, what the industry top performance is, how large the gap is to the top performers within the industry, and create a stimulus for continuous improvement. Benchmarking programs seek to understand the relationship between outcomes and management practices. The Process Safety Management benchmarking program is a systematic study of process safety between companies that will allow sites/companies to quantitatively compare their management systems with the goal of improving the design and implementation of management practices that impact process safety.

- Question Rating is based on the Answer Score (capped at 1.0) times Question Weighting (ranging from 0 – 6.0)
- Section Rating is based on all valid Question Ratings
    - Σ Actual Question Ratings / Σ Maximum Question Ratings providing a normalized percentage score for section
- Overall Rating is based on Weight Average of the Section Ratings
    - Σ Section Weight * Section Rating

Value Proposition of Process Safety Benchmarking

Process Safety Management benchmarking can elevate a company's process safety performance through four distinct steps:

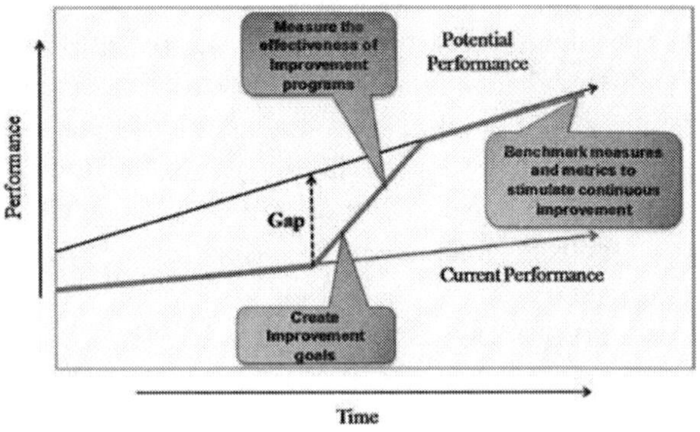

Figure 2: Value Proposition

1. Identify performance gaps

The identification of a company's performance in relation to Performance Quartiles will highlight key drivers for performance enhancement versus its peers. These quantifiable results replace perceptions with a data-based reality check.

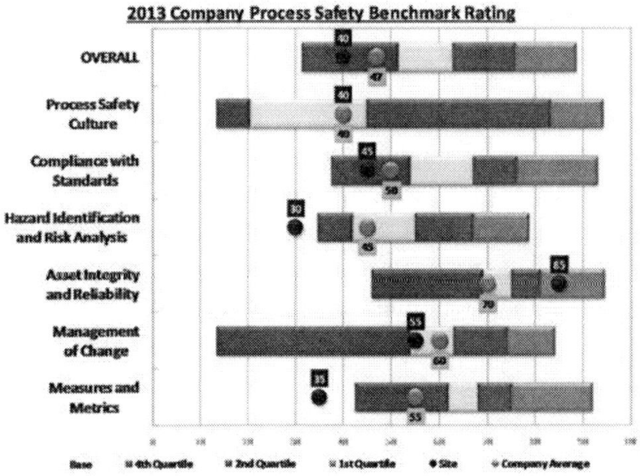

Figure 3: Performance Quartile Ranking

In addition, these results identify if the enterprise policies, practices, and procedures are operating as intended are achieving the desired results.

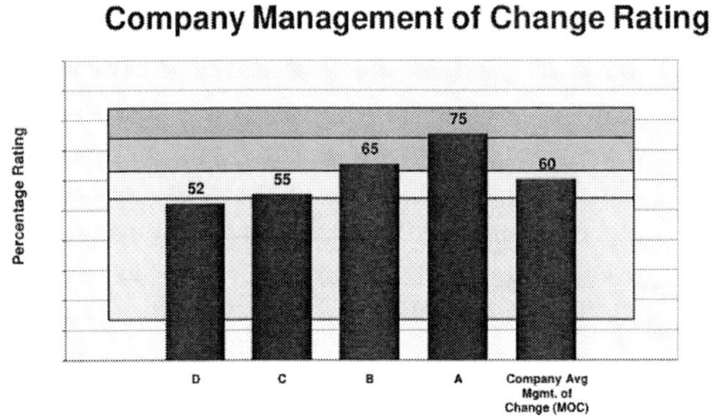

Figure 4: Company and site level comparisons

2. Create performance enhancement goals

Data driven information provides a measure of the enterprise effectiveness – emphasizing strengths, weaknesses and potential for improvement over time. This helps focus attention on the critical aspects of an enterprise that create risk. It also underscores what improvements should be made – providing quantifiable performance enhancement goals.

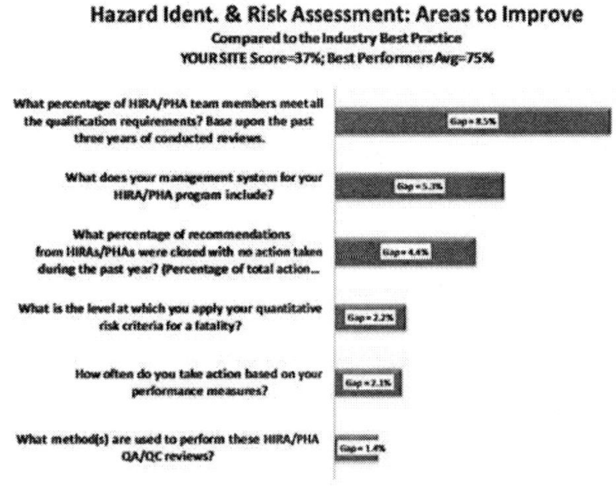

Figure 5: Measureable Opportunities Defined

3. Measure the effectiveness of improvement programs

Process Safety Management benchmarking can be a starting point for goal setting and performance initiatives aimed at boosting step change improvements in process safety. The clear outcome is that this will add significant shareholder value through reduced risks of catastrophic incidents.

Top Quartile Asset Integrity average rating 83% vs Your Company 63%

Areas representing largest opportunities for 20% gap

What percentage of time are your target inspection/testing intervals met?

Your Company:	75-90% of the time
Top performers:	98-99% of the time

%Δ Impact = 5.5%

How is metal thickness data from periodic inspections analyzed for trends?

Your Company:	No formal analysis of trends
Top performers:	Thickness measurement trends are analyzed, remai life projections calculated, and adjustments made to inspection interval

%Δ Impact = 5.0%

Figure 6: Impact of Improvement Initiatives

4. Maintain continuous improvement

The Process Safety Management benchmark helps maintain a stimulus for continuous improvement. It provides the necessary motivation to ensure consistent top performance results.

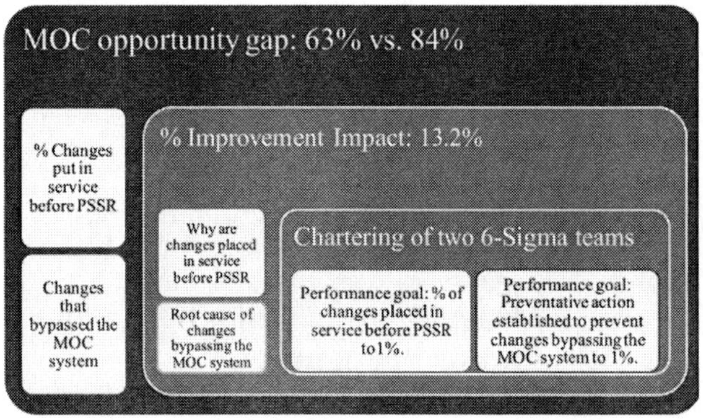

Figure 7: Leveraging Process Safety Excellence

Benchmarking Process

A peer group is established by PTAI of sites/companies agreeing to compare their process safety programs and performance in an effort to identify best practices and drive improvement. Ensuring consistency and protection of each participant's interests are critical to a successful benchmark. The process begins with clear and consistent definitions for the participating sites/companies being distributed. The use of a secure site over the internet ensures security of data and ease of use by participants as they are able to see all the survey questions and select their answers.

A kickoff discussion to align key stakeholders at each of the sites is critical. This can be done through a series of webinars between PTAI consultants and stakeholders at each of the participating sites. Following submission of the data, PTAI validates the data on a site and company basis and provides analysis of the results.

Statistical analysis of the data is performed by PTAI to evaluate various drivers of process safety using a Gap Analysis comparing performance to the top quartile of the peer group. The quantitative and qualitative results are distributed to participants with numerical scores for each element along with an overall site/company specific score. Strengths, as well as the main opportunities for improvement to close the gap towards the top performers are identified. A discussion of the study results and gap opportunities is held with the executive management in an executive management summary with specific individual gap analysis presentations to the top Line and HSE management of benchmarked unit.

Benefits of PSM Benchmarking

Management review is an integral part of moving from lessons learned to systematic improvement of Process Safety Management. Reviews of the performance gaps can identify potential problems and address them before an incident occurs; as well as, identify opportunities for improvement to move from current levels to the level of top performers. PSM benchmarking helps an organization prioritize the allocation of process safety resources, both financial and human, to focus on the highest identified risk areas.

Among the benefits of Process Safety Management benchmarking, there are ones that can have a bigger impact on an organization. There should be a marked increase in staff awareness of process safety issues. This awareness will expand the issues from the Health & Safety group to Engineering to Operations. Based on the organizational level of results discussion, it will potentially increase awareness from site management to impact those that are responsible for setting corporate strategy.

A key learning from the benchmarking survey has been the differences between the individual site scores when multiple units within the same company are benchmarked. Experience shows this can result in a changed perception by executive management. Process Safety Management benchmarking creates a common understanding and language on process safety matters. Benchmarking can also reduce risks by potentially providing an incident early warning signs.

Sometimes the opportunities for improvement can be very different from the perceived gaps in an organization. Identifying and gaining insight on perceived gaps versus actual gaps is greatly enhanced by an independent benchmark which effectively creates consistency across multiple sites within a company and with the others in the peer group.

Key Applications and Lessons Learned

Traditional process safety surveys have identified a list of both quantitative and qualitative issues. Translating the qualitative issues into measurable, actionable steps is a key issue for senior management decision making and allocation of resources. Benchmarking results must be measurable and use a scoring methodology that will not only identify the performance gaps, but also quantify the impact of selected actions. The Process Safety Management benchmarking process designed by CCPS® and PTAI includes a scoring methodology which enables participants to have measurable action plans. Benchmarking is a systematic way to make site-to-site company comparisons and view a company's performance level in relation to that of top performers.

The application of the benchmarking results to identify performance gaps provides meaningful insight for organizations as they develop goals and the associated improvement plans to reach those goals. Repeated benchmarking monitors the rate of improvement. One example of a lesson learned that applies to numerous organizations relates to Process Safety Culture. The findings from the Process Safety Culture section indicated the largest range of results compared to the other sections. The underlying cause of the low scores was a direct outcome of not monitoring the process safety culture through the use of an employee survey. Without a way to assess their work force's values, attitudes and behaviors pertaining to process safety, an organization will have no meaningful basis for making and measuring improvements.

The global process safety incidents have produced tremendous need for the use of Process Safety Management benchmark assessment to measure and improve process safety performance. Initially focused on fatality rates, manufacturers quickly learned that having strong Process Safety Management system and effective layers of protection were important to prevent any loss of containment scenarios.

Process Safety Management Benchmark assessment can help in identifying three main interrelated but distinct benefits: (a) Internal Benchmark of various sites, (b) learning through process of understanding the benchmark questionnaire, and (c) need for a detailed analysis of existing systems implementation.

Conclusion

Process safety is a mature, but continuously evolving concept with practices and procedures varying widely among the refining, chemical and chemical processing industry. Moving from a compliance model to a risk-based model requires a comprehensive approach to managing process safety. Substantial improvement opportunities exist for even the current top performers. Specifically, no site has made a perfect score on any of the risk-based elements of the Process Safety Management benchmark.

The Process Safety Management benchmarking program is a unique methodology to allow companies to identify how large the gap is to top performers and to "learn" how to prioritize resource allocations based on the greatest impact opportunity to improve their process safety management system. This shifts the understanding of current practices to the lessons learned that are measurable and actionable. The benchmarking process identifies top quartile management practices while highlighting not only a company's strengths, but also potential development opportunities. The program's scoring system provides a marker to track improvements and measure the gaps within the company sites and against external top performers.

Using a systematic Process Safety Management System benchmark drives lessons learned to strategic and organizational learning. Conducting periodical assessment of Process Safety Management through repeat benchmarking at regular intervals is a good value proposition to monitor trends of performance and to achieve pacesetter performance compared to industry peers.

References

[1] American Petroleum Institute, API RP 754, "Process Safety Performance Indicators for the Refining and Petrochemical Industries, First Edition," 2010.

[2] Center for Chemical Process Safety CCPS®, "Guidelines for Risk Based Process Safety," John Wiley & Sons, Inc., Hoboken, NJ, ISBN 978-0-470-16569-0: liii-liv, 2007.

[3] API Asset Integrity Key Performance Indicators Workshop; September, 2009

AUTHOR INDEX

AIChE
120 Wall Street, FL 23
New York, NY 10005-4020

ISBN 978-1-5108-5321-8

9 781510 853218

CURRAN ASSOCIATES INC.
proceedings.com